London Mathematical Society Lecture Note Series: 505

Surveys in Combinatorics 2026

Edited by

ANDREI GAGARIN
Cardiff University

ROGER BEHREND
Cardiff University

MATTHEW LETTINGTON
Cardiff University

ISKANDER ALIEV
Cardiff University

CAMBRIDGE
UNIVERSITY PRESS

CAMBRIDGE
UNIVERSITY PRESS

Shaftesbury Road, Cambridge CB2 8EA, United Kingdom

One Liberty Plaza, 20th Floor, New York, NY 10006, USA

477 Williamstown Road, Port Melbourne, VIC 3207, Australia

314–321, 3rd Floor, Plot 3, Splendor Forum, Jasola District Centre, New Delhi – 110025, India

Cambridge University Press is part of Cambridge University Press & Assessment, a department of the University of Cambridge.

We share the University's mission to contribute to society through the pursuit of education, learning and research at the highest international levels of excellence.

www.cambridge.org
Information on this title: www.cambridge.org/9781009766043
DOI: 10.1017/9781009766012

When citing this work, please include a reference to the DOI 10.1017/9781009766012

First published 2026

A catalogue record for this publication is available from the British Library

A Cataloging-in-Publication data record for this book is available from the Library of Congress

ISBN 978-1-009-76604-3 Paperback

LONDON MATHEMATICAL SOCIETY LECTURE NOTE SERIES

Managing Editor: Professor Colva Roney-Dougal, Mathematical Institute, University of St Andrews, North Haugh, St Andrews, KY16 9SS, United Kingdom

The titles below are available from booksellers, or from Cambridge University Press at www.cambridge.org/mathematics

395 How groups grow, A. MANN
396 Arithmetic differential operators over the p-adic integers, C.C. RALPH & S.R. SIMANCA
397 Hyperbolic geometry and applications in quantum chaos and cosmology, J. BOLTE & F. STEINER (eds)
398 Mathematical models in contact mechanics, M. SOFONEA & A. MATEI
399 Circuit double cover of graphs, C.-Q. ZHANG
400 Dense sphere packings: a blueprint for formal proofs, T. HALES
401 A double Hall algebra approach to affine quantum Schur–Weyl theory, B. DENG, J. DU & Q. FU
402 Mathematical aspects of fluid mechanics, J.C. ROBINSON, J.L. RODRIGO & W. SADOWSKI (eds)
403 Foundations of computational mathematics, Budapest 2011, F. CUCKER, T. KRICK, A. PINKUS & A. SZANTO (eds)
404 Operator methods for boundary value problems, S. HASSI, H.S.V. DE SNOO & F.H. SZAFRANIEC (eds)
405 Torsors, étale homotopy and applications to rational points, A.N. SKOROBOGATOV (ed)
406 Appalachian set theory, J. CUMMINGS & E. SCHIMMERLING (eds)
407 The maximal subgroups of the low-dimensional finite classical groups, J.N. BRAY, D.F. HOLT & C.M. RONEY-DOUGAL
408 Complexity science: the Warwick master's course, R. BALL, V. KOLOKOLTSOV & R.S. MACKAY (eds)
409 Surveys in combinatorics 2013, S.R. BLACKBURN, S. GERKE & M. WILDON (eds)
410 Representation theory and harmonic analysis of wreath products of finite groups, T. CECCHERINI-SILBERSTEIN, F. SCARABOTTI & F. TOLLI
411 Moduli spaces, L. BRAMBILA-PAZ, O. GARCÍA-PRADA, P. NEWSTEAD & R.P. THOMAS (eds)
412 Automorphisms and equivalence relations in topological dynamics, D.B. ELLIS & R. ELLIS
413 Optimal transportation, Y. OLLIVIER, H. PAJOT & C. VILLANI (eds)
414 Automorphic forms and Galois representations I, F. DIAMOND, P.L. KASSAEI & M. KIM (eds)
415 Automorphic forms and Galois representations II, F. DIAMOND, P.L. KASSAEI & M. KIM (eds)
416 Reversibility in dynamics and group theory, A.G. O'FARRELL & I. SHORT
417 Recent advances in algebraic geometry, C.D. HACON, M. MUSTAŢĂ & M. POPA (eds)
418 The Bloch–Kato conjecture for the Riemann zeta function, J. COATES, A. RAGHURAM, A. SAIKIA & R. SUJATHA (eds)
419 The Cauchy problem for non-Lipschitz semi-linear parabolic partial differential equations, J.C. MEYER & D.J. NEEDHAM
420 Arithmetic and geometry, L. DIEULEFAIT *et al* (eds)
421 O-minimality and Diophantine geometry, G.O. JONES & A.J. WILKIE (eds)
422 Groups St Andrews 2013, C.M. CAMPBELL *et al* (eds)
423 Inequalities for graph eigenvalues, Z. STANIĆ
424 Surveys in combinatorics 2015, A. CZUMAJ *et al* (eds)
425 Geometry, topology and dynamics in negative curvature, C.S. ARAVINDA, F.T. FARRELL & J.-F. LAFONT (eds)
426 Lectures on the theory of water waves, T. BRIDGES, M. GROVES & D. NICHOLLS (eds)
427 Recent advances in Hodge theory, M. KERR & G. PEARLSTEIN (eds)
428 Geometry in a Fréchet context, C.T.J. DODSON, G. GALANIS & E. VASSILIOU
429 Sheaves and functions modulo p, L. TAELMAN
430 Recent progress in the theory of the Euler and Navier–Stokes equations, J.C. ROBINSON, J.L. RODRIGO, W. SADOWSKI & A. VIDAL-LÓPEZ (eds)
431 Harmonic and subharmonic function theory on the real hyperbolic ball, M. STOLL
432 Topics in graph automorphisms and reconstruction (2nd Edition), J. LAURI & R. SCAPELLATO
433 Regular and irregular holonomic D-modules, M. KASHIWARA & P. SCHAPIRA
434 Analytic semigroups and semilinear initial boundary value problems (2nd Edition), K. TAIRA
435 Graded rings and graded Grothendieck groups, R. HAZRAT
436 Groups, graphs and random walks, T. CECCHERINI-SILBERSTEIN, M. SALVATORI & E. SAVA-HUSS (eds)
437 Dynamics and analytic number theory, D. BADZIAHIN, A. GORODNIK & N. PEYERIMHOFF (eds)
438 Random walks and heat kernels on graphs, M.T. BARLOW
439 Evolution equations, K. AMMARI & S. GERBI (eds)
440 Surveys in combinatorics 2017, A. CLAESSON *et al* (eds)
441 Polynomials and the mod 2 Steenrod algebra I, G. WALKER & R.M.W. WOOD
442 Polynomials and the mod 2 Steenrod algebra II, G. WALKER & R.M.W. WOOD
443 Asymptotic analysis in general relativity, T. DAUDÉ, D. HÄFNER & J.-P. NICOLAS (eds)
444 Geometric and cohomological group theory, P.H. KROPHOLLER, I.J. LEARY, C. MARTÍNEZ-PÉREZ & B.E.A. NUCINKIS (eds)

445 Introduction to hidden semi-Markov models, J. VAN DER HOEK & R.J. ELLIOTT
446 Advances in two-dimensional homotopy and combinatorial group theory, W. METZLER & S. ROSEBROCK (eds)
447 New directions in locally compact groups, P.-E. CAPRACE & N. MONOD (eds)
448 Synthetic differential topology, M.C. BUNGE, F. GAGO & A.M. SAN LUIS
449 Permutation groups and cartesian decompositions, C.E. PRAEGER & C. SCHNEIDER
450 Partial differential equations arising from physics and geometry, M. BEN AYED *et al* (eds)
451 Topological methods in group theory, N. BROADDUS, M. DAVIS, J.-F. LAFONT & I. ORTIZ (eds)
452 Partial differential equations in fluid mechanics, C.L. FEFFERMAN, J.C. ROBINSON & J.L. RODRIGO (eds)
453 Stochastic stability of differential equations in abstract spaces, K. LIU
454 Beyond hyperbolicity, M. HAGEN, R. WEBB & H. WILTON (eds)
455 Groups St Andrews 2017 in Birmingham, C.M. CAMPBELL *et al* (eds)
456 Surveys in combinatorics 2019, A. LO, R. MYCROFT, G. PERARNAU & A. TREGLOWN (eds)
457 Shimura varieties, T. HAINES & M. HARRIS (eds)
458 Integrable systems and algebraic geometry I, R. DONAGI & T. SHASKA (eds)
459 Integrable systems and algebraic geometry II, R. DONAGI & T. SHASKA (eds)
460 Wigner-type theorems for Hilbert Grassmannians, M. PANKOV
461 Analysis and geometry on graphs and manifolds, M. KELLER, D. LENZ & R.K. WOJCIECHOWSKI
462 Zeta and L-functions of varieties and motives, B. KAHN
463 Differential geometry in the large, O. DEARRICOTT *et al* (eds)
464 Lectures on orthogonal polynomials and special functions, H.S. COHL & M.E.H. ISMAIL (eds)
465 Constrained Willmore surfaces, Á.C. QUINTINO
466 Invariance of modules under automorphisms of their envelopes and covers, A.K. SRIVASTAVA, A. TUGANBAEV & P.A. GUIL ASENSIO
467 The genesis of the Langlands program, J. MUELLER & F. SHAHIDI
468 (Co)end calculus, F. LOREGIAN
469 Computational cryptography, J.W. BOS & M. STAM (eds)
470 Surveys in combinatorics 2021, K.K. DABROWSKI *et al* (eds)
471 Matrix analysis and entrywise positivity preservers, A. KHARE
472 Facets of algebraic geometry I, P. ALUFFI *et al* (eds)
473 Facets of algebraic geometry II, P. ALUFFI *et al* (eds)
474 Equivariant topology and derived algebra, S. BALCHIN, D. BARNES, M. KĘDZIOREK & M. SZYMIK (eds)
475 Effective results and methods for Diophantine equations over finitely generated domains, J.-H. EVERTSE & K. GYŐRY
476 An indefinite excursion in operator theory, A. GHEONDEA
477 Elliptic regularity theory by approximation methods, E.A. PIMENTEL
478 Recent developments in algebraic geometry, H. ABBAN, G. BROWN, A. KASPRZYK & S. MORI (eds)
479 Bounded cohomology and simplicial volume, C. CAMPAGNOLO, F. FOURNIER-FACIO, N. HEUER & M. MORASCHINI (eds)
480 Stacks Project Expository Collection (SPEC), P. BELMANS, W. HO & A.J. DE JONG (eds)
481 Surveys in combinatorics 2022, A. NIXON & S. PRENDIVILLE (eds)
482 The logical approach to automatic sequences, J. SHALLIT
483 Rectifiability: a survey, P. MATTILA
484 Discrete quantum walks on graphs and digraphs, C. GODSIL & H. ZHAN
485 The Calabi problem for Fano threefolds, C. ARAUJO *et al*
486 Modern trends in algebra and representation theory, D. JORDAN, N. MAZZA & S. SCHROLL (eds)
487 Algebraic combinatorics and the Monster group, A.A. IVANOV (ed)
488 Maurer–Cartan methods in deformation theory, V. DOTSENKO, S. SHADRIN & B. VALLETTE
489 Higher dimensional algebraic geometry, C. HACON & C. XU (eds)
490 C^{∞}-algebraic geometry with corners, K. FRANCIS-STAITE & D. JOYCE
491 Groups and graphs, designs and dynamics, R.A. BAILEY, P.J. CAMERON & Y. WU (eds)
492 Homotopy theory of enriched Mackey functors, N. JOHNSON & D. YAU
493 Surveys in combinatorics 2024, F. FISCHER & R. JOHNSON (eds)
494 K-theory and representation theory, R. PLYMEN & M.H. ŞENGÜN (eds)
495 Polygraphs: from rewriting to higher categories, D. ARA *et al*
496 Groups St Andrews 2022 in Newcastle, C.M. CAMPBELL *et al* (eds)
497 Proof complexity generators, J. KRAJÍČEK
498 Polynomial functors, N. NIU & D.I. SPIVAK
499 Moduli, motives and bundles, P.L. DEL ÁNGEL R., F. NEUMANN & A.H.W. SCHMITT (eds)
500 The Toda lattice and universality for the computation of the eigenvalues of a random matrix, P. DEIFT, G. DUBACH, C. TOMEI & T. TROGDON
501 Topics in number theory, J. MORALES
502 The geometry of jet bundles, D. J. SAUNDERS
503 Multifunctorial equivariant algebraic K-theory, D. YAU
504 Spectra of signed graphs, Z. STANIĆ

Contents

Preface

The 31st British Combinatorial Conference takes place at Cardiff University in Wales, from Monday 6th to Friday 10th of July 2026. The British Combinatorial Committee invited nine distinguished combinatorialists to give survey lectures in their areas of expertise. This volume contains the survey articles on which eight of these lectures are based. We hope this collection of surveys will be a valuable resource to bring researchers and graduate students up-to-date with the latest developments in some key areas of combinatorics and graph theory.

In compiling this volume, we are grateful to the authors for preparing their survey articles accurately and professionally, and to the referees for their careful reading of the manuscripts and constructive feedback. We would also like to thank Holly Paveling and Anna Scriven at Cambridge University Press for their assistance through different stages of preparing and editing this volume. Finally, we thank the editors of previous volumes of *Surveys in Combinatorics* and the British Combinatorial Committee for helpful guidance.

The 31st British Combinatorial Conference is organised in partnership with the Heilbronn Institute for Mathematical Research through the UKRI/EPSRC Additional Funding Programme for Mathematical Sciences, the Institute of Combinatorics and its Applications, the Combinatorics Consortium, the London Mathematical Society, Cardiff University, and the British Combinatorial Committee. We thank each of these organisations for their involvement and support.

Andrei Gagarin
Roger Behrend
Matthew Lettington
Iskander Aliev
School of Mathematics, Cardiff University
January 2026

Phase Transitions in Random Planar Graphs

Mihyun Kang

Abstract

We discuss classical results on phase transitions in Erdős–Rényi random graphs and related recent results on random planar graphs and random graphs embeddable on other orientable surfaces of constant genus. The main focus is on how imposing the planarity constraint (or more generally, the genus constraint) affects the global and local structure of random graphs, such as their component structure, local limits, and maximum degree.

1 Introduction

Erdős and Rényi laid the foundations for the theory of random graphs in their seminal work [43, 44]. Since then, various random graph models, at the intersection of different disciplines, including combinatorics, discrete probability, computer science, statistical physics, and network sciences, have been introduced and extensively studied. By now there are several monographs devoted solely to random graphs, e.g., [24, 47, 57, 101, 102].

In the context of random graphs, phase transitions were first discussed by Erdős and Rényi, who discovered in [44] that the random graph undergoes a drastic change in the size and structure of its components. Typically there are many 'small' components – each of which contains at most one cycle – if the average degree is smaller than one, while there is a 'giant' component (that is, a component whose size is much larger than the size of any other component) – which contains at least two cycles – if the average degree is larger than one. Erdős and Rényi showed in [44] that in the 'evolution' of a uniform random graph with n vertices and $N(n)$ edges, the typical size of the largest component changes from logarithmic order $\Theta(\ln n)$ to sublinear order $\Theta(n^{2/3})$ and then to linear order $\Theta(n)$, when the average degree increases from a constant smaller than one, to a constant equal to one, and then to a constant larger than one. Erdős and Rényi wrote in [44]:

> *'This double "jump" of the size of the largest component when $N(n)/n$ passes the value 1/2 is one of the most striking facts concerning random graphs'.*

This suggests the existence of a *'discontinuous' phase transition* in the evolution of the random graph when the average degree passes through the critical value one. Bollobás [23] and Łuczak [73] however showed that the Erdős–Rényi random graph undergoes not a discontinuous, but a *continuous* phase transition.

These fascinating results have been generalised to various other random graph models and strengthened in several ways. One random graph model that has received considerable attention in recent years is a random planar graph, whose study was pioneered by McDiarmid, Steger, and Welsh [77]. Imposing the planarity constraint on random graphs typically makes their investigation more challenging, but many exciting results have been obtained. It is well known that in the very sparse regime when the average degree is smaller than one, a random planar graph typically behaves similarly to the Erdős–Rényi random graph, because typically a very sparse Erdős–Rényi random graph is planar. However, different properties have been found to emerge when the average degree is larger than one. For example, Łuczak and Kang showed in [59] that surprisingly there are *two critical periods* in the evolution of a random planar graph, in stark contrast to the Erdős–Rényi random graph, which features only a *single* critical period. In a random planar graph, the *first critical point* is when the average degree equals *one*, where the unique giant component emerges, analogously to the Erdős–Rényi random graph, but the asymptotic size of the giant component in a random planar graph is roughly half the asymptotic size of the giant component in the Erdős–Rényi random graph. The *second critical point* that a random planar graph exhibits is when the average degree equals *two*, where the giant component contains almost all vertices. Such similarities and differences observed in phase transitions in the Erdős–Rényi random graph and a random planar graph form the basis of this paper.

Starting from the work by McDiarmid [75], many results on random planar graphs have now been extended to random graphs embeddable on other orientable surfaces of constant genus. It is well known that random graphs on orientable surfaces of constant genus enjoy *universal properties*. Such properties, especially in terms of component structures, local weak limits, and the maximum degree, will be discussed in Sections 3, 4, and 5, respectively.

2 Preliminaries

2.1 Erdős–Rényi random graphs

The most popular random graph model is arguably the binomial random graph $G_{n,p}$, i.e., a graph with vertex set $[n] := \{1, \dots, n\}$, where each unordered pair of distinct vertices is joined by an edge independently with probability $p \in [0,1]$. Another well-studied model is the uniform random graph introduced by Erdős–Rényi in [43]. To be more precise, let $\mathcal{G}(n,m)$ denote the set of all (vertex-labelled simple undirected) graphs with vertex set $[n]$ and with exactly m edges and let $G(n,m)$ be a graph chosen uniformly at random from $\mathcal{G}(n,m)$. Note that $G_{n,p}$ is essentially equivalent to $G(n,m)$ when $\binom{n}{2}p \sim m$ (see e.g., [47, Section 1.1] or [57, Section 1.4]).

Throughout the paper we let $G(n,m) \in_u \mathcal{G}(n,m)$ denote a graph chosen uniformly at random from $\mathcal{G}(n,m)$ and call it the Erdős–Rényi random graph $G(n,m)$.

2.2 Random planar graphs

The extensive study of planar graphs, planar maps (i.e., graphs that are *embedded* in the plane), and related families of graphs goes back to Tutte [100], leading to a plethora of impressive results, including results about asymptotic enumeration and bijections [9, 10, 14, 15, 26, 27, 29, 38, 41, 58, 72, 94, 99, 100], scaling and local limits [1, 4, 63, 66, 67, 68, 69, 70, 71, 78, 97, 98], random sampling and generation [13, 16, 17, 18, 48, 89, 95], and structural properties [12, 19, 28, 32, 33, 34, 36, 39, 40, 42, 46, 49, 50, 51, 52, 53, 59, 60, 61, 62, 63, 64, 65, 75, 76, 77, 80, 81, 82, 83, 84, 85, 86]. In particular, Noy [80] provided an excellent survey on the enumeration of planar graphs and asymptotic properties of random planar graphs. Many of these results are largely based on generating functions associated with decompositions of graphs into parts with appropriate connectivity properties and the methods from analytic combinatorics, whose general framework and theory can be found in [37] and [45].

In view of the main themes of the paper, which are the component structure, local limits, and the maximum degree of a random graph embeddable on an orientable surface of constant genus, the most important results concerning the component structure are obtained in [27, 46, 53, 59, 61, 64, 75, 77], the local limits in [63, 97], and the maximum degree in [34, 36, 39, 40, 60, 76, 86].

These intensive studies revealed rich and complex behaviour of random graphs on orientable surfaces. But, most of them dealt with random graph models *without a specification of the number of edges*, therefore not providing evolutionary viewpoints, or with random graph models in *dense regimes* corresponding to *late stable stages* in the evolution of random graphs.

The primary topic of this paper is *phase transition phenomena* in *sparse* random planar graphs (or more generally, *sparse* random graphs on orientable surfaces), which are unveiled only if the number of edges is taken into account in the model and if it is not too large, so that *early unstable* stages in the evolution of random graphs are properly detected, especially in the light of the phase transition phenomena discovered by Erdős and Rényi in [44].

Given $g \in \mathbf{N} \cup \{0\}$, let $\mathbb{S}_g$ denote the orientable surface of genus g. A graph H is called *embeddable* on $\mathbb{S}_g$ if H can be drawn on $\mathbb{S}_g$ without crossing edges. A graph embeddable on $\mathbb{S}_g$ will simply be called a graph on $\mathbb{S}_g$. We let $\mathcal{S}_g(n,m)$ denote the set of all (vertex-labelled simple undirected) graphs on $\mathbb{S}_g$ with vertex set $[n]$ and with exactly m edges. Note that the case when the genus $g = 0$ corresponds to planar graphs and

$$\mathcal{S}_0(n,m) \subset \cdots \subset \mathcal{S}_g(n,m) \subset \mathcal{S}_{g+1}(n,m) \subset \cdots \subset \mathcal{G}(n,m).$$

Throughout the paper we assume $g \in \mathbf{N} \cup \{0\}$ is a fixed constant. We let $S_g(n,m) \in_u \mathcal{S}_g(n,m)$ denote a graph chosen uniformly at random from $\mathcal{S}_g(n,m)$ and call it the random graph $S_g(n,m)$ on $\mathbb{S}_g$.

2.3 Local weak limit

The concept of *local weak convergence* was formally introduced by Benjamini and Schramm [11] and independently by Aldous and Steele [3]. For an overview of the subject, see the book of van der Hofstad [102, Chapter 2].

A (possibly infinite) graph H is *locally finite* if every vertex of H has finite degree. A *rooted graph* is a pair (H,r), where H is a locally finite graph and r is a vertex of H. Two rooted graphs (H_1,r_1) and (H_2,r_2) are *isomorphic*, which is denoted by

$$(H_1,r_1) \cong (H_2,r_2),$$

if there exists an isomorphism from H_1 to H_2 that takes r_1 to r_2. Given $\ell \in \mathbf{N}$ and a rooted graph (H,r), the ball $B_\ell(H,r)$ of radius ℓ in H with centre r is the subgraph of H induced on the vertices with distance at most ℓ from r and it is considered as a rooted graph with root r.

Given two rooted graphs (H_1, r_1) and (H_2, r_2), we define the distance between them by

$$d\Big((H_1, r_1), (H_2, r_2)\Big) \ := \ 2^{-k},$$

where k is the supremum of all $\ell \in \mathbf{N}$ satisfying

$$B_\ell(H_1, r_1) \cong B_\ell(H_2, r_2).$$

It is not difficult to verify that d is a metric on the space $\mathcal{G}$ of isomorphism classes of *connected* rooted graphs. Using the fact that the law of a random connected rooted graph is a probability measure on $\mathcal{G}$, we can define a 'limiting graph' for a sequence $\big((G_n, r_n)\big)_{n\in\mathbf{N}}$ of random connected rooted graphs (G_n, r_n) as follows. A random connected rooted graph (G_*, r_*) is the (Benjamini–Schramm) *local weak limit* of (G_n, r_n), which is denoted by

$$(G_n, r_n) \ \xrightarrow{d} \ (G_*, r_*),$$

if the law of (G_n, r_n) converges weakly to the law of (G_*, r_*). This is equivalent to

$$\mathbb{P}\left[B_\ell\left(G_n, r_n\right) \cong B_\ell(H, r_H)\right] \quad \to \quad \mathbb{P}\left[B_\ell(G_*, r_*) \cong B_\ell(H, r_H)\right] \quad \text{as} \quad n \to \infty$$

for each $\ell \in \mathbf{N}$ and each fixed rooted graph (H, r_H).

Roughly speaking, the local weak limit of (G_n, r_n) depends only on the local structure of G_n around the root r_n. For not necessarily connected rooted graphs (G_n, r_n) we define the local weak limit as the local weak limit of the connected rooted graphs (G_{r_n}, r_n) where G_{r_n} denotes the component of G_n that contains the root r_n.

Given a constant $a \in [0, 1]$ and two random rooted graphs (H_1, r_1) and (H_2, r_2), we denote by

$$a \cdot (H_1, r_1) + (1 - a) \cdot (H_2, r_2)$$

the random rooted graph (G, r) that satisfies

$$\mathbb{P}\left[(G, r) \cong (H, r_H)\right] \ = \ a \cdot \mathbb{P}\left[(H_1, r_1) \cong (H, r_H)\right] + (1 - a) \cdot \mathbb{P}\left[(H_2, r_2) \cong (H, r_H)\right]$$

for each fixed rooted graph (H, r_H).

2.4 Notations and notions

Throughout the paper all asymptotics are taken as $n \to \infty$. We say a property holds *with high probability*, or *whp* for short, if it holds with probability tending to one as $n \to \infty$. In addition to the standard Landau notation we use the following asymptotic notions. Given a sequence $(X_n)_{n\in\mathbf{N}}$ of random variables and a function $f\colon \mathbf{N} \to \mathbf{R}_{\geq 0}$, we say that

- $X_n = O_p(f(n))$, if for every $\delta > 0$, there exist $C_0 = C_0(\delta) > 0$ and $N_\delta \in \mathbf{N}$ such that with probability at least $1 - \delta$, we have $|X_n| \leq C_0 f(n)$ for every $n \geq N_\delta$;
- $X_n = \Theta_p(f(n))$, if for every $\delta > 0$, there exist $C_1 = C_1(\delta) > 0, C_2 = C_2(\delta) > 0$, and $N_\delta \in \mathbf{N}$ such that with probability at least $1-\delta$, we have $C_1 f(n) \leq |X_n| \leq C_2 f(n)$ for every $n \geq N_\delta$.

Given a graph H with vertex set $[n]$ and with m edges, we call

$$\frac{2m}{n} = \frac{1}{n}\sum_{v\in[n]} d(v)$$

the *average degree* of H. We denote by $\Delta(H)$ the *maximum degree* of H. A *component* of H is a maximal connected subgraph of H. Given a component $C(H)$ of H, its *size* $|C(H)|$ is the number of vertices in $C(H)$. For $i \in \mathbf{N}$, let $L_i(H)$ denote the i-th largest component of H. A graph $R(H) := H \setminus L_1(H)$, which is a subgraph of H obtained by deleting $L_1(H)$, is called the *fragment* of H. A component of H is called a *tree* if it contains no cycle, *unicyclic* if it contains exactly one cycle, and *complex* if it contains at least two cycles, respectively.

A *bare* path is a path whose internal vertices all have degree exactly two. Given a graph H, the *2-core* of H is the maximal subgraph of H whose minimum degree is at least two, and the *kernel* of H is a multigraph obtained from the 2-core of H by replacing each bare path by an edge.

The genus $g = g(H)$ of a graph H is defined as the minimum number of handles that must be attached to a sphere in order to be able to embed the graph H without any crossing edges.

3 Component structure

Among many features of the Erdős–Rényi random graph $G(n,m)$, the most prominent one is the *phase transition phenomenon* in the size and structure of its components, when the average degree $2m/n$ goes through the critical value one. At this point the graph typically changes in structure from being *planar* with relatively small components (of logarithmic size) to being *non-planar* with a unique giant component (of linear size). This result became a benchmark in the theory of random graphs.

In this section we will first overview the most important results in this regard. We then continue with corresponding results on the random graph $S_g(n,m)$ on $\mathbb{S}_g$.

3.1 Component structure in the Erdős–Rényi random graph

We begin with the classic results on the phase transition in the component structure of $G(n,m)$ that Erdős and Rényi discovered in [44]. (Some assertions in [44] relied on flawed arguments, but they were subsequently proved and improved by other authors – for a summary of the history, see e.g., [55, the two paragraphs before Theorem 8].) The following results are presented in a modified form to use unified notation throughout the paper for easy comparison.

Theorem 3.1 *Let $G = G(n,m) \in_u \mathcal{G}(n,m)$. Assume $2m/n$ tends to a constant $c \in [0,\infty)$.*

(a) If $c < 1$ (in the subcritical phase*), then whp every component of G is either a tree or unicyclic. In addition, whp G is planar and*

$$|L_1(G)| = O(\ln n).$$

(b) *If* $c = 1$ *(at the* critical point*), then*

$$|L_1(G)| = \Theta_p(n^{2/3}).$$

The probability that G is planar is bounded away from 0 and 1.

(c) *If* $c > 1$ *(in the* supercritical phase*), then whp G is not planar and*

$$|L_1(G)| = \Theta(n).$$

In short, the Erdős–Rényi random graph $G(n, m)$ undergoes a phase transition when the average degree is equal to one. This is indeed not surprising, in view of the *universality principle* in percolation theory, which says that the critical threshold for the emergence of the giant cluster is determined by the local structure of the graph, for example, the expected degree of a random vertex being equal to one.

In fact, Theorem 3.1 (b) holds only when $2m/n$ tends to the critical value one 'fast enough'. Bollobás [23] and Łuczak [73] strengthened Theorem 3.1 (b) when the average degree $2m/n$ is 'close' to 1. One of the main difficulties is to quantify how fast $2m/n$ converges to 1, in order to be able to capture meaningful changes in the size and structure of components. Bollobás [23] and Łuczak [73] found appropriate parametrisations of the average degree, determining the correct width of the critical window to be $O(n^{2/3})$. In addition to the sizes of components, they also studied whether these components are *trees*, *unicyclic*, or *complex*. Their results are further strengthened, e.g., by Janson, Knuth, Łuczak, and Pittel [55]. Revealing only the tip of the iceberg, we oversimplify their results as follows.

Theorem 3.2 *Let* $G = G(n, m) \in_u \mathcal{G}(n, m)$ *and* $R(G) := G \setminus L_1(G)$*. Assume*

$$2m/n = 1 + \epsilon \quad \textit{for} \quad \epsilon = \epsilon(n) = o(1).$$

(a) *If* $\epsilon n^{1/3} \to -\infty$ *(in the* weakly subcritical phase*), then whp every component of G is either a tree or unicyclic (and thus whp G is planar). For every* $i \in \mathbf{N}$*, whp* $L_i(G)$ *is a tree and*

$$|L_i(G)| = (1 + o(1))2|\epsilon|^{-2} \ln(|\epsilon|^3 n).$$

(b) *If* $\epsilon n^{1/3} \to a$ *for a constant* $a \in \mathbf{R}$ *(in the* critical phase*), then the probability that G has complex components is bounded away from 0 and 1. In addition, the probability that G is planar is bounded away from 0 and 1. For every* $i \in \mathbf{N}$*,*

$$|L_i(G)| = \Theta_p(n^{2/3}).$$

(c) *If* $\epsilon n^{1/3} \to +\infty$ *(in the* weakly supercritical phase*), then whp G is not planar,* $L_1(G)$ *is complex, and*

$$|L_1(G)| = (1 + o(1))2\epsilon n.$$

In addition, whp every component of $R(G)$ *is either a tree or unicyclic. For* $i \geq 2$*, whp* $L_i(G)$ *is a tree and*

$$|L_i(G)| = (1 + o(1))2\epsilon^{-2} \ln(\epsilon^3 n).$$

The behaviour of the *critical* Erdős–Rényi random graph $G(n,m)$ is more complex than as stated in Theorem 3.2 (b). For example, Aldous [2] showed that the ordered scaled sizes of the largest components converge to a multiplicative coalescent process, which is equivalent to ordered excursion lengths of the associated Brownian excursions.

The genus of a graph is one of the most fundamental topological characteristics of a graph, and thus it is intriguing to investigate the genus of a random graph. In Theorem 3.1 and Theorem 3.2 we discussed the planarity of the Erdős–Rényi random graph. In the *critical phase* we have stronger results than Theorem 3.2 (b). Łuczak, Pittel, and Wierman [74] showed that there exists a function $F : \mathbf{R} \to (0,1)$, $a \mapsto F(a)$, such that if $m = n/2 + a\,n^{2/3}$ for a constant $a \in \mathbf{R}$, then

$$\lim_{n\to\infty} \mathbb{P}[G(n,m) \text{ is planar}] \;=\; F(a), \tag{3.1}$$

where $\lim_{a\to-\infty} F(a) = 1$ and $\lim_{a\to+\infty} F(a) = 0$. Noy, Ravelomanana, and Rué [82] derived the analytic expression for $F(a)$.

The genus of the Erdős–Rényi random graph $G(n,m)$, or more precisely, the genus of $G_{n,p}$, was first studied by Archdeacon and Grable [6], who showed among other results that whp the genus of $G_{n,p}$ is $(1+o(1))pn^2/12$, if $p^2(1-p^2) \ge 8(\ln n)^4/n$. As noted in [6], results for the genus of $G_{n,p}$ can be transferred into analogous results for the genus of $G = G(n,m)$. Rödl and Thomas considered dense cases when m is superlinear in n and showed in [92] that the typical genus of G is superlinear in n. More precisely, whp $g(G) = (1+o(1))\,m/6$ when $m = \Theta(n^2)$ and $g(G) = (1+o(1))\,k\,m/(2k+4)$ when $m = \omega(n^{(k+2)/(k+1)})$ but $m = o(n^{(k+1)/k})$ for $k \in \mathbb{N}$. Complementing these results, Dowden, Kang, and Krivelevich [35] determined the genus of G in sparse regimes. For example, if $2m/n = 1+\epsilon$ for $\epsilon = \epsilon(n) = o(1)$ and $\epsilon n^{1/3} \to +\infty$ (in the *weakly supercritical phase*), then whp

$$g(G) \;=\; (1+o(1))\epsilon^3 n/3, \tag{3.2}$$

which is roughly one quarter of the asymptotic size of the *kernel* of G.

What else can we say about the genus $g(G)$ of $G = G(n,m)$ near the critical point? We might perhaps be able to guess the behaviour of $g(G)$ from what we already know about the genus in the weakly subcritical and supercritical phases. To this end, assume $2m/n = 1+\epsilon$ for $\epsilon = \epsilon(n) = o(1)$. Note that Theorem 3.2 (a) says that if $\epsilon n^{1/3} \to -\infty$ (in the *weakly subcritical phase*), then whp $g(G) = 0$. On the other hand, (3.2) says that if $\epsilon n^{1/3} \to +\infty$ (in the *weakly supercritical phase*), then whp $g(G) = \Theta(\epsilon^3 n) = \omega(1)$. Summing up, $g(G)$ changes from zero to $\omega(1)$, as $\epsilon n^{1/3}$ increases from $-\infty$ to $+\infty$ through the critical phase. A natural question is whether we can describe this change more precisely. In view of the planarity result (3.1) by Łuczak, Pittel, and Wierman [74], we are interested in the analogous result for a fixed constant $g \in \mathbf{N}$ and are curious whether there exists a function $F_g : \mathbf{R} \to (0,1)$, $a \mapsto F_g(a)$, such that if $m = n/2 + a n^{2/3}$ for a constant $a \in \mathbf{R}$, then

$$\lim_{n\to\infty} \mathbb{P}[G(n,m) \text{ has genus } g] \;=\; F_g(a).$$

If such a function F_g exists, it would be interesting to determine the behaviour of F_g. For example, is it continuous and increasing?

3.2 Component structure in a random planar graph

Early work on random planar graphs, or more generally, random graphs on surfaces, has been restricted to random graph models *without a specification of the number of edges*, but more recent work includes the study of random graphs on surfaces with given numbers of vertices and edges. Below we discuss the most important results about the component structure in random graph models both with and without a specification of the number of edges.

Given $g \in \mathbf{N} \cup \{0\}$, let $\mathcal{S}_g(n)$ denote the set of all (vertex-labelled simple undirected) graphs on $\mathbb{S}_g$ with vertex set $[n]$ and let $S_g(n) \in_u \mathcal{S}_g(n)$ be a random graph which is chosen uniformly at random from $\mathcal{S}_g(n)$. Note that Euler's formula implies that if $m > 3n - 6 + 6g$, then $\mathcal{S}_g(n,m) = \emptyset$. Thus we have

$$\mathcal{S}_g(n) \;=\; \bigcup_{0 \le m \le 3n-6+6g} \mathcal{S}_g(n,m).$$

When $2m/n$ tends to a constant $c \in [0,1)$, it follows from Theorem 3.1 (a) that whp $G(n,m)$ is planar, implying that

$$\frac{|\mathcal{S}_0(n,m)|}{|\mathcal{G}(n,m)|} \;\to\; 1 \quad \text{if} \quad 2m/n \;\to\; c \in [0,1)$$

and thus any property that holds whp in $G(n,m)$ will also hold whp in the random graph $S_g(n,m)$ on $\mathbb{S}_g$ for every $g \in \mathbf{N} \cup \{0\}$. As a consequence, we have, e.g., that whp

$$|L_1(S_g(n,m))| \;=\; O(\ln n) \quad \text{if} \quad 2m/n \;\to\; c \in [0,1).$$

The asymptotic behaviour of $|\mathcal{S}_0(n)|$ and of $|\mathcal{S}_0(n,m)|$ when $2m/n = c$ for a constant $c \in (2,6)$ were determined by Giménez and Noy [53], using a sophisticated singularity analysis of generating functions, based on prior work by Bender, Gao, and Wormald [10]. This work was extended by Bender and Gao [9] and independently by Chapuy, Fusy, Giménez, Mohar, and Noy [27] to arbitrary constant $g \in \mathbf{N}$. Among other interesting results, Chapuy, Fusy, Giménez, Mohar, and Noy showed in [27] that whp the size of the largest component in $S_g(n)$ is $n - O(1)$ and that the number of edges in $S_g(n)$ is asymptotically normally distributed with mean αn and variance βn, where $\alpha \approx 2.2133$ and $\beta \approx 0.4303$ are constants (which are independent of g). Their results imply that the average degree of $S_g(n)$ is very close to 4.42 and that $S_g(n)$ and $S_g(n,m)$ with $2m/n = c$ for a constant $c \in (2,6)$ behave like a *dense* Erdős–Rényi random graph $G(n,m)$ where the giant component containing nearly all but a bounded number of vertices appears only when $2m/n$ is close to the connectivity threshold, i.e., when $m \sim n \ln n$.

Thus, when studying the random graph $S_g(n,m)$ on $\mathbb{S}_g$ in the light of the phase transition phenomenon in the size and structure of components in the Erdős–Rényi random graph $G(n,m)$, the most interesting range to focus on is the *sparse* regime when the average degree $2m/n$ tends to a constant $c \in [1,2]$. Łuczak and Kang [59] showed that such a sparse random planar graph undergoes *two critical phases*, in comparison to the single critical phase that the Erdős–Rényi random graph undergoes. In the random planar graph $S_0 = S_0(n,m)$, the first critical point $c = 1$ is when the unique giant component $L_1(S_0)$ emerges, and the second critical point $c = 2$ is when $L_1(S_0)$ covers $n - \Theta(n^{3/5})$ vertices. In the *intercritical phase* when $c \in (1,2)$, whp $|L_1(S_0)| \sim (c-1)n$. This work has been extended by Kang, Moßhammer, and Sprüssel [64] to $S_g(n,m)$ for any arbitrary constant $g \in \mathbf{N} \cup \{0\}$. Below we discuss these results more precisely.

Analogously to the Erdős–Rényi random graph $G(n,m)$ (see Theorem 3.2), let us look closer into the *first critical phase* of the random graph $S_g(n,m)$ on $\mathbb{S}_g$. Łuczak and Kang [59] (for $g = 0$) and Kang, Moßhammer, and Sprüssel [64] (for arbitrary constant $g \in \mathbf{N} \cup \{0\}$) determined the correct width of the first critical window to be $O(n^{2/3})$ and investigated the size and structure of the largest components. Among other results, they proved the following.

Theorem 3.3 *Let $g \in \mathbf{N} \cup \{0\}$. Let $S_g = S_g(n,m) \in_u \mathcal{S}_g(n,m)$ and $R(S_g) := S_g \setminus L_1(S_g)$. Assume*

$$2m/n \;=\; 1+\epsilon \quad \textit{for} \quad \epsilon = \epsilon(n) = o(1).$$

(a) If $\epsilon n^{1/3} \to -\infty$ (in the first weakly subcritical phase*), then for every $i \in \mathbf{N}$, whp $L_i(S_g)$ is a tree and*

$$|L_i(S_g)| \;=\; (1+o(1))2|\epsilon|^{-2}\ln(|\epsilon|^3 n).$$

(b) If $\epsilon n^{1/3} \to a$ for a constant $a \in \mathbf{R}$ (in the first critical phase*), then the probability that S_g contains a complex component is bounded away from 0 and 1. For every $i \in \mathbf{N}$,*

$$|L_i(S_g)| \;=\; \Theta_p(n^{2/3}).$$

(c) If $\epsilon n^{1/3} \to +\infty$ (in the first weakly supercritical phase*), then whp $L_1(S_g)$ has genus g and it is complex. Furthermore,*

$$|L_1(S_g)| \;=\; (1+o(1))\epsilon n + O_p(n^{2/3}).$$

In addition, whp $R(S_g)$ is planar and it has $O_p(1)$ many complex components, each of which has size $O_p(n^{2/3})$. For $i \geq 2$,

$$|L_i(S_g)| \;=\; \Theta_p(n^{2/3}).$$

Theorem 3.3 (a) and (b) follow from Theorem 3.2 (a) and (b), respectively. But Theorem 3.3 (c) requires quite involved counting arguments based on the *core-kernel approach*, which will be discussed in Section 3.4.

It is worth mentioning that in the first weakly supercritical phase in $S_g(n,m)$, whp the genus of the giant component is g, while the genus of the fragment is zero, in other words, typically the giant component contains the full information on the genus of the whole graph.

From the first parts of Theorem 3.3 (c) and Theorem 3.2 (c) we have that the asymptotic size of the giant component in $S_g(n,m)$ is about ϵn, while the size of the giant component in $G(n,m)$ is about $2\epsilon n$.

The second parts of Theorem 3.3 (c) and Theorem 3.2 (c) tell us that whp the fragment in $S_g(n,m)$ contains (a bounded number of) *complex* components, but whp the fragment in $G(n, m)$ does not contain a complex component. In addition, for $i \geq 2$, the size of the i-th largest component in $S_g(n, m)$ is $\Theta_p(n^{2/3})$, while the counterpart in $G(n, m)$ is of order $\Theta(\epsilon^{-2} \ln(\epsilon^3 n)) = o(n^{2/3})$.

Summing up, Theorem 3.3 (c) clearly shows that in the first weakly supercritical phase, (the size and structure of) the largest components of $S_g(n, m)$ behave quite differently from those of $G(n, m)$. Roughly speaking, imposing the planarity or the genus constraint on a random graph makes its edges spread *more evenly* over different components, in comparison to the 'Rich-Get-Richer' principle that $G(n, m)$ enjoys. As a consequence, the asymptotic size of the giant component in $S_g(n, m)$ is only the half the asymptotic size of the giant component in $G(n, m)$, while the asymptotic size of the second largest component in $S_g(n, m)$ is much larger than the counterpart in $G(n, m)$. In addition, typically the fragment of $S_g(n, m)$ contains several components with at least two cycles, but every component in the fragment of $G(n, m)$ contains at most one cycle.

The existence of the *second critical phase* around the critical value two is a special aspect of sparse random planar graphs, or more generally, sparse random graphs on $\mathbb{S}_g$. Needless to say, such a phase does not exist in the Erdős–Rényi random graph.

Based on the phase transition results by Łuczak and Kang [59] (for $g = 0$) and Kang, Moßhammer, and Sprüssel [64] (for arbitrary constant $g \in \mathbf{N} \cup \{0\}$), the next two theorems summarise the results about the size and structure of the giant component $L_1(S_g)$ of $S_g = S_g(n, m)$ and the fragment $R(S_g) = S_g \setminus L_1(S_g)$ in the intercritical and the second critical phases.

In the *intercritical phase*, typically the rescaled asymptotic size of the giant component $L_1(S_g)$ grows *linearly* in $c - 1$ and the fragment $R(S_g)$ is planar (cf. [64, Theorem 1.6]).

Theorem 3.4 *Let $g \in \mathbf{N} \cup \{0\}$. Let $S_g = S_g(n, m) \in_u \mathcal{S}_g(n, m)$ and $R(S_g) := S_g \setminus L_1(S_g)$. Assume*

$$2m/n \ \to \ c \in (1, 2).$$

Then whp $L_1(S_g)$ has genus g and it is complex. Furthermore,

$$|L_1(S_g)| \ = \ (c - 1)n + O_p(n^{2/3}).$$

In addition, whp $R(S_g)$ is planar. For $i \geq 2$,

$$|L_i(S_g)| \ = \ \Theta_p(n^{2/3}).$$

Next we take a closer look into the *second critical phase* when the the average degree $2m/n$ tends to the critical value two. The following result (cf. [64, Theorem 1.5]) indicates that the correct width of the second critical window is $O(n^{3/5})$.

Theorem 3.5 *Let $g \in \mathbf{N} \cup \{0\}$. Let $S_g = S_g(n, m) \in_u \mathcal{S}_g(n, m)$ and $R(S_g) := S_g \setminus L_1(S_g)$. Assume*

$$2m/n \ = \ 2 + \eta \quad for \quad \eta = \eta(n) = o(1).$$

Then whp $L_1(S_g)$ has genus g and it is complex. Furthermore, whp $R(S_g)$ is planar.

(a) If $\eta n^{2/5} \to -\infty$ *(in the* second weakly subcritical phase*), then whp*

$$|R(S_g)| = (1+o(1))|\eta|\, n.$$

(b) If $\eta n^{2/5} \to a$ *for a constant* $a \in \mathbf{R}$ *(in the* second critical phase*), then whp*

$$|R(S_g)| = \Theta(n^{3/5}).$$

(c) If $\eta n^{2/5} \to +\infty$ *(in the* second weakly supercritical phase*) but* $\eta = o(\ln^{-2/3} n)$*, then whp*

$$|R(S_g)| = \Theta(\eta^{-3/2}).$$

Observe that Theorem 3.5 determines the width of the second critical window, up to the condition on $\eta = o(\ln^{-2/3} n)$. We believe that this gap (in the second weakly supercritical phase) is due to our proof techniques, and that Theorem 3.5 (c) should hold for any $\eta > 0$ satisfying $\eta = o(1)$ and $\eta n^{2/5} \to +\infty$, meaning the correct width of the second critical window is $O(n^{3/5})$.

Finally, Kang, Moßhammer, and Sprüssel [64, Theorem 1.7] discussed the number of vertices and edges in the fragment in all the ranges covered in Theorem 3.3 (b) and (c), Theorem 3.4, and Theorem 3.5.

Theorem 3.6 *Let* $g \in \mathbf{N} \cup \{0\}$*. Let* $S_g = S_g(n,m) \in_u \mathcal{S}_g(n,m)$ *and* $R = R(S_g) := S_g \setminus L_1(S_g)$*. Assume*

$$1+\epsilon \;\le\; 2m/n \;\le\; 2+\eta$$

where $\epsilon = \epsilon(n) = o(1)$ *satisfies* $\epsilon n^{1/3} \to a$ *for a constant* $a \in \mathbf{R}$ *and* $\eta = \eta(n) = o(1)$ *satisfies* $\eta n^{2/5} \to \infty$*, but* $\eta = o(\ln^{-2/3} n)$*. We denote by* n_R *and* m_R *the number of vertices and edges in* R*, respectively. Then*

$$m_R = \frac{n_R}{2} + O_p(n_R^{2/3}).$$

Theorem 3.6 tells us that ranging from the first critical phase to the second weakly supercritical phase, we have $m_R = n_R/2 + O_p(n_R^{2/3})$, meaning that the average degree $2m_R/n_R$ of the fragment R falls into the *critical phase* of the Erdős–Rényi random graph (cf. Theorem 3.2 (b)).

3.3 Comparison of component structures

Let us discuss two major differences in terms of the component structure in the Erdős–Rényi random graph $G(n,m)$ and the random graph $S_g(n,m)$ on $\mathbb{S}_g$.

The first one is the size of the largest component. As for $G = G(n,m)$, there is a standard approach to couple the component exploration process based on the Breadth-First Search algorithm with the Galton–Watson branching process, leading to the asymptotic size of the largest component of G. For example, if $2m/n$ tends to a constant $c > 1$ (in the *supercritical phase*), then whp

$$|L_1(G)| = \rho n + o(n), \tag{3.3}$$

where $\rho = \rho(c) \in (0,1)$ is the unique positive solution of the equation

$$1 - \rho = \exp(-c\rho) \tag{3.4}$$

and is equal to the survival probability of the Galton–Watson branching process with offspring distribution $\mathrm{Po}(c)$.

On the other hand, in the *intercritical phase* when $c \in (1,2)$, a typical size of the largest component of $S_g = S_g(n,m)$ is fairly simple and the asymptotic fraction of vertices in the largest component grows linearly in $c-1$. More precisely, whp

$$|L_1(S_g)| \;=\; (c-1)n + O_p(n^{2/3}). \tag{3.5}$$

It is worth noting that the rescaled limit

$$\rho \;=\; \lim_{n\to\infty} \frac{|L_1(G)|}{n}$$

obtained from (3.3) satisfies (3.4), and, if considered as a function

$$\rho : \mathbf{R}_{>1} \to (0,1), c \mapsto \rho(c),$$

it is a *concave* and increasing function in c and satisfies $\lim_{c\to 1^+} \rho(c) = 0$ and $\lim_{c\to+\infty} \rho(c) = 1$. On the other hand, the rescaled limit

$$c-1 \;=\; \lim_{n\to\infty} \frac{|L_1(S_g)|}{n}$$

obtained from (3.5) is a *linear* and increasing function in $c \in (1,2)$, and it satisfies trivially $\lim_{c\to 1^+}(c-1) = 0$ and $\lim_{c\to 2^-}(c-1) = 1$.

As we will discuss in Section 4, a coupling argument for the component exploration process in $G(n,m)$ and the Galton–Watson branching process shows that the local structure of $G(n,m)$ is closely related to the Galton–Watson tree generated by the Galton–Watson branching process. However, there is no such simple coupling approach to the component exploration process in $S_g(n,m)$. Instead, $S_g(n,m)$ features more *involved local structures* – this will be discussed in Section 4.2 (see Theorem 4.3–Theorem 4.6).

There is yet another significant difference between $G = G(n,m)$ and $S_g = S_g(n,m)$. An interesting property that G enjoys is the so-called *symmetry rule* or *duality*, in the sense that when $2m/n$ tends to a constant $c > 1$, typically the fragment $R = R(G) := G \setminus L_1(G)$ behaves like $G(n_R, m_R)$ with $2m_R/n_R = 1 + \epsilon_R$ for constant $\epsilon_R < 0$. That is, the fragment $R(G)$ behaves like the *subcritical* Erdős–Rényi random graph.

But, such a symmetry rule does *not* hold for $S_g = S_g(n,m)$. In fact, the fragment $R(S_g)$ behaves like the *critical* Erdős–Rényi random graph (throughout the ranges covered by Theorem 3.4 (b)–Theorem 3.5, as described in Theorem 3.6). Roughly speaking, the fraction of edges that could not be 'inserted' into the giant component of S_g, due to the planarity or the genus constraint, causes yet another interesting phenomenon outside the giant component, resulting in making the fragment $R(S_g)$ the *critical* Erdős–Rényi random graph.

3.4 Key proof ingredients

Most of the above results on the size and structure of components are largely based on counting arguments which stem from the fundamental work of Erdős and

Rényi [44], of Bollobás [23], and of Łuczak [73]. We derive estimates for various quantities, the most important one being the number of connected graphs with a fixed number of vertices and edges.

One of the accompanying tools is the *core-kernel approach*, which allows for fine control over the internal structure of the largest component in a *sparse* random graph. We decompose a graph into the simple part (in which every component contains at most one cycle) and the complex part (in which every component contains at least two cycles). Then we decompose the complex part into its 2-core (which is the maximal subgraph of the complex part whose minimum degree is at least two) and then into its kernel, a structure obtained from the 2-core by replacing each bare path by an edge. Conversely, each graph comprises its simple part and complex part, and the complex part can be uniquely constructed from the kernel by first subdividing the edges of the kernel – thereby obtaining the 2-core, then replacing vertices of the 2-core with rooted trees.

The reverse procedure in the core-kernel approach can be converted, for example, into a counting formula for the number of complex graphs (in which every component contains at least two cycles). Using a combinatorial variant of Laplace's method (see e.g., the part on *Laplace's method for sums* [45, Appendix B.6]), we can determine where the main contributions to the sum come from, including typical sizes of the 2-core and the kernel. For example, in the first weakly supercritical phase, i.e., when $2m/n = 1+\epsilon$ for $\epsilon = o(1)$ and $\epsilon n^{1/3} \to +\infty$, Bollobás [23] and Łuczak [73] showed that in $G = G(n,m)$, whp

$$|\mathrm{kernel}(G)| = \Theta(\epsilon^3 n) \quad \text{and} \quad |2\text{-core}(G)| = \Theta(\epsilon^2 n)$$

and Kang, Moßhammer, and Sprüssel [64] showed that in $S_g = S_g(n,m)$, whp

$$|\mathrm{kernel}(S_g)| = \Theta(\epsilon n^{1/3}) \quad \text{and} \quad |2\text{-core}(S_g)| = \Theta(\epsilon n^{2/3}).$$

These results show in particular that typically the asymptotic size of the kernel in the Erdős–Rényi random graph $G(n,m)$ is the *third power* of the asymptotic size of the kernel in the random graph $S_g(n,m)$ on $\mathbb{S}_g$, up to a multiplicative constant – yet another notable discrepancy between the two random graph models.

4 Local weak limits

Let us change our attention from *global* to *local* structures. When the average degree $2m/n$ tends to a constant $c \in [0,\infty)$, the Erdős–Rényi random graph $G(n,m)$ behaves locally like the *Galton–Watson tree* $\mathrm{GWT}(c)$ which is a rooted tree generated by the Galton–Watson branching process with offspring distribution $\mathrm{Po}(c)$. This can be formalised nicely in the language of *local weak convergence* (see Theorem 4.1). However, such a Galton–Watson tree is not always a local weak limit of the random graph $S_g(n,m)$ on $\mathbb{S}_g$ (see Theorem 4.5 and Theorem 4.6), especially after the first critical point.

4.1 Local weak limit of the Erdős–Rényi random graph

Consider the Erdős–Rényi random graph $G = G(n,m)$ as a random rooted graph (G,r) by assuming that given the realisation H of G, the root r is chosen uniformly at random from the vertices of H – we denote this by $r \in_u V(G)$. Using the concept and notation of local weak convergence in Section 2, we can describe the local structure of G.

Theorem 4.1 *Let $G = G(n,m) \in_u \mathcal{G}(n,m)$ and $r = r(n) \in_u V(G)$. Assume $2m/n$ tends to a constant $c \in [0,\infty)$. Then*

$$(G,r) \xrightarrow{d} GWT(c).$$

Another random graph model whose local structure is well understood is the random tree $T(n) \in_u \mathcal{T}(n)$ which is a graph chosen uniformly at random from the set $\mathcal{T}(n)$ of all (vertex-labelled simple undirected) connected acyclic graphs with vertex set $[n]$. Note that $\mathcal{T}(n) \subset \mathcal{S}_g(n)$ for any $g \in \mathbf{N} \cup \{0\}$.

Grimmett [54] showed that the local weak limit of $T(n)$ is the *Skeleton tree* T_∞ which is a rooted tree obtained from a rooted infinite directed path by replacing each vertex of the path by an independent critical Galton–Watson tree GWT(1). Alternatively, the Skeleton tree T_∞ can be obtained by connecting independent critical Galton–Watson trees $(T_1,r_1),(T_2,r_2),(T_3,r_3),\ldots$ via new edges $r_1r_2, r_2r_3,\ldots$ and by considering r_1 as the root of T_∞.

Theorem 4.2 *Let $T = T(n) \in_u \mathcal{T}(n)$ and $r = r(n) \in_u V(T)$. Then*

$$(T,r) \xrightarrow{d} T_\infty.$$

4.2 Local weak limits of random planar graphs

A natural generalisation of the random tree $T(n)$ is the random *planar* graph $S_0(n)$ (i.e., the random graph on $\mathbb{S}_0$ with vertex set $[n]$, but *without a specification of the number of edges*). Stufler [97] determined the local limit of $S_0(n)$, using big-jump phenomena of random walks and hidden branching processes. He established a novel uniform infinite planar graph as the local limit of $S_0(n)$.

In the light of the local weak limit of the Erdős–Rényi random graph $G(n,m)$ (see Theorem 4.1), Missethan and Kang [63] investigated the local weak limit of $S_g(n,m)$ in the sparse phase when the average degree $2m/n$ is smaller than two. They distinguished four cases depending on how large the average degree $2m/n$ is.

The first phase is when $m \leq n/2+O(n^{2/3})$, where $S_g(n,m)$ behaves very similarly to $G(n,m)$ and has the Galton–Watson tree GWT(c) as its local weak limit.

Theorem 4.3 *Let $g \in \mathbf{N} \cup \{0\}$. Let $S_g = S_g(n,m) \in_u \mathcal{S}_g(n,m)$ and $r = r(n) \in_u V(S_g)$. Assume $m = m(n) \leq n/2 + O(n^{2/3})$ such that $2m/n$ tends to a constant $c \in [0,1]$. Then*

$$(S_g,r) \xrightarrow{d} GWT(c).$$

Next we consider the case when $m \geq n/2 + \omega(n^{2/3})$, but we are still in the first weakly supercritical phase.

Theorem 4.4 *Let $g \in \mathbf{N} \cup \{0\}$. Let $S_g = S_g(n,m) \in_u \mathcal{S}_g(n,m)$ and $r = r(n) \in_u V(S_g)$. Assume $2m/n = 1 + \epsilon$ for $\epsilon = \epsilon(n) = o(1)$ and $\epsilon n^{1/3} \to +\infty$ (in the* first weakly supercritical phase*). Then*

$$(S_g,r) \xrightarrow{d} GWT(1).$$

Next, we consider the intercritical phase when the average degree $2m/n$ tends to a constant $c \in (1,2)$. Then the local weak limit of $S_g(n,m)$ is a 'mixture' of the *critical* Galton–Watson tree GWT(1) and the Skeleton tree T_∞.

Theorem 4.5 *Let $g \in \mathbf{N} \cup \{0\}$. Let $S_g = S_g(n,m) \in_u \mathcal{S}_g(n,m)$ and $r = r(n) \in_u V(S_g)$. Assume $2m/n$ tends to a constant $c \in (1,2)$ (in the* intercritical phase*). Then*

$$(S_g, r) \xrightarrow{d} (2-c) \cdot GWT(1) + (c-1) \cdot T_\infty.$$

Finally, we consider the case that the average degree $2m/n$ tends to the *second critical point.*

Theorem 4.6 *Let $g \in \mathbf{N} \cup \{0\}$. Let $S_g = S_g(n,m) \in_u \mathcal{S}_g(n,m)$ and $r = r(n) \in_u V(S_g)$. Assume $m = m(n)$ is such that $m = n + o(n)$ and $m \leq n + o(n(\log n)^{-2/3})$. Then*

$$(S_g, r) \xrightarrow{d} T_\infty.$$

4.3 Comparison of local weak limits

We saw in Theorem 4.1 that when the average degree $2m/n$ converges to a constant $c \in [0,\infty)$, the local weak limit of the Erdős–Rényi random graph $G(n,m)$ is the Galton–Watson tree GWT(c). Observe that the 'type' of the local weak limit of $G(n,m)$ is the same independent of the value of c (it is always a Galton–Watson tree).

In contrast, Theorem 4.3–Theorem 4.6 show that when $2m/n \to c \in [0,2]$, even the 'type' of the local weak limit of the random graph $S_g = S_g(n,m)$ on $\mathbb{S}_g$ changes *smoothly* from the Galton–Watson tree GWT(c) to the Skeleton tree T_∞ through a *'linear combination'* of the *critical* Galton–Watson tree GWT(1) and the Skeleton tree T_∞, as c increases from zero to two. Summing up, the local weak limit of S_g undergoes a *smooth* phase transition from GWT(c) for $c \in [0,1]$ to $(2-c) \cdot \text{GWT}(1) + (c-1) \cdot T_\infty$ for $c \in (1,2)$, and then to T_∞ for $c = 2$.

4.4 Main proof ideas for the intercritical phase

For simplicity let us discuss the most interesting and challenging *intercritical phase*, i.e., when the average degree is between one and two, and sketch the main ideas of the proof of Theorem 4.5.

The local structure of the random graph $S_g = S_g(n,m)$ on $\mathbb{S}_g$ is essentially determined by the *global, local, and internal* structure of its components.

First we observe a simple consequence of the *global* component structure of S_g (see Theorem 3.4) – typically S_g consists of the unique giant component $L_1(S_g)$ and the fragment $R(S_g) = S_g \setminus L_1(S_g)$ that satisfy

$$|L_1(S_g)| \sim (c-1)n \quad \text{and} \quad |R(S_g)| \sim (2-c)n.$$

Because the root $r = r(n) \in_u V(S_g)$ is chosen uniformly at random from $V(S_g)$, with asymptotic probability $c-1$ (which is equal to the asymptotic fraction of vertices in the giant component $L_1(S_g)$), the root r will be contained in $L_1(S_g)$, but with asymptotic probability $2 - c$ (which is equal to the asymptotic fraction of vertices in the fragment $R(S_g)$), the root r will be part of $R(S_g)$.

Next we use the *global and local* structure of the fragment $R = R(S_g)$. It follows from Theorem 3.6 that typically the fragment R behaves like the *critical* Erdős–Rényi random graph. Thus, conditioned on the event that the root r is contained in R (which happens with asymptotic probability $2 - c$), the local weak limit of (R, r) is the critical Galton–Watson tree GWT(1). In short, with asymptotic probability $2 - c$, we have $r \in R$ and

$$(R, r) \quad \xrightarrow{d} \quad \mathrm{GWT}(1).$$

Finally, we make use of the *internal* structure of the giant component $L_1 = L_1(S_g)$. Note that L_1 can be obtained from the 2-core $C = C(L_1)$ of L_1 (i.e., the maximal subgraph of L_1 whose minimum degree is at least two) by replacing each vertex in the 2-core C by a rooted tree. If we delete the *edges* of the 2-core C, we obtain a forest $F = F(L_1)$ such that each component of F is a rooted tree whose root is identical to a vertex in the 2-core C. Note that such a tree component in F may consist of a single vertex. Let $V(C)$ and $V(F)$ denote the set of vertices in C and F, respectively. We can embed the forest F in the giant component L_1 in such a way that the roots of tree components of F are identified with distinct vertices in the 2-core C, the number of tree components in F is equal to $|V(C)|$, and $|V(F)| = |L_1|$. Another important fact is that whp

$$|V(C)| \;=\; \Theta(n^{2/3}) \qquad \text{but} \qquad |V(F)| \;=\; \Theta(n).$$

Therefore, conditioned on the event that the root $r = r(n) \in_u V(S_g)$ is contained in L_1, typically the root r will be not part of $V(C)$, but part of $V(F) \setminus V(C)$. Let T_r be a tree component in F which contains the root r and let $v_T \in V(C)$ denote the vertex with which the root of T_r is identified in L_1. Then, we can show that whp the root r is quite far away from the vertex v_T. Observing that T_r can be any tree among the trees with the same number of vertices, we can show that the rooted tree (T_r, r) locally looks like the Skeleton tree T_∞. Given all of this, we can show that conditioned on the event that the root $r = r(n) \in_u V(S_g)$ is contained in L_1 (which happens with asymptotic probability $c - 1$), the local weak limit of (L_1, r) is T_∞. In short, with asymptotic probability $c - 1$, we have $r \in L_1$ and

$$(L_1, r) \quad \xrightarrow{d} \quad T_\infty.$$

Putting all these together, we have

$$(S_g, r) \quad \xrightarrow{d} \quad (2 - c) \cdot \mathrm{GWT}(1) \;+\; (c - 1) \cdot T_\infty.$$

5 Maximum degree

The maximum degree is a central topic in (random) graph theory, both as a matter of great interest in its own right and for its connections to colouring algorithms. In this section we present most important classical results on the maximum degree of the Erdős–Rényi random graph $G(n, m)$ and recent results on the maximum degree of the random graph $S_g(n, m)$ on $\mathbb{S}_g$.

5.1 Maximum degree of the Erdős–Rényi random graph

Erdős and Rényi [43] were the first to study the maximum degree of a random graph. Their results were strengthened in many ways (see e.g., [20, 21, 22, 91]), and Chapter 3 of [24] provides extensive discussions on the degree sequence of the Erdős–Rényi random graph and especially on the maximum degree. For example, the following oversimplified version of Theorem 3.7 of [24] gives the asymptotic order of the maximum degree of the Erdős–Rényi random graph, when the average degree is constant.

Theorem 5.1 *Let $G = G(n,m) \in_u \mathcal{G}(n,m)$. Assume $2m/n$ tends to a constant $c \in (0,\infty)$. Then whp*

$$\Delta(G) \;=\; \Theta\left(\frac{\ln n}{\ln\ln n}\right).$$

Bollobás [22] discovered a particularly interesting result on the two-point concentration of the maximum degree of the Erdős–Rényi random graph $G(n,m)$, showing that the connectivity threshold $m \sim n\ln n$ is also the threshold for the two-point concentration of $\Delta(G(n,m))$. More precisely, whp $\Delta(G(n,m))$ takes one of two values when $m = o(n\ln n)$, while $\Delta(G(n,m))$ is not concentrated on any subset of $[n]$ with bounded size when $m = \omega(n\ln n)$ and $m = o(n^2)$.

Theorem 5.2 *Let $G = G(n,m) \in_u \mathcal{G}(n,m)$. Assume $m = m(n) = o(n\ln n)$. Then there exists $D = D(n) \in \mathbf{N}$ such that whp*

$$\Delta(G) \;\in\; \{D, D+1\}.$$

Theorem 5.3 *Let $G = G(n,m) \in_u \mathcal{G}(n,m)$. Assume $m = m(n) = \omega(n\ln n)$ and $m = o(n^2)$. If $I = I(n) \subseteq [n]$ is such that whp $\Delta(G) \in I$, then $|I| = \omega(1)$.*

In fact, Bollobás [21, 22] considered the binomial random graph $G_{n,p}$. But by using standard tools relating $G(n,m)$ and $G_{n,p}$, we can translate his results as stated in Theorem 5.1, Theorem 5.2, and Theorem 5.3.

Missethan and Kang studied in [60, Proposition 1.9] how much the maximum degree of $G(n,m)$ increases when the average degree $2m/n$ grows.

Theorem 5.4 *For $i = 1,2$, assume $2m_i/n$ tends to a constant c_i, with $1 < c_1 < c_2$. Assume that $G_1 = G(n,m_1) \in_u \mathcal{G}(n,m_1)$ and $G_2 = G(n,m_2) \in_u \mathcal{G}(n,m_2)$ are chosen independently from each other. Then whp*

$$0 \;<\; \Delta(G_2) - \Delta(G_1) \;=\; \Theta\left(\ln n\,(\ln\ln n)^{-2}\right).$$

In addition, Missethan and Kang showed in [60, Proposition 1.9] that in the *supercritical* Erdős–Rényi random graph $G(n,m)$, the maximum degree of the giant component is typically much larger than the maximum degree of the fragment.

Theorem 5.5 *Let $G = G(n,m) \in_u \mathcal{G}(n,m)$ and $R(G) := G \setminus L_1(G)$. Assume $2m/n$ tends to a constant $c \in (1,\infty)$. Then whp*

$$0 \;<\; \Delta(L_1(G)) - \Delta(R(G)) \;=\; \Theta\left(\ln n\,(\ln\ln n)^{-2}\right).$$

5.2 Maximum degree of a random planar graph

The maximum degrees of random planar graphs and of random graphs on surfaces have also been extensively studied (see e.g., [34, 36, 39, 40, 60, 76, 86]).

The most prominent work is due to McDiarmid and Reed who determined the correct order of the maximum degree of $S_g(n) \in_u \mathcal{S}_g(n)$ (i.e., the random graph on $\mathbb{S}_g$ with vertex set $[n]$, but *without a specification of the number of edges*). They proved in [76] that there exist constants $0 < C_1 < C_2$ such that whp

$$C_1 \ln n \;\leq\; \Delta(S_g(n)) \;\leq\; C_2 \ln n,$$

using clever double counting arguments and the so-called component-degree of a vertex. Drmota, Giménez, Noy, Panagiotou, and Steger determined a more precise asymptotic behaviour of the maximum degree of the random *planar* graph $S_0(n) \in_u \mathcal{S}_0(n)$, using tools from analytic combinatorics and Boltzmann sampling techniques. They showed in [40] that whp

$$|\Delta(S_0(n)) \;-\; \gamma \ln n| \;=\; O(\ln \ln n),$$

where $\gamma \approx 2.52946$ is a constant.

Regarding $S_g(n,m) \in_u \mathcal{S}_g(n,m)$ (i.e., the random graph on $\mathbb{S}_g$ with vertex set $[n]$ and with exactly m edges), its maximum degree has not been much explored, until Dowden, Kang, and Sprüssel [36] derived bounds on $\Delta(S_g(n,m))$.

Recall from Theorem 3.1 (a) that when $2m/n$ tends to a constant $c \in [0,1)$, whp $G(n,m)$ is planar and therefore it is embeddable on $\mathbb{S}_g$ for any $g \in \mathbf{N} \cup \{0\}$. As a consequence, whp the maximum degree of $S_g(n,m)$ is equal to the maximum degree of $G(n,m)$ and it follows Theorem 5.1 that whp

$$\Delta(S_g(n,m)) \;=\; \Delta(G(n,m)) \;=\; \Theta\left(\frac{\ln n}{\ln \ln n}\right) \quad \text{if} \quad 2m/n \;\to\; c \in (0,1).$$

In the complementary case when $2m/n$ tends to a constant $c \in [1,6)$, Dowden, Kang, and Sprüssel used double counting arguments to obtain the following bounds on $\Delta(S_g(n,m))$, depending on whether the average degree is below the first critical point, in the intercritical phase, or above the second critical point.

Theorem 5.6 *Let $g \in \mathbf{N} \cup \{0\}$ and $S_g = S_g(n,m) \in_u \mathcal{S}_g(n,m)$. Assume $2m/n$ tends to a constant $c \in (0,6)$.*

(a) If $c \in (0,1)$, then whp $\Delta(S_g) = \Theta\left(\frac{\ln n}{\ln \ln n}\right)$.

(b) If $c \in [1,2]$, then whp $\Delta(S_g) = O(\ln n)$.

(c) If $c \in (2,6)$, then whp $\Delta(S_g) = \Theta(\ln n)$.

When $2m/n$ tends to a constant $c \in [1,2]$, Theorem 5.6 (b) provides only the asymptotic upper bound on $\Delta(S_g(n,m))$. In [60], Missethan and Kang obtained a stronger result on $\Delta(S_g(n,m))$, using Prüfer sequences and the balls-into-bins model (see Section 5.4).

Theorem 5.7 *Let $g \in \mathbf{N} \cup \{0\}$ and $S_g = S_g(n,m) \in_u \mathcal{S}_g(n,m)$. Assume $m = m(n)$ is such that $\liminf_{n\to\infty} m/n > 0$ and $m \le n + n^{1-\delta}$ for some $\delta > 0$. Then whp*

$$\Delta(S_g) = (1+o(1))\frac{\ln n}{\ln\ln n}.$$

Note that Theorem 5.7 covers the ranges from the first subcritical phase to the second critical point and shows in particular that whp

$$\Delta(S_g(n,m)) = (1+o(1))\frac{\ln n}{\ln\ln n} \quad \text{if} \quad 2m/n \to c \in (0,2].$$

Turning our attention to the two-point concentration of the maximum degree, let us recall from Theorem 5.2 that the two-point concentration of $\Delta(G(n,m))$ holds all the way below the connectivity threshold, i.e., when $m = o(n\ln n)$. In contrast, Missethan and Kang showed in [60] that the two-point concentration of $\Delta(S_g(n,m))$ holds only *below the second critical point.*

To be more precise, consider the first subcritical phase. When $2m/n$ tends to a constant $c \in (0,1)$, the two-point concentration of $\Delta(S_g(n,m))$ follows from Theorem 3.1 (a) and Theorem 5.2. More precisely, there exists $D = D(n) \in \mathbf{N}$ such that whp

$$\Delta(S_g(n,m)) = \Delta(G(n,m)) \in \{D, D+1\} \quad \text{if} \quad 2m/n \to c \in (0,1).$$

Furthermore, we know from Theorem 3.2 (a) that if $2m/n \le 1 - \omega(1)n^{-1/3}$ (in the weakly subcritical phase), then whp $G(n,m)$ is planar. In addition, Theorem 3.2 (b) says that when $m = n/2 + O(n^{2/3})$, the probability that $G(n,m)$ is planar is bounded away from zero and one. These imply that when $m \le n/2 + O(n^{2/3})$, typically $S_g(n,m)$ behaves like $G(n,m)$. However, when $m \ge n/2 + \omega(n^{2/3})$, this is not the case, because whp $G(n,m)$ is not planar (see Theorem 3.2 (b) and (c)). Thus, it is natural to study the maximum degree of $S_g(n,m)$ in two different phases separately when $m = n/2 + O(n^{2/3})$ and when $m \ge n/2 + \omega(n^{2/3})$.

First consider the case when $m \le n/2 + O(n^{2/3})$ (see [60, Theorem 1.4]).

Theorem 5.8 *Let $g \in \mathbf{N} \cup \{0\}$ and $S_g = S_g(n,m) \in_u \mathcal{S}_g(n,m)$. Assume $m = m(n)$ is such that $m \le n/2 + O(n^{2/3})$. Then there exists $D = D(n)$ such that whp*

$$\Delta(S_g) \in \{D, D+1\}.$$

In the complementary case when $m \ge n/2 + \omega(n^{2/3})$, Missethan and Kang (see [60, Theorem 1.6–Corollary 1.8]) obtained various results on $\Delta(S_g(n,m))$. For example, in the intercritical phase they proved the two-point concentration of the maximum degree of $S_g(n,m)$, that of the giant component, and that of the fragment (see [60, Corollary 1.8]).

Theorem 5.9 *Let $g \in \mathbf{N} \cup \{0\}$. Let $S_g = S_g(n,m) \in_u \mathcal{S}_g(n,m)$ and $R(S_g) := S_g \setminus L_1(S_g)$. Then there exists $\hat{D} = \hat{D}(n)$ such that for all $m = m(n)$ where $2m/n$ tends to a constant $c \in (1,2)$ (i.e.,* throughout the intercritical phase*), whp*

$$\Delta(S_g) \in \left\{\hat{D}, \hat{D}+1\right\},$$

$$\Delta(L_1(S_g)) \in \left\{\hat{D}, \hat{D}+1\right\},$$

$$\Delta(R(S_g)) \in \left\{\hat{D}-1, \hat{D}\right\}.$$

Note that $\hat{D}$ in Theorem 5.9 is the same throughout the intercritical phase, *independently* of the value of $c \in (1,2)$.

As a consequence of Theorem 5.9, we have that in the intercritical regime, typically the difference between the maximum degree of the giant component and that of the fragment is at most two.

Corollary 5.10 *Let $g \in \mathbf{N} \cup \{0\}$. Let $S_g = S_g(n,m) \in_u \mathcal{S}_g(n,m)$ and $R(S_g) := S_g \setminus L_1(S_g)$. Assume $2m/n$ tends to a constant $c \in (1,2)$. Then whp*

$$0 \leq \Delta(L_1(S_g)) - \Delta(R(S_g)) \leq 2.$$

Finally, Missethan and Kang considered the dense case *above the second critical point*. They showed in [60, Theorem 1.10] that $\Delta(S_g(n,m))$ is not concentrated on any subset of $[n]$ with bounded size when $2m/n$ tends to a constant $c \in (2,6)$, in the sprit of the non-concentration result on $\Delta(G(n,m))$ when $m = \omega(n \ln n)$ and $m = o(n^2)$ (cf. Theorem 5.3).

Theorem 5.11 *Let $g \in \mathbf{N} \cup \{0\}$ and $S_g = S_g(n,m) \in_u \mathcal{S}_g(n,m)$. Assume $2m/n$ tends to a constant $c \in (2,6)$. If $I = I(n) \subseteq [n]$ is such that whp $\Delta(S_g) \in I$, then $|I| = \omega(1)$.*

5.3 Comparison of maximum degrees

Let us discuss major differences between the maximum degree of the Erdős–Rényi random graph $G = G(n,m)$ and the maximum degree of the random graph $S_g = S_g(n,m)$ on $\mathbb{S}_g$.

In the Erdős–Rényi random graph G, whp the maximum degree $\Delta(G)$ is of the same order $\Theta\left(\frac{\ln n}{\ln \ln n}\right)$, independent of the value of c, when the average degree $2m/n$ tends to a constant $c \in [0, \infty)$ (see Theorem 5.1). On the other hand, in the random graph S_g, whp the maximum degree $\Delta(S_g)$ changes from the order $(1+o(1))\frac{\ln n}{\ln \ln n}$ to the order $\Theta(\ln n)$, when the average degree passes through the *second critical point* (see Theorem 5.7 and Theorem 5.6 (c)).

Next, Theorem 5.4 shows that $\Delta(G)$ features a large positive gap of order $\Theta\left(\ln n\,(\ln \ln n)^{-2}\right)$, when the average degree increases from c_1 to c_2 for any $1 < c_1 < c_2$. In comparison to this large gap in $\Delta(G)$, the first part of Theorem 5.9 shows that *throughout the intercritical phase*, whp $\Delta(S_g)$ takes one of the two values, which is the same *independent* of the value of $c \in (1,2)$.

Note further that in the supercritical Erdős–Rényi random graph, whp the difference between the maximum degree of the giant component $L_1(G)$ and that of the fragment $R(G)$ is of order $\Theta\left(\ln n\,(\ln \ln n)^{-2}\right)$ (see Theorem 5.5), while the corresponding difference in S_g is at most two (see Corollary 5.10).

Finally, the transition from the two-point concentration to the non-concentration of the maximum degree occurs *much earlier* in S_g than in G, namely when the average degree $2m/n$ passes through the *second critical point* (see Theorem 5.8, Theorem 5.9, and Theorem 5.11), instead of at the connectivity threshold $m \sim n \ln n$ in G (see Theorem 5.2 and Theorem 5.3). It is worth noting that in S_g, the threshold where the size of the fragment drops from linear to sublinear is around the second critical point (see Theorem 3.4 and Theorem 3.5), while in G, it is around the connectivity threshold.

5.4 Key proof techniques in the intercritical phase

The maximum degree of the random graph $S_g(n,m)$ on $\mathbb{S}_g$ is significantly affected by the *global and internal* structure of the giant component $L_1 = L_1(S_g(n,m))$. For simplicity we focus on the two-point concentration of the maximum degree of L_1 (see the middle part of Theorem 5.9).

As discussed in Section 4.4, deleting the edges of the 2-core $C = C(L_1)$ of the giant component L_1 results in a forest $F = F(L_1)$ which is embedded in L_1 in such a way that the roots of tree components in F are identified with distinct vertices in the 2-core C, the number of components in F is equal to $|V(C)|$ (which is whp of order $\Theta(n^{2/3})$), and $|V(F)| = |L_1|$ (which is whp of order $\Theta(n)$). Clearly, the maximum degree of L_1 is closely related to the maximum degree of F. Below we sketch how to determine $\Delta(F)$ and $\Delta(L_1)$.

Let $\mathcal{F}(n,t)$ denote the set of all forests with vertex set $[n]$ which consist of exactly t tree components such that the vertices $1,\ldots,t$ are the roots of the tree components and let

$$\mathcal{P}(n,t) \;:=\; \left\{(x_1,\ldots,x_{n-t}) : x_1,\ldots,x_{n-t-1} \in [n] \text{ and } x_{n-t} \in [t]\right\}.$$

Similarly to Prüfer sequences for trees, we can define Prüfer sequences for forests in $\mathcal{F}(n,t)$ in such a way that there is a bijection

$$\Psi : \mathcal{F}(n,t) \to \mathcal{P}(n,t),\; F \mapsto \Psi(F),$$

and for every $F \in \mathcal{F}(n,t)$ and $v \in [n]$, the degree of v in F is equal to the number of occurrences of v in $\Psi(F)$ (up to one, depending on whether $v \in [t]$ or not).

One crucial observation is that the maximum number of occurrences of $v \in [n]$ in the Prüfer sequence $\Psi(F) \in \mathcal{P}(n,t)$ is equal to the maximum number $M(n,t)$ of balls in a bin, when $n-t$ balls are assigned to n bins by choosing a bin for each ball, independently and uniformly at random. Using the first and second moments, we can show that if $t = O(n)$, then $M(n,t)$ and $\Delta(F)$ are strongly concentrated around their means. Putting all necessary ingredients together, Missethan and Kang [60, Definitions 1.3 and 1.5, Theorem 1.6, and Corollary 1.7] showed that for every $\epsilon > 0$, whp

$$\lfloor \nu(n) - \epsilon \rfloor + 1 \;\le\; \Delta(L_1) \;\le\; \lfloor \nu(n) + \epsilon \rfloor + 1,$$

where $\nu : \mathbf{N} \to \mathbf{R}^+$ is a function such that $\nu(n)$ is the unique positive zero of

$$f_n(x) \;:=\; x - (x + 1/2)\ln x + \ln n.$$

They also showed that whp

$$\Delta(L_1) \;\in\; \left\{\hat{D}, \hat{D}+1\right\},$$

where

$$\hat{D} \;:=\; \max\{\lfloor \nu(n) - 1/3 \rfloor, \lfloor \nu(n) + 2/3 \rfloor\}$$

and

$$\nu(n) \;=\; (1+o(1))\,\frac{\ln n}{\ln\ln n}.$$

6 Discussions and open problems

The results in Sections 3.2, 4.2, and 5.2 exhibit the *universal behaviour* of sparse random graphs on orientable surfaces of constant genus, in terms of their component structure, local weak limits, and maximum degree.

Summing up, for any constant $g \in \mathbf{N} \cup \{0\}$, the random graph $S_g = S_g(n,m)$ on the orientable surface $\mathbb{S}_g$ undergoes two critical periods. The first critical point is determined by the emergence of the giant component of linear size, while the second critical point is characterised by a phase transition in the size of the fragment. The local weak limit of S_g undergoes a smooth phase transition from the Galton–Watson tree $\mathrm{GWT}(c)$ for any $c \in [0,1]$ to the Skeleton tree T_∞ at the second critical point $c = 2$, via their linear combination in the intercritical phase when $c \in (1,2)$. Finally, the maximum degree of S_g features yet another phase transition around the second critical point, below which $\Delta(S_g)$ is concentrated on two points, but above which it is not concentrated on any subset of $[n]$ of bounded size.

In the study of random graphs on surfaces, the most unexplored and challenging area of research is about the behaviour *'near' the second critical point.* Below we present the most interesting open problems.

The first open problem concerns the limit distribution of the size of the giant component. As for the Erdős–Rényi random graph $G = G(n,m)$, when the average degree $2m/n$ tends to a constant $c > 1$ (in the supercritical phase), Theorem 3.1 provides a typical value of $|L_1(G)|$ up to fluctuations of order $o(n)$. On the other hand, *central limit theorems* and *local limit theorems* for the size of the giant component provide stronger concentration results. In the case of the Erdős–Rényi random graph, such limit theorems were obtained, e.g., via counting techniques [88, 96], two-round exposure and Martingale techniques [7], Fourier analysis [8], and random walks [25]. But, in the case of random graphs on surfaces, no such limit theorems have been obtained so far.

In the light of the current state of the art, we are interested in the limit distribution of the size of the giant component in $S_g = S_g(n,m)$. Recall that when $2m/n$ tends to a constant $c \in (1,2)$ (in the *intercritical phase*), Theorem 3.4 gives that whp

$$|L_1(S_g)| = (c-1)n + O_p(n^{2/3}),$$

and thus the typical value of $|L_1(S_g)|$ can be determined up to fluctuations of order $O_p(n^{2/3})$. It is natural to ask whether it can be more precisely characterised in terms of the limit distribution.

Question 6.1 *Let $g \in \mathbf{N} \cup \{0\}$ and $S_g = S_g(n,m) \in_u \mathcal{S}_g(n,m)$. Assume $2m/n$ tends to a constant $c \in (1,2)$. What is the limit distribution of $|L_1(S_g)|$?*

The second open problem concerns the *giant* k-core (i.e., the unique k-core of *linear* size). For $k \geq 2$, the k-core of a graph H is the maximal subgraph of H whose minimum degree is at least k and it can be determined algorithmically by the so-called *peeling process* which removes an arbitrary vertex of degree less than k recursively one after another, if one exists. The k-core has been a subject of much study, partly because it is a graph parameter that plays an essential role, e.g., in k-colouring and random k-SAT problems.

By now we have numerous results on the k-core of the Erdős–Rényi random graph $G(n, m)$, including central and local limit theorems [5, 30, 31, 56, 79, 87, 90, 93]. It is well known that the giant 2-core of $G(n, m)$ emerges smoothly together with the giant component when the average degree becomes *one*. But for larger k, the picture is very different. For $k \geq 3$, there exists a threshold c_k, below which whp the giant k-core of $G(n, m)$ is empty, and above which whp it is of linear size, exhibiting a sudden growth.

Regarding the random graph $S_g(n, m)$ on $\mathbb{S}_g$, we have hardly any results on the giant k-core, and even on the giant 2-core. Based upon preliminary work, we conjecture that the giant 2-core emerges smoothly in $S_g(n, m)$ when the average degree becomes *two*, in contrast to the fact that whp the giant 2-core of $G(n, m)$ appears when the average degree becomes *one*.

In fact, whp the giant 2-core of $S_g = S_g(n, m)$ does not exist even in the second weakly supercritical phase. Kang, Moßhammer, and Sprüssel [64] showed that if $2m/n = 2 + \eta$ where $\eta = o(1)$, $\eta n^{2/5} \to +\infty$, but $\eta = o(\ln^{-2/3} n)$ (in the second weakly supercritical phase), then whp

$$|\text{2-core}(S_g)| \;=\; \Theta(\eta^{1/2} n). \tag{6.1}$$

Note that $|\text{2-core}(S_g)| = \Theta(\eta^{1/2} n)$ is much larger that $n^{4/5}$, but much smaller than n, when $\eta n^{2/5} \to +\infty$, but $\eta = o(1)$. If we could extend the result (6.1) smoothly until η tends to a small positive constant, then the size of the 2-core would become linear in n from the point when the average degree becomes barely larger than two, meaning the threshold for the emergence of the giant 2-core in $S_g(n, m)$ would be equal to the second critical point (i.e., when the average degree converges to *two*).

Question 6.2 *What is the threshold for the giant 2-core in $S_g(n, m)$?*

The third open problem is about the local weak limit of $S_g(n, m)$ *beyond* the second supercritical phase.

Question 6.3 *What is the local weak limit of $S_g(n, m)$ when the average degree $2m/n$ converges to a constant $c \in (2, 6]$, especially when $c = 2 + \eta$ for some positive but very small constant η?*

The last open problem is inspired by Theorem 5.7 and Theorem 5.6 (c). The fact that whp $\Delta(S_g(n, m))$ increases from $(1 + o(1)) \ln n / (\ln \ln n)$ to $\Theta(\ln n)$ when the average degree goes through two suggests a possible 'jump' in $\Delta(S_g(n, m))$.

Question 6.4 *Does the maximum degree of $S_g(n, m)$ increase smoothly or discontinously when average degree passes through two?*

Acknowledgements

The author thanks Vincent Pfenninger, Ronen Wdowinski, and the anonymous referee for their very helpful comments and suggestions. This survey was supported in part by the Austrian Science Fund (FWF) [10.55776/F1002]. For the purpose of open access, the author has applied a CC-BY public copyright licence to any Author Accepted Manuscript version arising from this submission.

References

[1] Marie Albenque, Éric Fusy, and Thomas Lehéricy, *Random cubic planar graphs converge to the Brownian sphere*, Electron. J. Probab. **28** (2023), Paper No. 28, 54 pages.

[2] David Aldous, *Brownian excursions, critical random graphs and the multiplicative coalescent*, Ann. Probab. **25** (1997), no. 2, 812–854.

[3] David Aldous and J. Michael Steele, *The objective method: probabilistic combinatorial optimization and local weak convergence*, Probability on discrete structures, Encyclopaedia Math. Sci., vol. 110, Springer, Berlin, 2004, pp. 1–72.

[4] Daniel Amankwah, Jakob Björnberg, Sigurdur Ö. Stefánsson, Benedikt Stufler, and Joonas Turunen, *The scaling limit of random 2-connected series-parallel maps*, submitted for publication. arXiv: 2503.19705, 2025.

[5] Michael Anastos, Joshua Erde, Mihyun Kang, and Vincent Pfenninger, *A short proof of a central limit theorem for the order of the giant component and k-core*, submitted for publication. arXiv: 2506.11651, 2025.

[6] Dan Archdeacon and David A. Grable, *The genus of a random graph*, Discrete Math. **142** (1995), no. 1-3, 21–37.

[7] Michael Behrisch, Amin Coja-Oghlan, and Mihyun Kang, *The order of the giant component of random hypergraphs*, Random Structures Algorithms **36** (2010), no. 2, 149–184.

[8] ———, *Local limit theorems for the giant component of random hypergraphs*, Combin. Probab. Comput. **23** (2014), no. 3, 331–366.

[9] Edward A. Bender and Zhicheng Gao, *Asymptotic enumeration of labelled graphs by genus*, Electron. J. Combin. **18** (2011), no. 1, Paper 13, 28 pages.

[10] Edward A. Bender, Zhicheng Gao, and Nicholas C. Wormald, *The number of labeled 2-connected planar graphs*, Electron. J. Combin. **9** (2002), no. 1, Research Paper 43, 13 pages.

[11] Itai Benjamini and Oded Schramm, *Recurrence of distributional limits of finite planar graphs*, Electron. J. Probab. **6** (2001), no. 23, 13 pages.

[12] Nicla Bernasconi, Konstantinos Panagiotou, and Angelika Steger, *On properties of random dissections and triangulations*, Combinatorica **30** (2010), no. 6, 627–654.

[13] Manuel Bodirsky, Éric Fusy, Mihyun Kang, and Stefan Vigerske, *Boltzmann samplers, Pólya theory, and cycle pointing*, SIAM J. Comput. **40** (2011), no. 3, 721–769.

[14] Manuel Bodirsky, Omer Giménez, Mihyun Kang, and Marc Noy, *Enumeration and limit laws for series-parallel graphs*, European J. Combin. **28** (2007), no. 8, 2091–2105.

[15] Manuel Bodirsky, Clemens Gröpl, Daniel Johannsen, and Mihyun Kang, *A direct decomposition of 3-connected planar graphs*, Sém. Lothar. Combin. **54A** (2005), Art. B54Ak, 15 pages.

[16] Manuel Bodirsky, Clemens Gröpl, and Mihyun Kang, *Sampling unlabeled biconnected planar graphs*, Algorithms and computation, Lecture Notes in Comput. Sci., vol. 3827, Springer, Berlin, 2005, pp. 593–603.

[17] ———, *Generating labeled planar graphs uniformly at random*, Theoret. Comput. Sci. **379** (2007), no. 3, 377–386.

[18] ———, *Generating unlabeled connected cubic planar graphs uniformly at random*, Random Structures Algorithms **32** (2008), no. 2, 157–180.

[19] Manuel Bodirsky, Mihyun Kang, Mike Löffler, and Colin McDiarmid, *Random cubic planar graphs*, Random Structures Algorithms **30** (2007), no. 1-2, 78–94.

[20] Béla Bollobás, *The distribution of the maximum degree of a random graph*, Discrete Math. **32** (1980), no. 2, 201–203.

[21] Béla Bollobás, *Degree sequences of random graphs*, Discrete Math. **33** (1981), no. 1, 1–19.

[22] ———, *Vertices of given degree in a random graph*, J. Graph Theory **6** (1982), no. 2, 147–155.

[23] ———, *The evolution of random graphs*, Trans. Amer. Math. Soc. **286** (1984), no. 1, 257–274.

[24] ———, *Random graphs*, second ed., Cambridge Studies in Advanced Mathematics, vol. 73, Cambridge University Press, Cambridge, 2001.

[25] Béla Bollobás and Oliver Riordan, *Asymptotic normality of the size of the giant component via a random walk*, J. Combin. Theory Ser. B **102** (2012), no. 1, 53–61.

[26] Jérémie Bouttier, Philippe Di Francesco, and Emmanuel Guitter, *Planar maps as labeled mobiles*, Electron. J. Combin. **11** (2004), no. 1, Research Paper 69, 27 pages.

[27] Guillaume Chapuy, Éric Fusy, Omer Giménez, Bojan Mohar, and Marc Noy, *Asymptotic enumeration and limit laws for graphs of fixed genus*, J. Combin. Theory Ser. A **118** (2011), no. 3, 748–777.

[28] Guillaume Chapuy, Éric Fusy, Omer Giménez, and Marc Noy, *On the diameter of random planar graphs*, Combin. Probab. Comput. **24** (2015), no. 1, 145–178.

[29] Guillaume Chapuy, Éric Fusy, Mihyun Kang, and Bilyana Shoilekova, *A complete grammar for decomposing a family of graphs into 3-connected components*, Electron. J. Combin. **15** (2008), no. 1, Research Paper 148, 39 pages.

[30] Amin Coja-Oghlan, Oliver Cooley, Mihyun Kang, and Kathrin Skubch, *How does the core sit inside the mantle?*, Random Structures Algorithms **51** (2017), no. 3, 459–482.

[31] ———, *Core forging and local limit theorems for the k-core of random graphs*, J. Combin. Theory Ser. B **137** (2019), 178–231.

[32] Gwendal Collet, Michael Drmota, and Lukas D. Klausner, *Limit laws of planar maps with prescribed vertex degrees*, Combin. Probab. Comput. **28** (2019), no. 4, 519–541.

[33] Chris Dowden, *The evolution of uniform random planar graphs*, Electron. J. Combin. **17** (2010), no. 1, Research Paper 7, 20 pages.

[34] ———, *Random planar graphs with bounds on the maximum and minimum degrees*, Graphs Combin. **27** (2011), no. 1, 87–107.

[35] Chris Dowden, Mihyun Kang, and Michael Krivelevich, *The genus of the Erdős-Rényi random graph and the fragile genus property*, Random Structures Algorithms **56** (2020), no. 1, 97–121.

[36] Chris Dowden, Mihyun Kang, and Philipp Sprüssel, *The evolution of random graphs on surfaces*, SIAM J. Discrete Math. **32** (2018), no. 1, 695–727.

[37] Michael Drmota, *Random trees: an interplay between combinatorics and probability*, Springer, Vienna, 2009.

[38] Michael Drmota, Éric Fusy, Mihyun Kang, Veronika Kraus, and Juanjo Rué, *Asymptotic study of subcritical graph classes*, SIAM J. Discrete Math. **25** (2011), no. 4, 1615–1651.

[39] Michael Drmota, Omer Giménez, and Marc Noy, *Degree distribution in random planar graphs*, J. Combin. Theory Ser. A **118** (2011), no. 7, 2102–2130.

[40] Michael Drmota, Omer Giménez, Marc Noy, Konstantinos Panagiotou, and Angelika Steger, *The maximum degree of random planar graphs*, Proc. Lond. Math. Soc. (3) **109** (2014), no. 4, 892–920.

[41] Michael Drmota, Marc Noy, Clément Requilé, and Juanjo Rué, *Random cubic planar maps*, Electron. J. Combin. **30** (2023), no. 2, Paper No. 2.51, 41 pages.

[42] Michael Drmota, Marc Noy, and Benedikt Stufler, *Cut vertices in random planar maps*, Electron. J. Combin. **30** (2023), Paper No. 3.32, 32 pages.

[43] Paul Erdős and Alfréd Rényi, *On random graphs. I*, Publ. Math. Debrecen **6** (1959), 290–297.

[44] ———, *On the evolution of random graphs*, Magyar Tud. Akad. Mat. Kutató Int. Közl. **5** (1960), 17–61.

[45] Philippe Flajolet and Robert Sedgewick, *Analytic combinatorics*, Cambridge University Press, Cambridge, 2009.

[46] Nikolaos Fountoulakis and Konstantinos Panagiotou, *3-connected cores in random planar graphs*, Combin. Probab. Comput. **20** (2011), no. 3, 381–412.

[47] Alan Frieze and Michał Karoński, *Introduction to random graphs*, Cambridge University Press, 2015.

[48] Éric Fusy, *Uniform random sampling of planar graphs in linear time*, Random Structures Algorithms **35** (2009), no. 4, 464–522.

[49] Stefanie Gerke and Colin McDiarmid, *On the number of edges in random planar graphs*, Combin. Probab. Comput. **13** (2004), no. 2, 165–183.

[50] Stefanie Gerke, Colin McDiarmid, Angelika Steger, and Andreas Weißl, *Random planar graphs with n nodes and a fixed number of edges*, Proceedings of the Sixteenth Annual ACM-SIAM Symposium on Discrete Algorithms, ACM, New York, 2005, pp. 999–1007.

[51] ———, *Random planar graphs with given average degree*, Combinatorics, complexity, and chance, Oxford Lecture Ser. Math. Appl., vol. 34, Oxford Univ. Press, Oxford, 2007, pp. 83–102.

[52] Stefanie Gerke, Dirk Schlatter, Angelika Steger, and Anusch Taraz, *The random planar graph process*, Random Structures Algorithms **32** (2008), no. 2, 236–261.

[53] Omer Giménez and Marc Noy, *Asymptotic enumeration and limit laws of planar graphs*, J. Amer. Math. Soc. **22** (2009), no. 2, 309–329.

[54] Geoffrey R. Grimmett, *Random labelled trees and their branching networks*, J. Austral. Math. Soc. Ser. A **30** (1980/81), no. 2, 229–237.

[55] Svante Janson, Donald E. Knuth, Tomasz Łuczak, and Boris Pittel, *The birth of the giant component*, Random Structures Algorithms **4** (1993), no. 3, 231–358, With an introduction by the editors.

[56] Svante Janson and Malwina J. Luczak, *Asymptotic normality of the k-core in random graphs*, Ann. Appl. Probab. **18** (2008), no. 3, 1085–1137.

[57] Svante Janson, Tomasz Łuczak, and Andrzej Rucinski, *Random graphs*, Wiley-Interscience Series in Discrete Mathematics and Optimization, Wiley-Interscience, New York, 2000.

[58] Mihyun Kang and Martin Loebl, *The enumeration of planar graphs via Wick's theorem*, Adv. Math. **221** (2009), no. 5, 1703–1724.

[59] Mihyun Kang and Tomasz Łuczak, *Two critical periods in the evolution of random planar graphs*, Trans. Amer. Math. Soc. **364** (2012), no. 8, 4239–4265.

[60] Mihyun Kang and Michael Missethan, *Concentration of maximum degree in random planar graphs*, J. Combin. Theory Ser. B **156** (2022), 310–342.

[61] ———, *Longest and shortest cycles in random planar graphs*, Random Structures Algorithms **60** (2022), no. 3, 462–505.

[62] ———, *The early evolution of the random graph process in planar graphs and related classes*, SIAM J. Discrete Math. **37** (2023), no. 1, 146–162.

[63] ———, *Benjamini-Schramm local weak limit of sparse random planar graphs*, submitted for publication. arXiv: 2101.11910, 2025+.

[64] Mihyun Kang, Michael Moßhammer, and Philipp Sprüssel, *Phase transitions in graphs on orientable surfaces*, Random Structures Algorithms **56** (2020), no. 4, 1117–1170.

[65] Mihyun Kang and Philipp Sprüssel, *Symmetries of unlabelled planar triangulations*, Electron. J. Combin. **25** (2018), no. 1, Paper No. 1.34, 38 pages.

[66] Igor Kortchemski and Cyril Marzouk, *Large deviation local limit theorems and limits of biconditioned planar maps*, Ann. Appl. Probab. **33** (2023), no. 5, 3755–3802.

[67] Jean-François Le Gall, *The topological structure of scaling limits of large planar maps*, Invent. Math. **169** (2007), no. 3, 621–670.

[68] ———, *Large random planar maps and their scaling limits*, European Congress of Mathematics, Eur. Math. Soc., Zürich, 2010, pp. 253–276.

[69] ———, *The Brownian map: a universal limit for random planar maps*, XVIIth International Congress on Mathematical Physics, World Sci. Publ., Hackensack, NJ, 2014, pp. 420–428.

[70] Jean-François Le Gall and Grégory Miermont, *Scaling limits of random planar maps with large faces*, Ann. Probab. **39** (2011), no. 1, 1–69.

[71] ———, *Scaling limits of random trees and planar maps*, Probability and statistical physics in two and more dimensions, Clay Math. Proc., vol. 15, Amer. Math. Soc., Providence, RI, 2012, pp. 155–211.

[72] Valery A. Liskovets and Timothy R. Walsh, *Ten steps to counting planar graphs*, Congr. Numer. **60** (1987), 269–277.

[73] Tomasz Łuczak, *Component behavior near the critical point of the random graph process*, Random Structures Algorithms **1** (1990), no. 3, 287–310.

[74] Tomasz Łuczak, Boris Pittel, and John C. Wierman, *The structure of a random graph at the point of the phase transition*, Trans. Amer. Math. Soc. **341** (1994), no. 2, 721–748.

[75] Colin McDiarmid, *Random graphs on surfaces*, J. Combin. Theory Ser. B **98** (2008), no. 4, 778–797.

[76] Colin McDiarmid and Bruce Reed, *On the maximum degree of a random planar graph*, Combin. Probab. Comput. **17** (2008), no. 4, 591–601.

[77] Colin McDiarmid, Angelika Steger, and Dominic J. A. Welsh, *Random planar graphs*, J. Combin. Theory Ser. B **93** (2005), no. 2, 187–205.

[78] Grégory Miermont, *The Brownian map is the scaling limit of uniform random plane quadrangulations*, Acta Math. **210** (2013), no. 2, 319–401.

[79] Michael Molloy, *Cores in random hypergraphs and Boolean formulas*, Random Structures Algorithms **27** (2005), no. 1, 124–135.

[80] Marc Noy, *Random planar graphs and beyond*, Proceedings of the International Congress of Mathematicians—Seoul 2014. Vol. IV, Kyung Moon Sa, Seoul, 2014, pp. 407–430.

[81] Marc Noy and Lander Ramos, *Random planar maps and graphs with minimum degree two and three*, Electron. J. Combin. **25** (2018), no. 4, Paper No. 4.5, 38 pages.

[82] Marc Noy, Vlady Ravelomanana, and Juanjo Rué, *On the probability of planarity of a random graph near the critical point*, Proc. Amer. Math. Soc. **143** (2015), no. 3, 925–936.

[83] Marc Noy, Clément Requilé, and Juanjo Rué, *Further results on random cubic planar graphs*, Random Structures Algorithms **56** (2020), no. 3, 892–924.

[84] Deryk Osthus, Hans Jürgen Prömel, and Anusch Taraz, *On random planar graphs, the number of planar graphs and their triangulations*, J. Combin. Theory Ser. B **88** (2003), no. 1, 119–134.

[85] Konstantinos Panagiotou and Angelika Steger, *Maximal biconnected subgraphs of random planar graphs*, ACM Trans. Algorithms **6** (2010), no. 2, Art. 31, 21 pages.

[86] ———, *On the degree distribution of random planar graphs*, Proceedings of the Twenty-Second Annual ACM-SIAM Symposium on Discrete Algorithms, SIAM, Philadelphia, PA, 2011, pp. 1198–1210.

[87] Boris Pittel, Joel Spencer, and Nicholas Wormald, *Sudden emergence of a giant k-core in a random graph*, J. Combin. Theory Ser. B **67** (1996), no. 1, 111–151.

[88] Boris Pittel and Nicholas C. Wormald, *Counting connected graphs inside-out*, J. Combin. Theory Ser. B **93** (2005), no. 2, 127–172.

[89] Dominique Poulalhon and Gilles Schaeffer, *Optimal coding and sampling of triangulations*, Algorithmica **46** (2006), no. 3-4, 505–527.

[90] Oliver Riordan, *The k-core and branching processes*, Combin. Probab. Comput. **17** (2008), no. 1, 111–136.

[91] Oliver Riordan and Alex Selby, *The maximum degree of a random graph*, Combin. Probab. Comput. **9** (2000), no. 6, 549–572.

[92] Vojtěch Rödl and Robin Thomas, *On the genus of a random graph*, Random Structures and Algorithms **6** (1995), no. 1, 1–12.

[93] Cristiane M. Sato, *On the robustness of random k-cores*, European J. Combin. **41** (2014), 163–182.

[94] Gilles Schaeffer, *Conjugaison d'arbres et cartes combinatoires aléatoires*, Ph.D. thesis, Bordeaux 1, 1998.

[95] ———, *Random sampling of large planar maps and convex polyhedra*, Annual ACM Symposium on Theory of Computing (Atlanta, GA, 1999), ACM, New York, 1999, pp. 760–769.

[96] Vadim E. Stepanov, *The probability of the connectedness of a random graph* $\mathcal{G}_m(t)$, Teor. Verojatnost. i Primenen. **15** (1970), 58–67.

[97] Benedikt Stufler, *Local convergence of random planar graphs*, J. Eur. Math. Soc. (JEMS) **25** (2023), no. 1, 1–73.

[98] ———, *The scaling limit of random cubic planar graphs*, J. Lond. Math. Soc. (2) **110** (2024), no. 5, Paper No. e70018, 46 pages.

[99] William T. Tutte, *A census of planar triangulations*, Canad. J. Math. **14** (1962), 21–38.

[100] ———, *A census of planar maps*, Canad. J. Math. **15** (1963), 249–271.

[101] Remco van der Hofstad, *Random graphs and complex networks. Vol. 1*, Cambridge Series in Statistical and Probabilistic Mathematics, vol. 43, Cambridge University Press, Cambridge, 2017.

[102] ———, *Random graphs and complex networks. Vol. 2*, Cambridge Series in Statistical and Probabilistic Mathematics, vol. 54, Cambridge University Press, Cambridge, 2024.

Institute of Discrete Mathematics
Graz University of Technology
Steyrergasse 30, 8010 Graz
Austria
kang@math.tugraz.at

Extremal Poset Theory

Maria Axenovich, Ryan R. Martin, Balázs Patkós

Abstract

A cornerstone of combinatorics is extremal theory, in which one asks for the optimal value of a parameter of a combinatorial object, given certain constraints. Extremal graph theory is one of the most robust areas in combinatorics, with classical results that has produced sophisticated techniques and a fundamental understanding of graphs.

Extremal questions have been asked about other combinatorial objects, and, in recent years, there has been a growing interest in extremal problems in the area of partially ordered sets (posets). In this survey, we will review some of the more well-known results in extremal poset theory, focusing on Turán-type and Ramsey-type problems. We will also present fundamental open questions that have motivated and continue to motivate this emerging area of combinatorics.

1 Introduction

An extremal problem in combinatorics is one that asks for the optimum value of a parameter, given a set of constraints. Arguably the most fundamental result in extremal graph theory is Turán's theorem:

Theorem 1.1 (Turán [119]) *A graph on n vertices with no copy of a clique on $k+1$ vertices has no more edges than that of the complete k-partite graph on n vertices, with each part of size $\lceil n/k \rceil$ or $\lfloor n/k \rfloor$.*

The case of $k = 2$ is attributed to Mantel [94].

A *Turán-type* problem in graph theory is one that asks for the maximum number of a particular type of subgraph (say the edge K_2) that occurs when a particular subgraph or a particular set of subgraphs is forbidden. In poset theory, the most basic Turán-type problem is one that asks for the maximum number of elements in a subposet of a Boolean lattice such that some particular subposet is forbidden. For example, perhaps the most persistent open problem in the theory is the maximum size of a subposet $\mathcal{F}$ of a Boolean lattice of dimension n such that $\mathcal{F}$ has no subposet that is a diamond. See Problem 3 below.

On the other hand, Ramsey's theorem established, in the words of Theodore Motzkin that "complete disorder is impossible". In the graph setting:

Theorem 1.2 (Ramsey [111]) *For every integers $k \geq 2$ and $\ell \geq 2$, there is an integer $R = R(k, \ell)$ such that if the edges of a complete graph on R vertices are colored red and blue, there will either be a clique of order k with all edges red or a clique of order ℓ with all edges blue.*

In posets, the most basic Ramsey problem is one that asks for the minimum dimension of a Boolean lattice such that if the elements are colored red and blue, there is either a monochromatic red poset P or a monochromatic blue poset Q. The earliest open problem in this area is the case in which both P and Q are fixed-dimension Boolean lattices. See Problem 11 below.

1.1 Notation and terminology

In this survey, we will use terminology that has become standard in the literature. For every positive integer n, $[n]$ will denote $\{1,\dots,n\}$. For a set S, 2^S will denote the set of subsets of S and $\binom{S}{k}$ will denote the set of all k-subsets of S. The notation $X \sqcup Y$ denotes the disjoint union of a set X and a set Y.

Definition 1.3 *A poset $P = (X, \leq)$ is a set X together with a partial order $\leq$ which satisfies the following properties for all $x, y, z \in X$:*

- *(reflexivity)* $x \leq x$;
- *(antisymmetry) if $x \leq y$ and $y \leq x$, then $x = y$;*
- *(transitivity) if $x \leq y$ and $y \leq z$, then $x \leq z$.*

If, in addition, P satisfies

- *for each $x, y \in X$, either $x \leq y$ or $y \leq x$,*

then P is called a total order or a linear order.
If $x \leq y$ and $x \neq y$, then we write $x < y$.

In this survey, all posets will be finite, unless otherwise indicated. A poset is *nontrivial* if it has at least two elements.

For a poset P, we will often just use P to represent the underlying set. An element of the underlying set will often be called a *vertex* and the *size* of P, denoted $|P|$ is the size of its underlying set. If there are multiple posets under discussion, we will use $\leqslant_P$ to denote the partial order with respect to poset P.

For convenience the notation $x_1, \dots, x_k \leq y_1, \dots, y_\ell$ will mean that $x_i \leq y_j$ for all $i \in [k]$ and all $j \in [\ell]$.

As in graphs, there is both the notion of a (weak) subposet and an induced subposet.

Definition 1.4 *Let $P = (P, \leqslant_P)$ and $Q = (Q, \leqslant_Q)$ be posets.*

- *A function $\varphi : P \to Q$, is an order-preserving map if it is an injection with the property that $p_1 \leqslant_P p_2$ implies $\varphi(p_1) \leqslant_Q \varphi(p_2)$.*
- *A function $\varphi : P \to Q$ is an order-embedding map if it is an injection with the property that $p_1 \leqslant_P p_2$ if and only if $\varphi(p_1) \leqslant_Q \varphi(p_2)$.*
- *The poset P is a (weak) subposet of Q if there is an order-preserving map from P to Q. If P is not a subposet of Q, then Q is P-free.*
- *The poset P is an induced subposet or a strong subposet of A if there is an order-embedding map from P to Q. If P is not an induced subposet of Q, then Q is induced-P-free.*
- *For a set of posets $\mathcal{P}$, if no member of $\mathcal{P}$ is a subposet of Q [induced subposet of Q], then Q is said to be $\mathcal{P}$-free [induced-$\mathcal{P}$-free].*
- *Two elements p_1, p_2 of a poset P are unrelated or parallel if neither $p_1 \leq p_2$ nor $p_2 \leq p_1$. In this case, we write $p_1 \parallel p_2$.*

- *Two subposets P_1, P_2 of a larger poset Q are parallel if no vertex of one poset is comparable to a vertex in another poset. In this case, we write $P_1 \parallel P_2$.*

- *A poset D is the dual of poset P if they have the same underlying set and obey the property that $p_1 \leqslant_P p_2$ if and only if $p_2 \leqslant_D p_1$.*

- *A height-two or bipartite poset is one in which the vertices are the union of minimal elements and maximal elements, and every nonequality relation is between a minimal element and a maximal element.*

The key features of a poset are the antichains and chains.

Definition 1.5 *Let P be a poset.*

- *An antichain is a (sub)poset in which no pair elements are related. An antichain with n elements is denoted $\mathcal{A}_n$.*

- *A chain is a (sub)poset in which every pair of elements is related. A chain with n elements is denoted $\mathcal{C}_n$.*

- *The size of the largest antichain of a poset is its width and the width of P is denoted $w(P)$.*

- *The size of the largest chain of a poset is its height and the height of P is denoted $h(P)$.*

We note that the *length* of a chain is $h(P) - 1$ but we will avoid this terminology whenever possible.

1.2 Glossary of posets and operations on posets

In this section, we define some well-studied common posets. See Figure 1.

- The *Boolean lattice of dimension n* is the poset of subsets of the set $[n]$, ordered by inclusion.[1] It is denoted $\mathcal{B}_n = (2^{[n]}, \subseteq)$. Moreover, for $k = 0, \ldots, n$, the kth *layer* of $\mathcal{B}_n$ is the family $\binom{[n]}{k}$.

- The *r-fork* $\mathcal{V}_r$ is a poset of $r+1$ elements which consists of an antichain of size r and another element that is less than all of them. The 2-fork is denoted V. Its dual is denoted Λ.

- The *diamond* is a poset of 4 elements a, b, c, d such that $a \leq b, c \leq d$ and $b \parallel c$. The diamond is denoted $\Diamond$.

- The *r-diamond* $\Diamond_r$ is a poset on $r+2$ elements consisting of an antichain on r vertices, a vertex smaller than all of these, as well as a vertex larger than all of these. The 2-diamond is the same poset as $\Diamond$.

- The *butterfly* is a poset of 4 elements a, b, c, and d such that $a, b \leq c, d$, $a \parallel b$, and $c \parallel d$. The butterfly is denoted $\bowtie$.

[1] In order to avoid the awkward term "set of subsets", we will call a $\mathcal{F} \subseteq 2^{[n]}$ a *family* and it will come with the partial order induced on it. Thus, we may also write $\mathcal{F} \subseteq \mathcal{B}_n$

- The poset $\mathcal{N}$ is a poset with four elements a, b, c, and c, such that $a, c \leq b$ and $c \leq d$, whereas all other pairs of elements are incomparable.

- For $s, t \geq 1$, the *complete bipartite poset* $\mathcal{K}_{s,t}$ is a poset with $s + t$ elements satisfying $a_1, \ldots, a_s \leq b_1, \ldots, b_t$ and both $\{a_1, \ldots, a_s\}$ and $\{b_1, \ldots, b_t\}$ are antichains.

- For $s_1, \ldots, s_k$, the poset $\mathcal{K}_{s_1,\ldots,s_k} = (X, \leq)$ is a poset with $s_1 + \cdots + s_k$ elements called a *complete multipartite poset* if $X = A_1 \sqcup \cdots \sqcup A_k$, where the A_i's are pairwise disjoint antichains, $|A_i| = s_i$ for all $i \in [k]$ and for any $x \in A_i$ and $y \in A_j$, $x \leq y$ iff $i < j$.

- The *parallel composition* of posets P_1 and P_2 is the poset $P_1 + P_2$ which has the vertex set $P_1 \sqcup P_2$ with all relations inherited from P_1 and P_2 and for every $p_1 \in P_1$ and every $p_2 \in P_2$, the pair p_1 and p_2 are incomparable. We write kP for $\underbrace{P + \cdots + P}_{k}$.

- The *gluing* of posets P_1 and P_2 is the poset $P_1 * P_2$ which has the vertex set $P_1 \sqcup P_2$ with all relations inherited from P_1 and P_2 and for every $p_1 \in P_1$ and every $p_2 \in P_2$, the inequality $p_1 <_{P_1 * P_2} p_2$ holds.

- If P is a poset with a largest element p, and Q is a poset with a smallest element q, then P^Q denotes the poset with elements from $(P \cup Q) \setminus \{q\}$ with relations inherited from P and Q and all $p' \in P$ being smaller than all $q' \in Q \setminus \{q\}$. One can think of identifying p with q. One such example is the *broom*, $C_k^{\mathcal{V}_r}$.

Definition 1.6 *Let $P = (X, \leq)$ be a poset and $x, y \in X$. We say y covers x and write $x \lessdot y$ if $x \leqslant y$ and $x \leqslant z \leqslant y$ implies either $x = z$ or $z = y$.*

The Hasse diagram of a poset is drawn as a graph in which the elements are designated as vertices and edges are drawn between vertices that satisfy a cover relation. If $x \lessdot y$ then x is drawn below y. See Figure 1 for several examples.

1.3 Other concepts

Perhaps the most fundamental theorem of poset theory is Dilworth's theorem [35]. There are many proofs, including one that follows from the König-Hall [83, 64] theorem on matchings in bipartite graphs. A *chain cover* of a poset is a collection of chains whose union is the entire poset.

Theorem 1.7 (Dilworth [35]) *For a finite poset P, the size of the largest antichain is the same as the size of the smallest chain cover.*

There are several definitions of dimension in poset theory. The most well-known is due to Dushnik and Miller [37].

Definition 1.8 *A linear extension of a poset P is a linear order which has P as a subposet. If $L_1, \ldots, L_d$ are posets on the same vertex set X, the poset $\bigcap_{i=1}^d L_i$ is the poset such that for all $x, y \in X$, then $x \leq y$ if and only if $x \leqslant_{L_i} y$ for all $i \in [d]$. A realizer of P is a set of linear extensions $\{L_1, \ldots, L_d\}$ of P such that $P = \bigcap_{i=1}^d L_i$.*

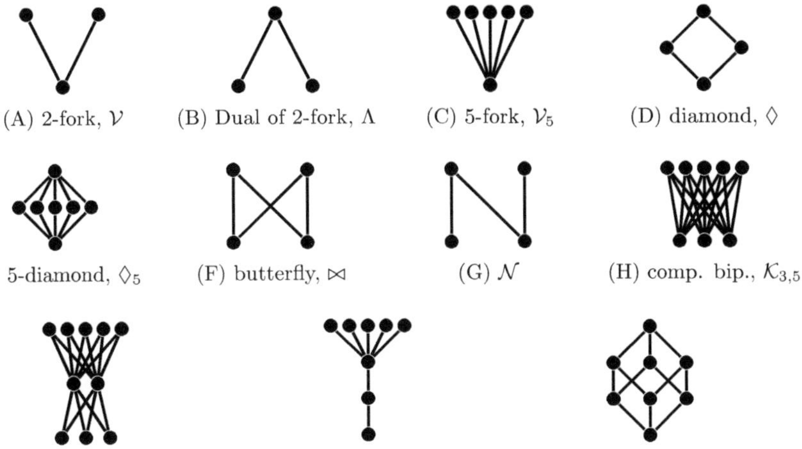

(A) 2-fork, $\mathcal{V}$ (B) Dual of 2-fork, Λ (C) 5-fork, $\mathcal{V}_5$ (D) diamond, $\Diamond$

(E) 5-diamond, $\Diamond_5$ (F) butterfly, $\bowtie$ (G) $\mathcal{N}$ (H) comp. bip., $\mathcal{K}_{3,5}$

(I) comp. multip., $\mathcal{K}_{3,2,5}$ (J) broom, $\mathcal{C}_3^{\mathcal{V}_5}$ (K) Boolean, $\mathcal{B}_3$

Figure 1: Glossary of posets

- *The (Dushnik-Miller) dimension of a poset P, denoted* $\dim(P)$ *is the size of its smallest realizer.*
- *The 2-dimension of a poset P, denoted* $\dim_2(P)$*, is the smallest integer k such that P is an induced subposet of $\mathcal{B}_k$.*

Both $\dim(P)$ and $\dim_2(P)$ are well-defined for finite posets P. For $\dim(P)$, it is because P is isomorphic to the intersection of its linear extensions and so $\dim(P) \leq |P|!$. For $\dim_2(P)$ it is because it can be shown that any P is an induced subposet of $\mathcal{B}_{|P|}$ by mapping $x \in P$ to the downset $\{y \leq x : y \in P\}$ and so $\dim_2(P) \leq |P|$. Note that these are two very different concepts. For example, if $\mathcal{C}_k$ is a chain with k elements, then $\dim(\mathcal{C}_k) = k$ but $\dim_2(\mathcal{C}_k) = k - 1$. Further, if $\mathcal{A}_k$ is an antichain with k elements, then $\dim(\mathcal{A}_k) = 2$ but $\dim_2(\mathcal{A}_k) = \mathrm{sp}(k)$ as defined below.

Definition 1.9 *The Sperner number* $\mathrm{sp}(k)$ *is the smallest integer j such that the Boolean lattice $\mathcal{B}_j$ has an antichain of size at least k. That is,* $\mathrm{sp}(k) = \min\{j \in \mathbf{N} : \binom{j}{\lfloor j/2 \rfloor} \geq k\}$.

The name "Sperner number" is inspired by Sperner's theorem (Theorem 2.1 below) and can be estimated accurately:

Proposition 1.10 (Habib, Nourine, Raynaud, Thierry [62]) *For all $k \geq 2$,*

$$\log_2 k + \tfrac{1}{2}\log_2\log_2 k < \mathrm{sp}(k) < \log_2 k + \tfrac{1}{2}\log_2\log_2 k + 2.$$

Hence, $\mathrm{sp}(k) \in \left\{\left\lfloor \log_2 k + \tfrac{1}{2}\log_2\log_2 k + 1\right\rfloor, \left\lfloor \log_2 k + \tfrac{1}{2}\log_2\log_2 k + 2\right\rfloor\right\}$.

1.4 Structure of the paper

This survey is laid out as follows: Section 2 addresses classical Turán-type problems and results. Section 3 addresses saturation problems and results. Section 4 addresses Ramsey problems and results. The appendices are tables of known results.

2 Turán-type poset theory

The study of extremal poset theory usually begins with Sperner's theorem from 1928.

Theorem 2.1 (Sperner [116]) *If $\mathcal{F} \subseteq 2^{[n]}$ is an antichain, then $|\mathcal{F}| \leq \binom{n}{\lfloor n/2 \rfloor}$. Equality holds if and only if either $\mathcal{F} = \binom{[n]}{\lfloor n/2 \rfloor}$ or $\mathcal{F} = \binom{[n]}{\lceil n/2 \rceil}$.*

This was later generalized by Erdős [40] in 1945. A subposet that does not contain a chain with $k+1$ elements is called a *k-Sperner* family. If $\mathcal{F}$ contains sets only of k possible set sizes, then $\mathcal{F}$ is clearly k-Sperner. Therefore, the union of the largest k layers of the Boolean lattice is k-Sperner, and thus the sum of the k largest binomial coefficients is a lower bound on the maximum size of a k-Sperner family.

We will denote the family of all sets of k largest layers of the Boolean lattice by $\mathcal{M}_{n,k}$. That is,

$$\mathcal{M}_{n,k} = \left\{ \bigcup_{i=0}^{k-1} \binom{[n]}{\lfloor \frac{n-k+1}{2} \rfloor + i}, \bigcup_{i=0}^{k-1} \binom{[n]}{\lceil \frac{n-k+1}{2} \rceil + i} \right\}. \tag{2.1}$$

Note that $|\mathcal{M}_{n,k}| = 1$ if $n+k$ is odd and $|\mathcal{M}_{n,k}| = 2$ if $n+k$ is even. The notation $\Sigma(n,k)$ is the size of a family in $\mathcal{M}_{n,k}$. That is,

$$\Sigma(n,k) = \sum_{i=0}^{k-1} \binom{n}{\lfloor \frac{n-k+1}{2} \rfloor + i}.$$

Theorem 2.2 (Erdős [40]) *If $\mathcal{F} \subseteq 2^{[n]}$ is k-Sperner, then $|\mathcal{F}| \leq \Sigma(n,k)$. Equality holds if and only if $\mathcal{F} \in \mathcal{M}_{n,k}$.*

In this section, we are interested in the maximum possible size, $\mathrm{La}(n,P)$ [$\mathrm{La}^*(n,P)$], that a [an induced] P-free family $\mathcal{F} \subseteq 2^{[n]}$ can have. Formally,

$$\begin{aligned} \mathrm{La}(n,P) &= \max\{\mathcal{F} \subseteq 2^{[n]} : \mathcal{F} \text{ is } P\text{-free}\} \\ \mathrm{La}^*(n,P) &= \max\{\mathcal{F} \subseteq 2^{[n]} : \mathcal{F} \text{ is induced-}P\text{-free}\}. \end{aligned}$$

Theorem 2.2 can be stated as $\mathrm{La}(n, \mathcal{C}_{k+1}) = \mathrm{La}^*(n, \mathcal{C}_{k+1}) = \Sigma(n,k)$, where $\mathcal{C}_{k+1}$ is the chain on $k+1$ elements. For any poset P, the chain $\mathcal{C}_{|P|}$ is a non-induced copy of P, therefore Theorem 2.2 implies $\mathrm{La}(n,P) \leq (|P|-1)\binom{n}{\lfloor n/2 \rfloor}$. Note that in this survey, we consider $\mathrm{La}(n,P)$ and $\mathrm{La}^*(n,P)$ only for fixed P. Hence, when considering asymptotic behavior, we fix P and let $n \to \infty$.

If P is not an antichain, then $\binom{[n]}{\lfloor n/2 \rfloor}$ is P-free, so for such posets we have $\mathrm{La}(n,P) = \Theta_P\left(\binom{n}{\lfloor n/2 \rfloor}\right)$. Similarly, it is true that $\mathrm{La}^*(n,P) = \Theta_P\left(\binom{n}{\lfloor n/2 \rfloor}\right)$ holds for each poset P that is not an antichain, but this requires a more involved proof that was found by Methuku and Pálvölgyi [102] only in 2017 (after partial results by Lu and Milans [90]). It is natural to conjecture that the limits $\lim_{n\to\infty} \frac{\mathrm{La}(n,P)}{\binom{n}{\lfloor n/2 \rfloor}}$ and $\lim_{n\to\infty} \frac{\mathrm{La}^*(n,P)}{\binom{n}{\lfloor n/2 \rfloor}}$ exist, but the existence has neither been shown in the induced case nor in the non-induced case.

The systematic study of Turán-type problems in posets was initiated by Gyula O.H. Katona and Tamás Tarján [78], who defined the P-free property and started investigating $\mathrm{La}(n,P)$ in 1983. In the next twenty years, sporadic exact and asymptotic results were obtained by Katona and some of his students, visitors and hosts on the value of $\mathrm{La}(n,P)$ for various posets. By the mid-2000s, a natural conjecture regarding the asymptotics of $\mathrm{La}(n,P)$ (which we will describe in Section 2.2) was formulated [21, 59] and a very intensive and fruitful decade of forbidden-subposet papers started thanks to the continued efforts of Katona and the fresh enthusiasm from Jerry Griggs to promote the area.

All results achieved during this period affirmed the conjecture, some of them for specific posets, some for more general poset classes, but the full conjecture remained open and seemed out of reach, so research shifted towards variants of the problem such as stability, supersaturation, and random versions, on which we will elaborate in later subsections. In 2024, Ellis, Ivan, and Leader [38] gave a very short and simple counterexample to the conjecture, reforming our understanding of the field.

2.1 Early results and observations

Katona and Tarján in their introductory paper [78] proved the exact result $\mathrm{La}(n,\{\mathcal{V},\Lambda\}) = 2\binom{n-1}{\lfloor (n-1)/2\rfloor}$ for the 2-fork, $\mathcal{V}$, and its dual, Λ. General r-fork posets were addressed by DeBonis and Katona [33] who proved

$$\mathrm{La}(n,\mathcal{V}_r) = \left(1+O_r(\frac{1}{n})\right)\binom{n}{\lfloor n/2\rfloor}.$$

The induced analog of the problem was considered by Caroll and Katona [22] who showed $\mathrm{La}^*(n,\mathcal{V}_r) = \left(1+O_r(\frac{1}{n})\right)\binom{n}{\lfloor n/2\rfloor}$.

Even earlier, a more general result, albeit with a worse error term, was established by Thanh [117] on broom posets: $\mathrm{La}\big(n,\mathcal{C}_k^{\mathcal{V}_r}\big) = \big(k+o(1)\big)\binom{n}{\lfloor n/2\rfloor}$. Using the cycle method (see Section 2.5.2) DeBonis, Katona, and Swanepoel [34] obtained the exact result for the butterfly $\mathrm{La}(n,\bowtie) = \Sigma(n,2)$, while Griggs and Katona [56] showed that $\mathrm{La}(n,\mathcal{N}) = \left(1+O(\frac{1}{n})\right)\binom{n}{\lfloor n/2\rfloor}$.

All of the above asymptotic results were later superseded by Theorem 2.3 by Bukh [21]. Yet, these early papers are very important as they introduced many of the tools and approaches that one still uses today to attack forbidden subposet problems. Let us mention two important observations that can be traced back to these papers.

First, as a consequence of a Chernoff-Hoeffding inequality, one can show that

$$\left|\left\{F\in 2^{[n]} : \left||F|-\frac{n}{2}\right| \ge \sqrt{n\log n}\right\}\right| = O\left(\frac{1}{n^2}\binom{n}{\lfloor n/2\rfloor}\right).$$

Therefore, if one is only interested in the asymptotics of $\mathrm{La}(n,P)$, then when considering a P-free family $\mathcal{F}$, one can assume that all sets in $\mathcal{F}$ are of size close to $n/2$.

A second observation is about the second order term of $\mathrm{La}(n,P)$ when it is not equal to $\Sigma(n,k)$ for any k. If $P=\mathcal{V}$, then one can add to $\binom{n}{\lfloor n/2\rfloor}$ any family $\mathcal{F}\subseteq\binom{n}{\lfloor n/2\rfloor+1}$ with the property that $|F\triangle F'|\ge 2$ for any $F,F'\in\mathcal{F}$ and obtain a $\mathcal{V}$-free family. So any $G \in \binom{n}{\lfloor n/2 \rfloor}$ is contained in at most one $F \in \mathcal{F}$, and since any F contains $\lfloor n/2 \rfloor + 1$ of the G's, we deduce $|\mathcal{F}| \leq \left(\frac{2}{n} + O(\frac{1}{n^2})\right)\binom{n}{\lfloor n/2 \rfloor}$ and thus the total size of such a construction is at most $\left(1 + \frac{2}{n} + O(\frac{1}{n^2})\right)\binom{n}{\lfloor n/2 \rfloor}$. This upper bound can be verified for any $\mathcal{V}$-free family.

The determination of the maximum size of such a family $\mathcal{F}$ above (called a 1-*error correcting code*) is a major open problem in coding theory. The best general known construction from Graham and Sloane [54] is as follows: for any $0 \leq i < n$, consider $\mathcal{F}_{n,i} = \left\{F \in \binom{[n]}{\lfloor n/2 \rfloor + 1} : \sum_{j \in F} j \equiv i \pmod{n}\right\}$. All $\mathcal{F}_{n,i}$'s have the required property, and they partition $\binom{[n]}{\lfloor n/2 \rfloor + 1}$, so for at least one value of i, we have $|\mathcal{F}_{n,i}| \geq \frac{1}{n}\binom{[n]}{\lfloor n/2 \rfloor + 1}$.

2.2 The conjecture that turned out to be false

Based on the results above, a general and very natural construction of P-free families can be obtained. Let $e(P)$ $[e^*(P)]$ denote the largest integer k such that for any n, the middle k layers of $\mathcal{B}_n$, i.e., a member of $\mathcal{M}_{n,k}$ (recall (2.1)) is [induced] P-free. By definition, we have

$$\Sigma\big(n, e(P)\big) \leq \mathrm{La}(n, P) \qquad \text{and} \qquad \Sigma\big(n, e^*(P)\big) \leq \mathrm{La}^*(n, P).$$

Griggs and Lu [59] and independently Bukh [21] conjectured in 2009 that for any poset P, $\mathrm{La}(n, P) = \big(e(P) + o(1)\big)\binom{n}{\lfloor n/2 \rfloor}$ and proved this conjecture for important classes of posets. (The analogous conjecture for $\mathrm{La}^*(n, P)$ was formulated only in 2015 [109].)

In order to describe Bukh's result, we define a *tree poset* T to be one such that its unoriented Hasse diagram is a tree. The *height* $h(P)$ of a poset P is the length of the longest chain in P. Observe that for any tree poset, it is the case that $e(P) = h(P) - 1$.

Theorem 2.3 (Bukh [21]) *For any tree poset T, we have* $\mathrm{La}(n, T) = \big(h(T) - 1 + O_T(\frac{1}{n})\big)\binom{n}{\lfloor n/2 \rfloor}$.

The same result for the asymptotics of $\mathrm{La}^*(n, T)$ is Theorem 2.12 by Boehnlein and Jiang [14] but their error term is $o_T(1)$ instead of $O_T(\frac{1}{n})$.

Now we consider the construction of Ellis, Ivan, and Leader [39] that serves as a counterexample to the conjecture that $\mathrm{La}(n, P) = \big(e(P) + o(1)\big)\binom{n}{\lfloor n/2 \rfloor}$.

In fact, this construction disproved the so-called *Daisy Conjecture* of Bollobás, Leader, and Malvenuto [18] and, in fact, disproved many (seemingly unrelated) conjectures in extremal set theory. The Daisy Conjecture was widely believed to be true and the Ellis-Ivan-Leader counterexample came as a surprise.

To formulate the statement, let us define an r-*daisy* to be the family of all six r-sets that contain some fixed $(r-2)$-set R (the *stem* of the daisy) and that are subsets of a fixed $(r+2)$-set $T \supset R$. Let $\mathrm{ex}(n, D_r)$ be the maximum number of r-sets in a family $\mathcal{F}$ of r-sets on vertex set $[n]$ such that $\mathcal{F}$ does not contain any r-daisies and let $\pi_r = \lim_{n \to \infty} \frac{\mathrm{ex}(n, D_r)}{\binom{n}{r}}$. The Daisy conjecture stated that π_r tends to 0 as r tends to infinity. Bollobás, Leader, and Malvenuto claimed that the Daisy conjecture is equivalent to the statement that $\mathrm{ex}(2n, D_n)/\binom{2n}{n}$ tends to 0 as n tends to infinity.

Lemma 2.4 (Bollobás, Leader, Malvenuto [18]) *For any $c \geq 0$,*

$$\liminf_{r\to\infty} \pi_r > c \qquad \Longleftrightarrow \qquad \liminf_{n\to\infty} \frac{\mathrm{ex}(2n, D_n)}{\binom{2n}{n}} > c.$$

The construction by Ellis, Ivan and Leader [39] establishes that there is a family in the middle layer of the Boolean lattice which has nontrivial density.

Theorem 2.5 (Ellis, Ivan, and Leader [39])

$$\liminf_{r\to\infty} \pi_r \geq \prod_{k=1}^{\infty}\left(1-2^{-k}\right) \geq 0.288787.$$

Consequently, $\left(0.288787 - o(1)\right)\binom{2n}{n} \leq \mathrm{ex}(2n, D_n) \leq \left(2/3 + o(1)\right)\binom{2n}{n}$.

Proof Here we only sketch the main points of the proof, leaving the details to the reader. First, define $\mathcal{F}_{r,2^r-1}$ as follows: fix a bijection f between the nonzero vectors of the vector field $\mathbb{F}_2^r$ and $[2^r-1]$, and let

$$\mathcal{F}_{r,2^r-1} = \left\{f(B) : B \text{ is a basis of } \mathbb{F}_2^r\right\}.$$

By counting the number of ordered bases of $\mathbb{F}_2^r$, one can obtain that $\left|\mathcal{F}_{r,2^r-1}\right|/\binom{n}{r} > \prod_{k=1}^{r}\left(1-2^{-k}\right)$.

The next goal is to show that any r-daisy in $\left[2^r-1\right]$ contains an r-set corresponding to a linearly dependent set of r vectors. Suppose for a contradiction that there exists some r-daisy F in $\left[2^r-1\right]$ whose r-sets all correspond to linearly independent sets of r-vectors. Thus, the stem of the daisy forms the basis of an $(r-2)$-dimensional subspace W. Let $\{u_1, u_2, u_3, u_4\}$ be the remaining vertices of the daisy. Modding out by W, $\mathbb{F}_2^r/W$ is a 2-dimensional subspace over $\mathbb{F}_2$. Since there is no $u_i \in W$, it must be the case that there are distinct u_i, u_j with $f^{-1}(u_i) + W = f^{-1}(u_j) + W$, contradicting the fact that $f^{-1}(u_i)$, $f^{-1}(u_j)$, and the vectors corresponding to the vertices of the stem form an independent set.

Next, let $n \geq 2^r - 1$ and the family $\mathcal{F}_{r,n}$ be the blow-up of $\mathcal{F}_{r,2^r-1}$. That is, we partition $[n]$ into 2^r-1 classes $X_1, X_2, \ldots, X_{2^r-1}$ as equally as possible. The index $i(x)$ of $x \in [n]$ is defined by $x \in X_{i(x)}$. Then let

$$\mathcal{F}_{r,n} = \left\{\left\{x_1, x_2, \ldots, x_r\right\} : \left\{i(x_1), i(x_2), \ldots, i(x_r)\right\} \in \mathcal{F}_{r,2^r-1}\right\}.$$

This is also r-daisy-free, and the lower bound on $\lim_{r\to\infty} \pi_r$ now follows.

Lemma 2.4 establishes the lower bound for $\mathrm{ex}(2n, D_n)$. For the upper bound, observe that for any $S \in \binom{[2n]}{n-2}$, the members of an n-daisy-free family that are supersets of S correspond to a K_4-free graph on $n+2$ vertices. Turán's theorem gives an edge density of $2/3 + o(1)$. Averaging gives the $2/3 + o(1)$ upper bound for the density of $\mathrm{ex}(2n, D_n)$. □

Using Theorem 2.5, there is a n-daisy-free $\mathcal{F}_n \subseteq \binom{[2n]}{n}$ with $|\mathcal{F}_n| \geq \big(0.288787 - o(1)\big)\binom{[2n]}{n}$. Then $\mathcal{G}_n = \binom{[2n]}{n-2} \cup \binom{[2n]}{n-1} \cup \mathcal{F} \cup \binom{[2n]}{n+1} \cup \binom{[2n]}{n+2}$ is $\mathcal{B}_4$-free with $|\mathcal{G}_n| \geq \left(4.288787 + o(1)\right)\binom{2n}{n}$. However, $e(\mathcal{B}_4) = 4$ and so $\mathrm{La}(2n, \mathcal{B}_4) \neq \left(e(\mathcal{B}_4) + o(1)\right)\binom{2n}{n}$.

Let us mention that Ellis, Ivan, and Leader gave more general constructions in their paper showing $\mathrm{La}(2n, \mathcal{B}_s) \neq \big(e(\mathcal{B}_s) + o(1)\big)\binom{2n}{n}$ for any $s \geq 4$.

The family $\mathcal{G}_n$ and the other constructions from [38] show that $\mathrm{La}(n, \Diamond_{14}) \neq \big(e(\Diamond_{14}) + o(1)\big)\binom{n}{\lfloor \frac{n}{2} \rfloor}$ and $\mathrm{La}^*(n, \Diamond_6) \neq \big(e^*(\Diamond_6) + o(1)\big)\binom{n}{\lfloor n/2 \rfloor}$. More generally, for any $s \geq 4$,

$$\mathrm{La}\Big(n, \Diamond_{2^s-2}\Big) \neq \Big(e(\Diamond_{2^s-2}) + o(1)\Big)\binom{n}{\lfloor n/2 \rfloor}$$

$$\mathrm{La}^*\Big(n, \Diamond_{\binom{s}{\lfloor s/2 \rfloor}}\Big) \neq \Big(e^*(\Diamond_{\binom{s}{\lfloor s/2 \rfloor}}) + o(1)\Big)\binom{n}{\lfloor n/2 \rfloor}.$$

On the other hand, in Theorem 2.18 below, Griggs, Li, and Lu [58] proved $\mathrm{La}(n, \Diamond_m) = \Sigma(n, s) = \big(e(\Diamond_m) + o(1)\big)\binom{n}{\lfloor n/2 \rfloor}$ for any m satisfying the inequalities $2^{s-1} - 1 \leq m \leq 2^s - \binom{s}{\lfloor s/2 \rfloor} - 1$, for some $s \geq 3$. Patkós [109] proved that for an infinite sequence of values of m (in fact for a set M that has density 1 in $\mathbb{N}$), one has $\mathrm{La}^*(n, \Diamond_m) = \big(e^*(\Diamond_m) + o(1)\big)\binom{n}{\lfloor n/2 \rfloor}$.

Problem 1 *Determine all the values m for which $\mathrm{La}(n, \Diamond_m) = \big(e(\Diamond_m)+o(1)\big)\binom{n}{\lfloor n/2 \rfloor}$ or $\mathrm{La}^*(n, \Diamond_m) = \big(e^*(\Diamond_m) + o(1)\big)\binom{n}{\lfloor n/2 \rfloor}$ hold.*

Finally, with the Ellis-Ivan-Leader construction in hand, it is quite unlikely that $\mathrm{La}(n, P) = \big(k + o(1)\big)\binom{n}{\lfloor n/2 \rfloor}$ for an integer k for all posets $P = \mathcal{B}_m$ or $P = \Diamond_m$. However, we are unaware of good upper bounds on $\mathrm{La}(n, \mathcal{B}_m)$ or on $\mathrm{La}(n, \Diamond_m)$. On the other hand, as we now discuss, there is an example of a poset P for which $\lim_{n\to\infty} \frac{\mathrm{La}(n,P)}{\binom{n}{\lfloor n/2 \rfloor}}$ is not an integer if it exists.

Consider the poset $\mathcal{B}_4^{1,2}$ that is induced by the rank 1 and rank 2 elements of $\mathcal{B}_4$. See Figure 2.

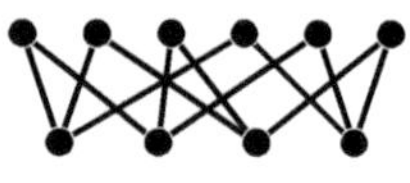

Figure 2: $\mathcal{B}_4^{1,2}$

The family $\mathcal{F}_n \cup \binom{[2n]}{n-1}$ shows $\mathrm{La}\big(2n, \mathcal{B}_4^{1,2}\big) \geq \big(1.288787 - o(1)\big)\binom{2n}{n}$. On the other hand, as pointed out by Gerbner and Patkós [51], Theorem 2.8 below yields $\mathrm{La}\big(n, \mathcal{B}_4^{1,2}\big) \leq \big(1 + \sqrt{2/3} + o(1)\big)\binom{n}{\lfloor n/2 \rfloor}$. Thus, if $\lim_{n\to\infty} \frac{\mathrm{La}(n,\mathcal{B}_4^{1,2})}{\binom{n}{\lfloor n/2 \rfloor}}$ exists, then it cannot be an integer.

Corollary 2.6 *We have*

$$\big(1.288787 - o(1)\big)\binom{n}{\lfloor n/2 \rfloor} \leq \mathrm{La}\big(n, \mathcal{B}_4^{1,2}\big) \leq \big(1.816497 + o(1)\big)\binom{n}{\lfloor n/2 \rfloor}.$$

Problem 2 *Determine whether $\lim_{n\to\infty} \mathrm{La}(n, P)/\binom{n}{\lfloor n/2 \rfloor}$ exists for all posets P.*

2.3 Diamonds, crowns and other open questions

Although the result of Griggs, Li, and Lu [58] establish the exact values of $\mathrm{La}(n, \Diamond_3)$ and $\mathrm{La}(n, \Diamond_4)$, even the asymptotic value of $\mathrm{La}(n, \Diamond)$ is still unknown.

It is easy to see that $\mathrm{La}(n, \Diamond) \geq \Sigma(n, 2)$. In the case when n is an even integer, $\Sigma(n, 2) = \left(2 - \frac{2}{n+2}\right)\binom{n}{\lfloor n/2 \rfloor}$ but Dove [36] gave a construction for $n \geq 2$ and $0 < k < n$ that shows $\mathrm{La}(n, \Diamond) \geq \left(2 - \frac{1}{n-1}\right)\binom{n}{\lfloor n/2 \rfloor}$ for even n. Dove's construction identifies two distinct $a, b \in [n]$ and is the following:

$$\left\{S \in \binom{[n]}{k-1} : |S \cap \{a,b\}| \in \{0,1\}\right\} \cup \left\{T \in \binom{[n]}{k} : |T \cap \{a,b\}| \in \{0,2\}\right\}$$
$$\cup \left\{U \in \binom{[n]}{k+1} : |U \cap \{a,b\}| \in \{1,2\}\right\}$$

and it has size $\left(2 - \frac{1}{n-1} + \frac{(n-2k)^2}{n(n-1)}\right)\binom{n}{k}$.

Finding the upper bound for $\mathrm{La}(n, \Diamond)$ has driven a lot of the research in poset Turán theory. As a $\Diamond$-free family is a 3-Sperner family, Theorem 2.2 implies that $\mathrm{La}(n, \Diamond) \leq \Sigma(n, 3)$. Griggs and Lu [59] established that $\mathrm{La}(n, \Diamond) \leq \left(2.3 + o(1)\right)\binom{n}{\lfloor n/2 \rfloor}$. Axenovich, Manske, and Martin [1] proved that $\mathrm{La}(n, \Diamond) \leq \left(2.284 + o(1)\right)\binom{n}{\lfloor n/2 \rfloor}$. Griggs, Li, and Lu [58] gave $\mathrm{La}(n, \Diamond) \leq \left(2.273 + o(1)\right)\binom{n}{\lfloor n/2 \rfloor}$. Using the method of flag algebras, Kramer, Martin, and Young [84] proved that $\mathrm{La}(n, \Diamond) \leq \left(2.25 + o(1)\right)\binom{n}{\lfloor n/2 \rfloor}$. The best current upper bound is due to Grósz, Methuku, and Tompkins [60].

Theorem 2.7 (Dove [36]; Grósz, Methuku, Tompkins [60])

$$\max\left\{\Sigma(n,2), \left(2 - \frac{1}{n-1}\right)\binom{n}{\lfloor n/2 \rfloor}\right\} \leq \mathrm{La}(n, \Diamond) \leq \left(\frac{3+\sqrt{2}}{2} + o(1)\right)\binom{n}{\lfloor n/2 \rfloor}$$
$$\approx (2.207107 + o(1))\binom{n}{\lfloor n/2 \rfloor}.$$

Because the upper bound for the size of $\Diamond$-free families seemed so out-of-reach, researchers attempted to find the largest $\Diamond$-free family in three consecutive layers of the Boolean lattice. That is, researchers looked to bound the size of the largest $\Diamond$-free family in $\binom{[n]}{k-1} \cup \binom{[n]}{k} \cup \binom{[n]}{k+1}$ for some k. We note that Dove's lower bound still holds because it is a subfamily of three consecutive layers.

For families in three consecutive layers, Axenovich, Manske, and Martin [1] establishes that a $\Diamond$-free family has size at most $\left(3/2 + \sqrt{2}/2 + o(1)\right)\binom{n}{\lfloor n/2 \rfloor} \approx \left(2.207107 + o(1)\right)\binom{n}{\lfloor n/2 \rfloor}$. This upper bound was improved by Manske and Shen [93] using a Cauchy-Schwarz argument to $\left(1 + 2/\sqrt{3} + o(1)\right)\binom{n}{\lfloor n/2 \rfloor} \approx \left(2.154701 + o(1)\right)\binom{n}{\lfloor n/2 \rfloor}$. Balogh, Hu, Lidický, and Liu [7] used flag algebras to give an upper bound of $\left(2.151210 + o(1)\right)\binom{n}{\lfloor n/2 \rfloor}$ for $\Diamond$-free families in three layers.

An exact result of $\mathrm{La}(n, \bowtie) = \Sigma(n, 2)$ was established by De Bonis, Katona, and Swanepoel [34]. The asymptotic value of $\mathrm{La}(n, \bowtie) = \left(2 + o(1)\right)\binom{n}{\lfloor n/2 \rfloor}$ comes from De Bonis and Katona [33]. Most of the other earlier asymptotic results for $\mathrm{La}(n, P)$ have been superceded by Bukh's result [21] since for most of them, the Hasse diagram of P is a tree.

The $2k$-*crown*, $\mathcal{O}_{2k}$ is a height-two poset whose Hasse diagram is a cycle on $2k$ vertices such that the vertices alternate between minimal and maximal elements. See Figure 3(A) for $\mathcal{O}_6$ and Figure 3(B) for $\mathcal{O}_{10}$.

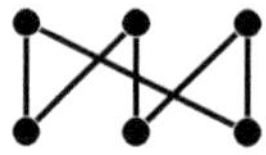

(A) 6-crown, $\mathcal{O}_6$

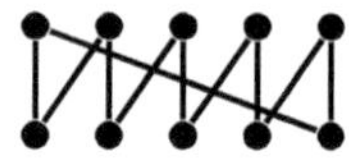

(B) 10-crown, $\mathcal{O}_{10}$

Figure 3: Crown posets

Griggs and Lu [59] have a general result for certain types of height-two posets.

Theorem 2.8 (Griggs, Lu [59]) *Let G be a graph and let $P(G)$ be the height-two poset in which each maximal element corresponds to an $e \in E(G)$ and each minimal element corresponds to a $v \in V(G)$ where $v <_{P(G)} e$ if and only if e is incident to v in G. Then*

$$\mathrm{La}(n, P(G)) \leq \left(1 + \sqrt{1 - \frac{1}{\chi(G) - 1}} + o(1)\right)\binom{n}{\lfloor n/2 \rfloor}.$$

As used in Corollary 2.6 above, $\mathcal{B}_4^{1,2}$ is $P(K_4)$. Since $\mathcal{O}_{2k}$ is $P(C_k)$, a corollary follows:

Corollary 2.9 (Griggs, Lu [59]) *If k is even, then*

$$\mathrm{La}(n, \mathcal{O}_{2k}) = (1 + o(1))\binom{n}{\lfloor n/2 \rfloor}.$$

Linyuan Lu [89] used Conlon's *k-partite representations* [29] to resolve all of the remaining crown cases except for $\mathcal{O}_6$ and $\mathcal{O}_{10}$.

Theorem 2.10 (Lu [89]) *If $k \geq 7$ is odd, then*

$$\mathrm{La}(n, \mathcal{O}_{2k}) = (1 + o(1))\binom{n}{\lfloor n/2 \rfloor}.$$

In fact, Lu [89] shows that the asymptotic value of $\binom{n}{\lfloor n/2 \rfloor}$ holds for any height-two poset that has a k-partite representation, for some $k \geq 2$.

Problem 3 *Improve any of the following bounds:*

$$\begin{aligned}
\max\left\{\Sigma(n,2), \left(2 - \frac{1}{n-1}\right)\binom{n}{\lfloor n/2 \rfloor}\right\} \leq \mathrm{La}(n, \Diamond) &\leq \left(\frac{3+\sqrt{2}}{2} + o(1)\right)\binom{n}{\lfloor n/2 \rfloor} \\
(1 + o(1))\binom{n}{\lfloor n/2 \rfloor} \leq \mathrm{La}(n, \mathcal{O}_6) &\leq \left(1 + \frac{\sqrt{2}}{2} + o(1)\right)\binom{n}{\lfloor n/2 \rfloor} \\
(1 + o(1))\binom{n}{\lfloor n/2 \rfloor} \leq \mathrm{La}(n, \mathcal{O}_{10}) &\leq \left(1 + \frac{\sqrt{2}}{2} + o(1)\right)\binom{n}{\lfloor n/2 \rfloor} \\
(1.288787 - o(1))\binom{n}{\lfloor n/2 \rfloor} \leq \mathrm{La}(n, \mathcal{B}_4^{1,2}) &\leq \left(1 + \sqrt{\frac{2}{3}} + o(1)\right)\binom{n}{\lfloor n/2 \rfloor}.
\end{aligned}$$

2.4 Induced Turán results

Recall that $\mathrm{La}^*(n,P)$ denotes the maximum size of an *induced-P-free* family in $2^{[n]}$. Katona and Tarján [78] proved $\mathrm{La}^*(n,\{\mathcal{V},\Lambda\}) = \mathrm{La}(n,\{\mathcal{V},\Lambda\}) = \binom{n-1}{\lfloor (n-1)/2\rfloor}$. Katona [77] also proved that $\mathrm{La}^*(n,\mathcal{V}_r) \le \big(1+\frac{2r}{n}+O(\frac{1}{n^2})\big)\binom{n}{\lfloor n/2\rfloor}$, for the case where P is the r-fork.

The *standard example* of dimension r, denoted $\mathcal{S}_r$, is the smallest poset which has Dushnik-Miller [37] dimension r and has $2r$ elements. The Hasse diagram is a complete bipartite graph which misses a perfect matching. See Figure 4.

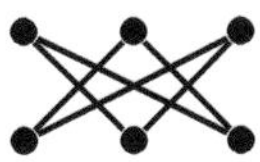

(A) Standard example $\mathcal{S}_3$, isomorphic to $\mathcal{O}_6$

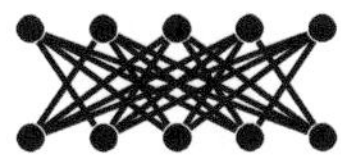

(B) Standard example $\mathcal{S}_5$

Figure 4: Standard examples

Theorem 2.11 (Lu, Milans [90])

$$(2.28+o(1))\binom{n}{\lfloor n/2\rfloor} \le \mathrm{La}^*(n,\Diamond) \le (2.58+o(1))\binom{n}{\lfloor n/2\rfloor}$$

$$\mathrm{La}^*(n,\mathcal{S}_r) = \big(r+O(\sqrt{r})\big)\binom{n}{\lfloor n/2\rfloor}.$$

While Erdős' Theorem 2.2 establishes that $\mathrm{La}(n,P) \le \Sigma(n,|P|-1)$, it is not obvious that $\mathrm{La}^*(n,P)/\binom{n}{\lfloor n/2\rfloor}$ is bounded for all posets P. Lu and Milans established that it was bounded for all height-two posets P.

Boehnlein and Jiang [14] proved an induced analogy to Bukh's Theorem 2.3.

Theorem 2.12 (Boehnlein, Jiang [14]) *For any tree poset T, recall that $e(T)$ is the least integer such that T is not a subposet of $e(T)$ layers of $\mathcal{B}_k$ for k sufficiently large. We have* $\mathrm{La}^*(n,T) = \big(e(T)+o(1)\big)\binom{n}{\lfloor n/2\rfloor}$.

Using the Klazar-Marcus [82] high-dimensional variant of the Marcus-Tardos [95] theorem, Methuku and Pálvölgyi [102] showed $\mathrm{La}^*(n,P)/\binom{n}{\lfloor n/2\rfloor}$ is bounded for all P. Later, Tomon [118] gave that C_P is polynomial in $|P|$.

Theorem 2.13 (Methuku, Pálvölgyi [102]) *For every P, there exists a C_P such that* $\mathrm{La}^*(n,P) \le C_P\binom{n}{\lfloor n/2\rfloor}$.

Theorem 2.14 (Tomon [118]) *For every poset P of height h, there exists a c_h such that* $\mathrm{La}^*(n,P) \le |P|^{c_h}\binom{n}{\lfloor n/2\rfloor}$.

Finding $\mathrm{La}^*(n,P)$ asymptotically seems to be quite difficult for posets P not covered by Theorem 2.12. Diamonds are, of course, of particular interest.

Problem 4 *Improve either of the following bounds:*

$$(2.28+o(1))\binom{n}{\lfloor n/2\rfloor} \le \mathrm{La}^*(n,\Diamond) \le (2.58+o(1))\binom{n}{\lfloor n/2\rfloor}.$$

Problem 5 *Determine whether* $\lim_{n\to\infty}\mathrm{La}^*(n,P)/\binom{n}{\lfloor n/2\rfloor}$ *exists for all posets P.*

2.5 The art of double counting

Every paper establishing an upper bound on $\mathrm{La}(n, P)$ or on $\mathrm{La}^*(n, P)$ uses some kind of a double counting argument. A couple of papers [7, 84] apply, explicitly or implicitly, the flag algebra method, but most proofs count pairs $(F, \mathcal{S})$, where F is a set from the P-free family $\mathcal{F} \subseteq 2^{[n]}$ we want to bound the size of, while $\mathcal{S} \subseteq \mathcal{B}_n$ is a family that is isomorphic to a (relatively) simple substructure of $\mathcal{B}_n$. The most natural substructure is a maximal chain $\emptyset = C_0 \subsetneq C_1 \subsetneq \dots C_{n-1} \subsetneq C_n = [n]$ on which we will elaborate later.

More complicated structures include

- *pairs of complement chains*, i.e., a maximal chain $\mathcal{C}$ and another one $\mathcal{C}' = \{[n] \setminus C : C \in \mathcal{C}\}$,
- *double-chains* $\mathcal{C} \cup \mathcal{C}^+$, where $\mathcal{C} = \{C_i : i = 0, 1, \dots, n\}$ is a maximal chain and $\mathcal{C}^+ = \{C_i \cup (C_{i+2} \setminus C_{i+1}) : i = 0, 1, \dots, n-2\}$,
- *intervals of a cycle*, that is for any cyclic permutation π of n, the family $\{I_{i,j} : 1 \leq i, j \leq n\} \cup \{\emptyset\}$, where $I_{i,j} = \{\pi(i), \pi(i+1), \dots, \pi(j)\}$ if $i \leq j$, while $I_{i,j} = \{\pi(i), \pi(i+1), \dots, \pi(n), \pi(1), \pi(2), \dots, \pi(j)\}$ if $j < i$,

and several other structures that we do not list here.

Let us go back to counting pairs $(F, \mathcal{C})$ with F running through members of a P-free family $\mathcal{F}$ and $\mathcal{C}$ being a maximal chain. We start by the famous inequality that is due to Lubell [92], Yamamoto [125], and Meshalkin [101] and is most often referred to as the *LYM inequality*, but it can be easily derived from a the more general "set-pairs" result of Bollobás [16]. So, in Hungary, it is sometimes called the *YBLM inequality*.[2] The simple observation is that if $\mathcal{F}$ has no chain with $k+1$ elements, then no maximal chain $\mathcal{C}$ can contain more than k sets from $\mathcal{F}$, so the number of pairs $(F, \mathcal{C})$ is at most $k \cdot n!$, while the number of maximal chains containing a fixed set F is $|F|!(n-|F|)!$. Adding these terms up for all members of $\mathcal{F}$ and rearranging yields the following inequality.

Theorem 2.15 (LYM inequality [92, 125, 101]) *If $\mathcal{F} \subseteq 2^{[n]}$ is a k-Sperner (i.e. C_{k+1}-free) family, then*

$$\lambda_n(\mathcal{F}) \stackrel{\text{def}}{=} \sum_{F \in \mathcal{F}} \binom{n}{|F|}^{-1} \leq k.$$

The quantity $\lambda_n(\mathcal{F})$ is called the *Lubell mass* of $\mathcal{F}$ is one of the major and most common used tools for establishing upper bounds on $\mathrm{La}(n, P)$. The connection between upper bounds on $\lambda_n(\mathcal{F})$ and the size of $\mathcal{F}$ is made in the following simple lemma. Together with Theorem 2.15, they immediately yield Theorem 2.2.

Lemma 2.16 (Griggs, Lu [59]) *If for some $\mathcal{F} \subseteq 2^{[n]}$ and positive real c, we have $\lambda_n(\mathcal{F}) \leq c$, then $|\mathcal{F}| \leq c\binom{n}{\lfloor n/2 \rfloor}$. Moreover, if $c = k$ is an integer, then $|\mathcal{F}| \leq \Sigma(n, k)$.*

In fact, Kramer, Martin, and Young's result [84] involved proving that, for any $\Diamond$-free family in $2^{[n]}$, $\lambda_n(\mathcal{F}) \leq 2.25 + o(1)$. This is best possible, up to the error term.

[2] Miklós Ybl (Ybl, M.) is a famous Hungarian architect whose many works in Budapest include the world-renowned Opera House.

As every P-free family is $\mathcal{C}_{|P|}$-free, another immediate corollary of Theorem 2.15 is that for any P-free family $\mathcal{F} \subseteq 2^{[n]}$, we have $\lambda_n(\mathcal{F}) \leq |P|-1$. An induced analog of this inequality resisted the attempts of researchers for long time, until Méroueh [100] proved the following.

Theorem 2.17 (Méroueh [100]) *For an P there exists a constant $C = C(P)$ such that for any induced-P-free family $\mathcal{F} \subseteq 2^{[n]}$ we have $\lambda_n(\mathcal{F}) \leq C$.*

Note that Theorem 2.17, via Lemma 2.16, establishes Methuku and Pálvölgyi's Theorem 2.13 that $\mathrm{La}^*(n, P) = O(\binom{n}{\lfloor n/2 \rfloor})$.

In probabilistic terminology, the Lubell mass is the expected value $\mathbb{E}[X]$ of the random variable $X = X_{\mathcal{F}}$ which counts the number of sets $F \in \mathcal{F}$ that a maximal chain picked uniformly at random contains. Using a bound on the average number of sets in $\mathcal{F} \cap \mathcal{C}$ was addressed from early on. Katona wrote several survey-like papers [76, 77] where he tried to systematically introduce methods for bounding $\lambda_n(\mathcal{F})$ albeit without using this notation. Katona's approach was to partition the set of maximal chains according to which *connected component of* $\mathcal{F}$ they intersect. The connected components of $\mathcal{F}$ are the connected components of the undirected Hasse diagram of $\mathcal{F}$ when considered as a poset ordered by inclusion. However, the method that is now known as the *chain partitioning method* was introduced by Griggs, Li, and Lu [58] and a detailed discussion appeared in the paper by Griggs and Li [57].

The method can be very simply described in general. Fix a family $\mathcal{F} \subseteq 2^{[n]}$. Let $\mathbf{C}$ denote the set of all maximal chains in $[n]$ and let $\mathbf{C}^1 \cup \mathbf{C}^2 \cup \cdots \cup \mathbf{C}^m$ be a partition of $\mathbf{C}$. Then one can introduce the corresponding random variables $X^1, X^2, \ldots, X^m$ with X^j being the number of sets $F \in \mathcal{F}$ that a maximal chain picked uniformly at random from $\mathbf{C}^j$ contains. Then clearly, the Lubell mass $\lambda_n(\mathcal{F})$ is the weighted average $\sum_{j=1}^{m} \frac{|\mathbf{C}^j|}{n!}\mathbb{E}(X^j)$. In particular, if $\mathbb{E}[X^j] \leq C$ holds for $1 \leq j \leq m$, then $\lambda_n(\mathcal{F}) \leq C$.

Commonly used partitions are

- the *min-partition* $\mathbf{C}_{\emptyset} \cup \bigcup_{F \in \mathcal{F}} \mathbf{C}_F^{\min}$, where $\mathbf{C}_F^{\min}$ is the set of those maximal chains C for which F is the smallest set in $\mathcal{F}$ that lies in C, and $\mathbf{C}_{\emptyset} = \{\mathcal{C} \in \mathbf{C} : \mathcal{C} \cap \mathcal{F} = \emptyset\}$,
- the *max-partition* $\mathbf{C}_{\emptyset} \cup \bigcup_{F \in \mathcal{F}} \mathbf{C}_F^{\max}$, where $\mathbf{C}_F^{\max}$ is the set of those maximal chains C for which F is the largest set in $\mathcal{F}$ that lies in C,
- the *min-max-partition* $\mathbf{C}_{\emptyset} \cup \bigcup_{F,F' \in \mathcal{F}, F \subseteq F'} \mathbf{C}_{F,F'}$, where $\mathbf{C}_{F,F'}$ is the set of those maximal chains C for which F is the smallest and F' is the largest element of $\mathcal{F}$ that each lie in C, in particular, if $F = F'$, then F is the only set in $\mathcal{F} \cap \mathcal{C}$.

2.5.1 Application of chain partitioning Let us give a short illustration of how the chain partitioning method works. Suppose $\mathcal{F}$ is a $\Diamond_r$-free or an induced-$\Diamond_r$-free family, and consider its min-max-partition. Then for any $F \subseteq F'$ in $\mathcal{F}$,

- $|\{F^* \in \mathcal{F} : F \subsetneq F^* \subsetneq F'\}| < r$ if $\mathcal{F}$ is $\Diamond_r$-free,
- $\{F^* \in \mathcal{F} : F \subsetneq F^* \subsetneq F'\}$ does not contain antichains of size r, in particular, $|\{F^* \in \mathcal{F} : F \subsetneq F^* \subsetneq F', |F^*| = i\}| < r$ for any $|F| < i < |F'|$.

Consider first $\Diamond_r$-free families. Set $m = m_r = \lceil \log_2(r+2) \rceil$. If $|F'| - |F| < m$, then $\big|\{F^* \in \mathcal{F} : F \subsetneq F^* \subsetneq F'\}\big| < r$, and so $\mathcal{M}_{n,m}$ is $\Diamond_r$-free. Griggs, Li, and Lu [58] could show that for some values of r, if $\mathcal{F} \subseteq 2^{[k]}$ and $|\mathcal{F}| < r+2$ (the plus 2 stands for the extra sets of $\emptyset$ and $[k]$), then $\lambda_k(\mathcal{F}) \leq m-2$. Together with Lemma 2.16, this implies the following theorem that was mentioned in Section 2.3.

Theorem 2.18 (Griggs, Li, Lu [58]) *If* $r \in \left[2^{m-1}-1, 2^m - \binom{m}{\lfloor m/2 \rfloor} - 1\right]$, *then* $\mathrm{La}(n, \Diamond_r) = \Sigma(n, m)$.

For induced-$\Diamond_r$-free families, set $m' = m'_r = \min\left\{m : r \leq \binom{m}{\lfloor m/2 \rfloor}\right\}$. If $|F'| - |F| < m$, then, by Theorem 2.2, any antichain in $\{F^* \in \mathcal{F} : F \subsetneq F^* \subsetneq F'\}$ has size less than r, and so $\mathcal{M}_{n,m}$ is induced-$\Diamond_r$-free. Patkós [109] showed that, for an infinite number of values of r, the maximum value of $\lambda_k(\mathcal{F})$ of a family not containing antichains of size r is m'. This yields $\mathrm{La}^*(n, \Diamond_r) = m'$ for these values of r. It was possible to extend both the non-induced and the induced results to complete multipartite posets $\mathcal{K}_{s,r_1,r_2,\ldots,r_\ell,t}$, whenever all r_i are values covered by Theorem 2.18 or the result of Patkós from [109].

2.5.2 The cycle method In order to bound the size of a family $\mathcal{F} \subseteq 2^{[n]}$, the cycle method counts pairs (F, π) such that $F \in \mathcal{F}$, π is a cyclic permutation of $[n]$, and F forms an interval in π. The set of cyclic permutations of n is denoted Σ_n°. For a simple example, let us reprove Theorem 2.15, the YBLM-inequality.

Let $\mathrm{int}(\pi)$ denote the intervals of a cyclic permutation π. Since any set F is in $|F|!(n-|F|)!$ cyclic permutations,

$$\sum_{F \in \mathcal{F}} \sum_{\pi \in \Sigma_n^\circ} \mathbf{1}_{F \in \mathrm{int}(\pi)} = \sum_{F \in \mathcal{F}} |F|!(n-|F|)!. \tag{2.2}$$

On the other hand, by Dilworth's theorem (Theorem 1.7), each antichain has at most n members of $\mathrm{int}(\pi)$ for any $\pi \in \Sigma_n^\circ$ because each must begin at a different element. Hence a k-Sperner family has $|\mathrm{int}(\pi)| \leq kn$. Moreover, if equality holds, then for each antichain, all of the n sets have the same size. Since $|\Sigma_n^\circ| = (n-1)!$,

$$\sum_{\pi \in \Sigma_n^\circ} \sum_{F \in \mathcal{F}} \mathbf{1}_{F \in \mathrm{int}(\pi)} \leq \sum_{\pi \in \Sigma_n^\circ} kn = kn!. \tag{2.3}$$

Putting (2.2) and (2.3) together, the result in Theorem 2.15 follows and a little extra work establishes that in the equality case, the sets must take on at most k distinct sizes.

In many applications of the method, one counts sets with a weight. The most often used are $|F|$, $|F|^{-1}$, $\binom{n}{|F|}$, and $\binom{n}{|F|}^{-1}$, but sometimes ad hoc weights are of use. As mentioned earlier, DeBonis, Katona, and Swanepoel [34] obtained $\mathrm{La}(n, \bowtie) = \Sigma(n, 2)$ via the cycle method. Other notable applications are results on k-Sperner families that satisfy some additional intersecting property [17, 55, 46, 75, 49, 8].

2.6 The number of P-free families and random versions of forbidden subposet problems

In 1973, Erdős, Kleitman, and Rothschild [41] proved that, for any graph F with chromatic number $\chi(F)$, there is a $c = c(F) = 1 - \frac{1}{\chi(F)-1}$ such that the number of n-

vertex graphs that are F-free is $2^{\mathrm{ex}(n,F)+o(n^2)}$. In 1995, Bollobás and Thomason [19] determined that if F is neither complete nor empty, there is a $c^* = c^*(F)$ such that the number of n-vertex graphs that are induced-F-free is $2^{(c^*+o(1))\binom{n}{2}}$.

In the poset setting, being P-free for some poset P is a monotone property and so a trivial lower bound on the number of P-free families is $2^{\mathrm{La}(n,P)}$. One might look to [41] and conjecture that the number of such families is $2^{(1+o(1))\mathrm{La}(n,P)}$. Broadly speaking, such a statement would be a strong stability result, as it states that the number of all subfamilies of non-extremal maximal P-free families is negligible compared to the trivial ones. Of course, since $\mathrm{La}(n,P)$ is exponential in n for all nontrivial posets P, an error term of $o\big(\mathrm{La}(n,P)\big)$ would still be substantial. Nevertheless, it is interesting to obtain the asymptotics of the exponent.

A problem related to the number of P-free families is a random version of the Turán problem. In a parallel to the Erdős-Rényi [42] random graph model, we let $\mathcal{P}(n,p)$ denote the probability space such that each member of $2^{[n]}$ is chosen independently with probability p. This is a process known as *percolation*. Hence, for all $\mathcal{F} \subseteq 2^{[n]}$ it is the case that $\mathbb{P}\big(\mathcal{P}(n,p) = \mathcal{F}\big) = p^{|\mathcal{F}|}(1-p)^{2^n-|\mathcal{F}|}$.

Given a $p \in [0,1]$, we ask for the size of the largest P-free [induced-P-free] family in $\mathcal{P}(n,p)$. A sequence of events $\{\mathcal{E}_n\}$ holds *with high probability* (whp for short) if $\mathbb{P}(\mathcal{E}_n)$ tends to one as n tends to infinity. Given a maximum-sized P-free family $\mathcal{F} \subseteq 2^{[n]}$, percolating that family with probability p yields a P-free family of size $(1+o(1))\cdot \mathrm{La}(n,P)$ whp, and this can be easily shown via a concentration inequality such as a Chebyshev or Chernoff bound.

Osthus [106] addressed this problem for the case of $P = \mathcal{C}_2$, which can be restated as the size of the largest antichain.

Theorem 2.19 (Osthus [106])

- *If* $p = \omega(\log n/n)$, *then the size of the largest antichain in* $\mathcal{P}(n,p)$ *is* $(1+o(1))\cdot p\binom{n}{\lfloor n/2\rfloor}$ whp.
- *If* $p = o(1/n)$, *then the size of the largest antichain in* $\mathcal{P}(n,p)$ *is at least* $(2-o(1))\cdot p\binom{n}{\lfloor n/2\rfloor}$ whp.

Furthermore, Osthus conjectured that the threshold for the property of $\mathcal{P}(n,p)$ having its maximum-sized antichain of size $(1+o(1))p\binom{n}{\lfloor n/2\rfloor}$ is $1/n$. We say that $p_0 = p_0(n)$ is a *threshold function* for a monotone property if the property holds whp for all $p = \omega(p_0)$ and fails to hold whp whenever $p = o(p_0)$.

Osthus' conjecture was verified by Balogh, Mycroft, and Treglown [10] and generalized by Collares-Neto and Morris [28] to $P = \mathcal{C}_{k+1}$ to show that the threshold for k-families is $1/n$ for every integer k.

Theorem 2.20 (Balogh, Mycroft, Treglown [10]; Collares-Neto, Morris [28]) *Let* $k \geq 1$.

- *If* $p = \omega(1/n)$, *then the size of the largest* k*-family in* $\mathcal{P}(n,p)$ *is* $(k+o(1))\cdot p\binom{n}{\lfloor n/2\rfloor}$ whp.
- *If* $p = o(1/n)$, *then the size of the largest* k*-family in* $\mathcal{P}(n,p)$ *is at least* $(k+1-o(1))\cdot p\binom{n}{\lfloor n/2\rfloor}$ whp.

The case of $p = o(1/n)$ follows from generalizing Osthus' observation that the expected number of copies of $\mathcal{C}_2$ in two consecutive layers of $2^{[n]}$ is $O\big(np \cdot p\binom{n}{\lfloor n/2 \rfloor}\big)$ and a concentration argument shows that the number of these copies goes to 0 whp.

Both Balogh, Mycroft, and Treglown [10] and Collares-Neto and Morris [28] use the container method by Balogh, Morris, Samotij [9] and Saxton and Thomason [113] to prove the case of $p = \omega(1/n)$. For $k = 1$, Balogh, Mycroft, and Treglown [10] use what is sometimes called the graph container method and for $k \geq 2$, Collares-Neto and Morris [28] uses the more complex hypergraph container method. Hogenson [65] used the graph container method to prove that the threshold for a $\mathcal{V}_r$-free family in $\mathcal{P}(n,p)$ to have size $(1+o(1)) \cdot p\binom{n}{\lfloor n/2 \rfloor}$ is also $1/n$. This was extended by Gerbner, Nagy, Patkós and Vizer [50] to induced-$\mathcal{V}_r$-free families.

Patkós and Treglown [110] derive similar results for tree posets with Hasse diagrams of radius at most two.

Further ideas led to the proof of the counting and random versions of Theorem 2.3. Theorem 2.21 was obtained first in the non-induced case by Balogh, Garcia, and Wigal [6] with a simpler proof.

Theorem 2.21 (Jiang, Longbrake, Spiro, Yepremyan [72]) *Let T be a tree poset of height k. The number of induced-T-free families in $2^{[n]}$ is $2^{(k-1+o(1))\binom{n}{\lfloor n/2 \rfloor}}$.*

Theorem 2.22 (Jiang, Longbrake, Spiro, Yepremyan [72]) *Let T be a tree poset of height k. There exists C such that if $p \geq C/n$, then the size of the largest induced-T-free family in $\mathcal{P}(n,p)$ is $(k-1+o(1)) \cdot p\binom{n}{\lfloor n/2 \rfloor}$ whp.*

This leads us to the general questions in this area:

Problem 6 *Let P be a poset. Let $N(n,P)$ denote the number of P-free families in $2^{[n]}$ and let $N^*(n,P)$ denote the number of induced-P-free families in $2^{[n]}$.*

- *Determine asymptotic bounds for $\log_2 N(n,P)/\binom{n}{\lfloor n/2 \rfloor}$.*
- *Determine asymptotic bounds for $\log_2 N^*(n,P)/\binom{n}{\lfloor n/2 \rfloor}$.*
- *Determine the values of p for which the largest P-free family in $\mathcal{P}(n,p)$ has size $(1+o(1)) \cdot p\mathrm{La}(n,P)$.*

In particular, since the expected number of copies of a $\bowtie$ in three consecutive middle layers of $2^{[n]}$ is $O\big(n^4p^3 \cdot p\binom{n}{\lfloor n/2 \rfloor}\big)$ and the expected number of copies of a $\Diamond$ in three consecutive middle layers of $2^{[n]}$ is $O\big(n^2p^3 \cdot p\binom{n}{\lfloor n/2 \rfloor}\big)$, we ask the following:

Problem 7

- *Determine if a threshold for the largest $\bowtie$-free family in $\mathcal{P}(n,p)$ being $\big(2 + o(1)\big) \cdot p\binom{n}{\lfloor n/2 \rfloor}$ exists. Is $n^{-4/3}$ such a threshold function?*
- *Determine if a threshold for the largest $\Diamond$-free family in $\mathcal{P}(n,p)$ being $\big(1+o(1)\big) \cdot p\mathrm{La}\big(n, \Diamond\big)$ exists. Is $n^{-2/3}$ such a threshold function?*

The recent breakthrough result of Park and Pham [108] (answering a conjecture of Kahn and Kalai [74]) relates the expectation threshold of a monotone property to the threshold of the existence of that property. Though it doesn't apply directly, some ideas in [108] may inspire a solution to Problem 7.

3 Saturation

Whereas Turán-type problems ask for the maximum size of a combinatorial object that has no forbidden substructure, saturation problems ask for the *minimum* size of a combinatorial object that has no forbidden substructure, but the addition of **any** element creates a copy of the forbidden substructure.

For posets, the definition of saturation naturally parallels that of saturation studied in graph theory, with the additional notion of induced saturation.

Definition 3.1 *A family $\mathcal{F} \subseteq 2^{[n]}$ is [induced] P-saturated if it is*

- *[induced-]P-free,*
- *for any $G \in 2^{[n]} \setminus \mathcal{F}$, the family $\mathcal{F} \cup \{G\}$ contains a [an induced] copy of P.*

The [induced] saturation number of P is minimum size of a [an induced] P-saturated family $\mathcal{F} \subseteq 2^{[n]}$ and is denoted by $\mathrm{sat}(n, P)$ *[*$\mathrm{sat}^*(n, P)$*].*

Saturation for posets was first investigated by Gerbner, Keszegh, Lemons, Palmer, Pálvölgyi, and Patkós [48] in 2013. Induced saturation was initiated by Ferrara, Kay, Kramer, Martin, Reiniger, Smith, and Sullivan [44] in 2017. It should be noted that induced saturation in graphs does not follow naturally from traditional extremal graph theory and that, when studied [97], it either cannot apply to all graphs or requires notions such as Chudnovski's "trigraph" [27]. In posets, however, induced saturation is well-defined and, indeed, well-studied.

There exist several variants of this problem that parallel notions from graph theory: Removing the first condition from Definition 3.1 gives the notion of *oversaturated* families, while if one can enumerate sets of $2^{[n]} \setminus \mathcal{F}$ as $G_1, G_2, \ldots, G_{2^n-|\mathcal{F}|}$ such that for any $1 \le j \le 2^n - |\mathcal{F}|$ the family $\mathcal{F} \cup \{G_i : 1 \le i \le j\}$ contains a copy $\mathcal{H}$ of P with $G_j \in \mathcal{H}$, then $\mathcal{F}$ is said to be *weakly P-saturated.* These notions are well-developed in graphs and Morrison, Noel, and Scott [103] discuss oversaturation in posets. For this manuscript, however, we will only consider P-saturated and induced-P-saturated families.

3.1 Saturation – chains and general bounds

Gerbner, Kesegh, Lemons, Palmer, Pálvölgyi, and Patkós [48] first considered the case $P = \mathcal{C}_{k+1}$, i.e., the saturation version of k-Sperner families. Note that chains are the only posets for which saturation and induced saturation are the same. The following natural construction shows that there is upper bound on $\mathrm{sat}(n, \mathcal{C}_{k+1})$ that is independent of n. Let

$$\mathcal{F}(n,k) = 2^{[k-2]} \cup \{[n] \setminus F : F \in 2^{[k-2]}\}.$$

It is easy to verify that $\mathcal{F}(n,k)$ is $\mathcal{C}_{k+1}$-saturated and has size 2^{k-1}.

Gerbner, et al. [48] gave a better bound for $k = 6$, showing that $\mathrm{sat}(6, \mathcal{C}_{6+1}) \le 30 = \frac{15}{16}2^{6-1}$ and the upper bound for higher values of k was extended to give the result $\mathrm{sat}(k, \mathcal{C}_{k+1}) \le \frac{15}{16}2^{k-1}$. Then Morrison, Noel, and Scott [103] used the 30-element construction of Gerbner, et al. [48] as a starting point along with several structural lemmas to obtain $\mathrm{sat}(n, \mathcal{C}_{k+1}) \le 2^{(1-\varepsilon)k}$ for $\varepsilon = 1 - \frac{1}{4}\log_2 15 \approx 0.023277$

for all $k \geq 6$. In fact, they were able to show that any construction that improves the upper bound for some fixed k will increase the value of ε for larger values of k.

Martin and Veldt [99] found a $\mathcal{C}_{7+1}$-free construction (i.e., a 7-Sperner family) on 56 elements, giving a better value of $\varepsilon = 1 - \frac{1}{5}\log_2 28 \approx 0.038529$. This was later improved by Charton, Ellenberg, Wagner, and Williamson [24] who used the method of machine learning in order to find a $\mathcal{C}_{8+1}$-free construction on 108 elements, giving a better value of $\varepsilon = 1 - \frac{1}{6}\log_2 54 \approx 0.0408521$.

A lower bound of $\mathrm{sat}(n, \mathcal{C}_{k+1}) \geq 2^{k/2-1.5}$ was given by Gerbner, et al. [48], which was improved by Martin and Veldt [99] to $\mathrm{sat}(n, \mathcal{C}_{k+1}) \geq \sqrt{k}\, 2^{k/2}$ for $k \geq 497$ and $\mathrm{sat}(n, \mathcal{C}_{k+1}) \geq \sqrt{k}\, 2^{k/2-1.66}$ for all k.

Problem 8 *Improve the following bounds for sufficiently large* k*:*

$$\sqrt{k}\, 2^{k/2} \leq \mathrm{sat}\big(n, \mathcal{C}_{k+1}\big) \leq 2^{(1-0.0408521)k}.$$

It turns out that the saturation number $\mathrm{sat}(n, P)$ is bounded by an absolute constant, independent of n.

Theorem 3.2 (Keszegh, Lemons, Martin, Pálvölgyi, Patkós [80]) *For any finite poset* P*, it is the case that* $\mathrm{sat}(n, P) \leq 2^{|P|-2}$.

The proof of this result uses a simple algorithm that generates a[n induced-] P-saturated family and so we present it here.

Proof Let $G_1, G_2, \ldots, G_{2^n}$ be an enumeration of the sets of $2^{[n]}$ and set $\mathcal{F}_0 = \emptyset$. For $j = 1, 2 \ldots, 2^n$, let $\mathcal{F}_j = \mathcal{F}_{j-1}$ if $\mathcal{F}_{j-1} \cup \{G_j\}$ contains a [an induced] copy of P, and let $\mathcal{F}_j = \mathcal{F}_{j-1} \cup \{G_j\}$ if $\mathcal{F}_{j-1} \cup \{G_j\}$ is [induced-]P-free. Finally, set $\mathcal{F} = \mathcal{F}_{2^n}$. It is clear that $\mathcal{F}$ is [induced-]P-saturated.

However, the size of the output family $\mathcal{F}$ depends very much on P, on induced or non-induced copies we are looking for, and most importantly on the enumeration of the sets. The enumeration is as follows: let $\leqslant_{colex}$ be the usual *colex* ordering of $2^{[m]}$, that is $F \leqslant_{colex} G$ if and only if the largest element of $(F \setminus G) \cup (G \setminus F)$ belongs to G. Then let $F_1, F_2, \ldots, F_{2^{n-1}}$ be the enumeration of sets of $2^{[n-1]}$ according to the colex ordering, and let the final ordering of $2^{[n]}$ be

$$F_1, [n] \setminus F_1, F_2, [n] \setminus F_2, F_3, [n] \setminus F_3, \ldots, F_{2^{n-1}}, [n] \setminus F_{2^{n-1}}.$$

Observe that the $\mathcal{C}_{k+1}$-saturated family $\mathcal{F}(n, k)$ is exactly the output of the greedy process with this ordering of the sets. It was shown by Keszegh, Lemons, Martin, Pálvolgyi, and Patkós [80] that, in the non-induced case, no matter what P is, the last set that can be possibly added to $\mathcal{F}$ is $[n] \setminus F_{2^{|P|-3}}$. □

3.2 Induced saturation – general bounds

Although $\mathrm{sat}(n, P)$ is bounded by a constant independent of n, this is not necessarily true of $\mathrm{sat}^*(n, P)$. For example, consider the two-element antichain $\mathcal{A}_2$. By the induced-$\mathcal{A}_2$-free property, an induced-$\mathcal{A}_2$-saturated family must be a chain, and by the definition of saturation, it has to be a maximal chain, and so any induced-$\mathcal{A}_2$-saturated family in $2^{[n]}$ has size $n + 1$. Hence, $\mathrm{sat}^*(n, \mathcal{A}_2) = n + 1$.

We will return to induced saturation of antichains, but let us overview the general results first. Keszegh, Lemons, Martin, Pálvölgyi, and Patkós [80] established that there is a dichotomy phenomenon with the behavior for $\mathrm{sat}^*(n,P)$. For a pair $x,y \in [m]$, a family $\mathcal{F} \subset 2^{[m]}$ is said to *not separate x and y* if $\left|F \cap \{x,y\}\right| \in \{0,2\}$ for every $F \in \mathcal{F}$.

Lemma 3.3 (Keszegh, Lemons, Martin, Patkós, Pálvölgyi [80]) *If there is an integer m and an induced-P-saturated family $\mathcal{F}_m \subseteq 2^{[m]}$ such that it does not separate a pair $x,y \in [m]$, then $\mathrm{sat}^*(n,P) \le \left|\mathcal{F}_m\right|$ for all $n \ge m$. Otherwise, $\mathrm{sat}^*(n,P) \ge \log_2 n$.*

Proof If $\mathcal{F}_m \subseteq 2^{[m]}$ is induced-P-saturated and does not separate x and y, then let $\mathcal{F}^i = \left\{F \in F_m : \left|F \cap \{x,y\}\right| = i\right\}$ for $i = 0,2$. Then for any $n \ge m$, the family

$$\mathcal{F}_n = \mathcal{F}^0 \cup \left\{F \cup ([n] \setminus [m]) : F \in \mathcal{F}^2\right\}$$

is induced-P-saturated and of size $|\mathcal{F}_m|$ for all $n \ge m$.

Indeed, construct $\mathcal{F}_n$ as the image of the function $f : \mathcal{F}_m \to 2^{[n]}$ where $f(F) = F$ for all $F \in \mathcal{F}^0$ and $f(F) = F \cup \big([n]\setminus[m]\big)$ for all $F \in \mathcal{F}^2$. The containment relationship between sets is preserved by f, this shows that $\mathcal{F}_m$ and $\mathcal{F}_n$ induce the same poset structure in $2^{[m]}$ and $2^{[n]}$, respectively. In particular, $\mathcal{F}_n$ is induced P-free.

We still need to show that for any $G \in 2^{[n]} \setminus \mathcal{F}_n$, the family $\mathcal{F}_n \cup \{G\}$ contains an induced copy of P. If $G \cap \big(\{x,y\} \cup ([n] \setminus [m])\big) = \emptyset$, then $G \in 2^{[m]} \setminus \mathcal{F}_m$, so $\{G\} \cup \mathcal{F}_m$ contains an induced copy $\{G\} \cup \mathcal{G}$ of P, and $\{G\} \cup f(\mathcal{G})$ is an induced copy of P in $\mathcal{F}_n \cup \{G\}$. If G contains $\{x,y\} \cup \big([n] \setminus [m]\big)$, then $G' = G \cap [m] \notin \mathcal{F}_m$ as otherwise $G = f(G') \in \mathcal{F}_n$. So $\{G'\} \cup \mathcal{F}_m$ contains an induced copy $\{G'\} \cup \mathcal{G}$ of P and again $\{G\} \cup f(\mathcal{G})$ is an induced copy of P in $\{G\} \cup \mathcal{F}_n$. Finally, if $G \cap \big(\{x,y\} \cup \big([n]\setminus[m]\big)\big) \ne \emptyset$ and $\big(\{x,y\} \cup \big([n] \setminus [m]\big)\big) \setminus G \ne \emptyset$, then $G' = \big(G \cap \big([m] \setminus \{x,y\}\big) \cup \{x\} \in 2^{[m]} \setminus \mathcal{F}_m$ (as $|G' \cap \{x,y\}| = 1$), so there is an induced copy $\mathcal{G}$ of P in $\{G'\} \cup \mathcal{F}_m$. Now $f(\mathcal{G} \setminus \{G'\}) \cup \{G\}$ is an induced copy of P in $\mathcal{F}_n \cup \{G\}$. So, $\mathcal{F}_n$ is indeed induced-P-saturated.

If $\mathcal{F} = \left\{F_1, F_2, \ldots, F_k\right\} \subseteq 2^{[n]}$ separates every $x,y \in [n]$, then construct $0-1$ vectors $v_1, v_2, \ldots, v_n$, each of length k, so that $v_i^j = 1$ if and only if $i \in F_j$. Since all vectors are distinct, $2^k \ge n$. □

Freschi, Piga, Sharifzadeh, and Treglown [47] proved a stronger lower bound of $2\sqrt{n}$ for $n \ge 11$:

Theorem 3.4 (Freschi, Piga, Sharifzadeh, Treglown [47]) *For any finite poset P either there exists a constant $C = C_P$ such that $\mathrm{sat}^*(n,P) \le C$ for all positive integers n, or we have $\mathrm{sat}^*(n,P) \ge \min\left\{2\sqrt{n}, n/2+1\right\}$.*

Note that $\lceil 2\sqrt{n}\rceil \le \lceil n/2+1\rceil$ for all $n \ge 11$, so we will assume $n \ge 11$ when using this bound.

We can easily show a weaker lower bound of $\sqrt{n}$ by adapting the above proof of Lemma 3.3 (and using an idea from Martin, Smith, and Walker [96]): Suppose for some $x \in [m]$ and an induced-P-saturated family $\mathcal{F}_m \in 2^{[m]}$, there is no pair

$F, F' \in \mathcal{F}_m$ with $F \setminus F' = \{x\}$. With $\mathcal{F}^- = \{F \in \mathcal{F}_m : x \notin F\}$ and $\mathcal{F}^+ = \{F \in \mathcal{F}_m : x \in F\}$, the family

$$\mathcal{F}_n = \mathcal{F}^- \cup \{F \cup ([n] \setminus [m]) : F \in \mathcal{F}^+\}$$

is induced-P-saturated and of size $|\mathcal{F}_m|$ for all $n \geq m$.

Hence, either $\text{sat}^*(n, P)$ is bounded by a function of P or, if $\mathcal{F}_n \subseteq 2^{[n]}$, then for every $x \in [n]$, there is a pair $F, F' \in \mathcal{F}_n$ such that $F \setminus F' = \{x\}$. Thus, $2\binom{|\mathcal{F}_n|}{2} \geq n$ and so $|\mathcal{F}_n| \geq \sqrt{n}$.

It is not clear whether the lower bound in Theorem 3.4 of $2\sqrt{n}$ for unbounded $\text{sat}^*(n, P)$ functions can be asymptotically improved. However, there have been some recent breakthroughs in improving square root lower bounds for $\text{sat}^*(n, P)$ to linear ones. Ivan and Jaffe [70] showed that $\text{sat}^*(n, \Diamond) \geq (n + 1)/5$ and Brownlie and Jaffe [20] showed that $\text{sat}^*(n, \mathcal{V}_3) \geq n/2$. There is no poset P known for which $\text{sat}^*(n, P)$ goes to infinity sublinearly.

No general upper bound on $\text{sat}^*(n, P)$ was known up until very recently. A polynomial upper bound was proved by Bastide, Groenland, Ivan, and Johnston [11] using a careful analysis of the greedy algorithm mentioned above. The enumeration that the authors used lists subsets of $[n]$ in increasing order of their sizes, applies the Sauer Lemma [112, 114] together with several newly introduced notions.

Theorem 3.5 (Bastide, Groenland, Ivan, Johnston [11]) *For every finite poset P and positive integer n, it is the case that* $\text{sat}^*(n, P) \leq n^{|P|^2/2}$.

This was later improved to $2n^{|P|-1}$ by Flídr, Ivan, and Jaffe [45] and independently by Bastide, Hodor, La, and Trotter [13].

Theorem 3.6 (Flídr, Ivan, Jaffe [45]; Bastide, Hodor, La, Trotter [13]) *For every finite poset P and positive integer n, it is the case that* $\text{sat}^*(n, P) \leq 2n^{|P|-1}$.

For all of the posets for which $\text{sat}^*(n, P)$ is known asymptotically, it is either bounded by a constant or grows linearly. See Appendix A.

Problem 9

- *Determine whether there a poset P for which* $\text{sat}^*(n, P)$ *goes to infinity but* $\text{sat}^*(n, P) = o(n)$.
- *Determine whether there a poset P for which* $\text{sat}^*(n, P) = \omega(n)$.

3.3 Poset properties that imply unbounded $\text{sat}^*(n, P)$

Although Theorem 3.4 establishes a strong duality for $\text{sat}^*(n, P)$, it is still not clear from the properties of P whether $\text{sat}^*(n, P)$ is bounded or increases with n.

Recall that for distinct $x, y \in P$, we say that y *covers* x if $x \leqslant_P z \leqslant_P y$ implies that either $x = z$ or $z = y$. A poset P is said to have the *unique cover twin property* (UCTP) if whenever an $x \in P$ has a unique cover $y \in P$, then it must be the case that y also covers another $z \in P \setminus \{x\}$.

Theorem 3.7 (Ferrara, Kay, Kramer, Martin, Reiniger, Smith, Sullivan [44]) *If P has the UCTP, then any P-saturated family $\mathcal{F} \subseteq 2^{[n]}$ is separating.*

Therefore, by Theorem 3.4, if P has the UCTP (and $n \geq 11$) then $\text{sat}^*(n, P) \geq 2\sqrt{n}$.

We remark that one can observe that P has the UCTP if and only if its dual $P^\perp$ has the UCTP. The dual $P^\perp$ of $P = (X, \leqslant_{P^\perp})$ is defined on the same set X of vertices and $x \leqslant_{P^\perp} y$ if and only if $y \leqslant_P x$.

A condition is presented by Freschi, Piga, Sharifzadeh, and Treglown [47] that implies $\text{sat}^*(n, P) \geq n + 1$. Two vertices x, y of P are said to be *two legs of P*, if x and y are incomparable and for any $z \in P \setminus \{x, y\}$ we have both $x < z$ and $y < z$. A *hip* of a poset P with two legs x, y is a vertex $h \in P \setminus \{x, y\}$ such that $h < z$ for any $z \in P \setminus \{x, y, h\}$.

Theorem 3.8 (Freschi, Piga, Sharifzadeh, Treglown [47]) *Let $n \geq 3$ and let P be a poset with two legs and a hip. Then* $\text{sat}^*(n, P) \geq n + 1$. *Moreover, if both P and $P^\perp$ have two legs and a hip, then* $\text{sat}^*(n, P) \geq 2n + 2$.

Theorem 3.8 was strengthened by Liu who proved that the condition about the existence of a hip can be dropped.

Theorem 3.9 (Liu [87]) *If a poset P has two legs, then* $\text{sat}^*(n, P) \geq n + 1$.

Ivan and Jaffe [68] considered the effect of the gluing operation on the poset saturation number. Recall that the poset $P_1 * P_2$ is the poset on $P_1 \cup P_2$ with all relations inherited from P_1 and P_2 and every $p_1 \in P_1$ being smaller than every $p_2 \in P_2$.

By applying Theorem 3.4, Ivan and Jaffe [68] proved the following result.

Theorem 3.10 (Ivan, Jaffe [68]) *Let P_1 and P_2 be two finite posets such that at least one of them has the UCTP and at least two elements. If $n \geq 11$, then* $\text{sat}^*(n, P_1 * P_2) \geq 2\sqrt{n}$.

3.4 $\text{sat}^*(n, P)$ for specific posets

In this subsection we enumerate results on the saturation number of particular posets or classes of posets.

Antichains $\mathcal{A}_{k+1}$ are among the most natural posets to consider and yet the solution turned out to be difficult. For fixed k, lower and upper bounds that are linear in n were obtained in the introductory paper by Ferrara, Kay, Kramer, Martin, Reiniger, Smith, and Sullivan [44]. The trivial upper bound of $k(n-1)+2$ is given by k maximum chains which share no common vertices except $\emptyset$ and $[n]$. This can be improved to $kn - \Omega(k \log k)$ by letting $k = \text{sp}(j)$ from Definition 1.9 and creating a family with a copy of the bottom half of $\mathcal{B}_j$ in the bottom $\lfloor j/2 \rfloor$ layers of $\mathcal{B}_n$, a copy of the top half of $\mathcal{B}_j$ in the top $\lfloor j/2 \rfloor$ layers and connect them in between with k disjoint maximal chains.

A lower bound of $\text{sat}^*(n, \mathcal{A}_{k+1}) \geq 3n-1$ was shown for $k \geq 3$ with the exact value $\text{sat}^*(n, \mathcal{A}_4) = 3n - 1$. The general lower bound was improved by Martin, Smith, and Walker [96] to $\text{sat}^*(n, \mathcal{A}_{k+1}) \geq \left(1 - \frac{1}{\log_2 k}\right) \frac{k}{\log_2 k} n$. Đanković and Ivan [32] determined $\text{sat}^*(n, \mathcal{A}_5) = 4n - 2$ and $\text{sat}^*(n, \mathcal{A}_6) = 5n - 5$.

Finally, Bastide, Groenland, Jacob, and Johnston [12] determined the exact value of $\text{sat}^*(n, \mathcal{A}_{k+1}) = kn - \Theta(k \log k)$. The upper bound improves on that of Ferrara, et al. [44] by making the $\lceil j/2 \rceil$ bottom and $\lceil j/2 \rceil$ top layers more efficient. The lower bound uses a result of Lehman and Ron [86].

The case of $\text{sat}^*(n, \mathcal{V}) = n + 1$ was established by Ferrara, et al. [44]. Note that $\text{sat}^*(n, P)$ is the same as $\text{sat}^*(n, P^\perp)$ where $P^\perp$ is the dual of P.

For any new variant of the Turán problem, the butterfly poset $\bowtie$ is a common "toy example". It does have the UCTP, so its saturation number is unbounded. It has two legs, so Theorem 3.9 implies $\text{sat}^*(n, \bowtie) \geq n+1$. This was earlier established by Ivan [66], and, in fact, her ideas were used by Liu [87] in order to obtain Theorem 3.9. An upper bound of $\text{sat}^*(n, \bowtie) \leq 6n - 10$, was shown by Keszegh, Lemons, Martin, Pálvölgyi, and Patkós [80], and so $\bowtie$ was one of the earliest posets with unbounded induced saturation number for which the order of magnitude was determined.

Once again the diamond poset, $\Diamond$, is an open problem. As it has the UCTP, it was stated by Ferrara, et al. [44] that its saturation number is unbounded. Before Theorem 3.4, the lower bound was improved by Martin, Smith, and Walker [96] to $\sqrt{n}$ and was improved later by Ivan [67] to $(4 - o(1))\sqrt{n}$. Recently, Ivan and Jaffe [70] established the best-known lower bound, $\text{sat}^*(n, \Diamond) \geq (n + 1)/5$, which is of particular interest because it is the first linear lower bound for the $\Diamond$.

An easy upper bound, stated by Ferrera, et al. [44], is $\text{sat}^*(n, \Diamond) \leq n + 1$. This is given by two nonisomorphic families: A maximum chain and the family consisting of the emptyset and all singletons. Moreover, it was proved by Ferrara, et al. [44] that any induced-$\Diamond$-saturated family in $2^{[n]}$ that contains either the element $\emptyset$ or the element $[n]$ has size at least $n+1$. Therefore, in order to verify that the upper bound for $\text{sat}^*(n, \Diamond)$ is correct, it suffices to show that every induced-$\Diamond$-saturated family in $2^{[n]}$ without $\emptyset$ or $[n]$ has at least $n + 1$ elements.

Consider $\mathcal{J}$, a poset on four vertices a, b, c, d with $a < b, c, d$ and $c < d$, i.e., the diamond with one of the upper relations removed (see Figure 5(A)). It is not difficult to see that the family $\emptyset, \{1\}, [n] - \{1\}, [n]$ forms a minimum-sized induced-$\mathcal{J}$-free family.

For $\text{sat}^*(n, \mathcal{N})$, the lower bound $\sqrt{n}$ was first obtained by Ivan [66], later superseded by Theorem 3.4. An easy construction can be given by the union of a maximal chain and all singletons, showing that $2\sqrt{n} \leq \text{sat}^*(n, \mathcal{N}) \leq 2n$.

The $\mathcal{X}$-poset has 5 vertices a, b, c, d, e with $a, b < c < d, e$ (see Figure 5(B)) whereas the $\mathcal{Y}$-poset has 4 vertices a, b, c, d with $a < b < c, d$ (see Figure 5(C)). In other notation, $\mathcal{X} = \Lambda^{\mathcal{V}}$ and $\mathcal{Y} = \mathcal{C}_2^{\mathcal{V}}$. Using Theorem 3.8, Freschi, Piga, Sharifzadeh, and Treglown proved [47] $\text{sat}^*(n, \mathcal{X}) = 2n + 2$ and $\text{sat}^*(n, \mathcal{Y}) = n + 2$.

(A) $\mathcal{J}$ (B) $\mathcal{X}$ (C) $\mathcal{Y}$ (D) $\mathcal{V}_3$

Figure 5: Posets with known values of $\text{sat}^*(n, P)$

Liu [87] considered complete bipartite posets (see, e.g., Figure 1(H)). Observe that $\mathcal{C}_2$, $\mathcal{V}_2$, and $\bowtie$ each belong to this class of posets. By a general construction, he obtained $\text{sat}^*(n, \mathcal{K}_{s,t}) = O_{s,t}(n)$ for all positive integers s and t. In addition, Theorem 3.9 implies a linear lower bound if one class has size 2, and thus

$\text{sat}^*(n, \mathcal{K}_{s,2}) = \Theta(n)$.

Ivan and Jaffe [69] managed to generalize the linear upper bound to complete multipartite posets (see, e.g., Figure 1(I)), showing $\text{sat}^*(n, K_{s_1,s_2,\ldots,s_k}) = O(n)$.

Later Ivan and Jaffe [69] showed a linear lower bound where, if P_1 and P_2 are (possibly empty) posets, then, as long as $k \geq 2$, $\text{sat}^*(n, P_1 * \mathcal{A}_k * P_2) = \Omega(n)$. Note that the case where both P_1 and P_2 are empty, this is a consequence of the result of Bastide, Groenland, Jacob, and Johnston [12]. As a consequence, the asymptotic case for multipartite posets has been settled.

Theorem 3.11 (Ivan, Jaffe [68, 69]) *Let k and $s_1, \ldots, s_k$ be positive integers. If $\max\{s_1, \ldots, s_k\} \geq 2$, then $\text{sat}^*(n, \mathcal{K}_{s_1,s_2,\ldots,s_k}) = \Theta(n)$. Otherwise $\mathcal{K}_{s_1,s_2,\ldots,s_k}$ is a chain and $\text{sat}^*(n, \mathcal{C}_k) = O(1)$.*

Recently, Brownlie and Jaffe [20] established a linear lower bound of $\text{sat}^*(n, \mathcal{V}_3) \geq n/2$ (see Figure 5(D)).

Ivan and Jaffe [68] also categorized a class of posets which have bounded saturation number which they called *special*. A poset P is special if $P = P' \cup \{p\}$ such that P' has a smallest and a largest vertex and p is incomparable to all vertices of P'. Note that all posets of the form $\mathcal{C}_k + \mathcal{C}_1$ are special.

This leads us to the last class of posets we consider here: posets that are formed by the union of pairwise incomparable chains $\mathcal{C}_{i_1} + \cdots + \mathcal{C}_{i_m}$. This investigation was already started by Keszegh, Lemons, Martin, Pálvölgyi, and Patkós [80] in which the colex greedy algorithm yielded Theorem 3.2 on several small posets. Whenever the algorithm's output was a non-separating saturated family, it implied that $\text{sat}^*(n, P)$ is bounded by a constant. This was the case for the following: $\mathcal{C}_2 + 2\mathcal{C}_1$, $\mathcal{C}_3 + \mathcal{C}_1$, $3\mathcal{C}_2$, $5\mathcal{C}_2$, $7\mathcal{C}_2$, $3\mathcal{C}_3$, $3\mathcal{C}_4$, $2\mathcal{C}_i + \mathcal{C}_j$ for $i \in \{3,4\}$, $j \in \{1,2,3\}$ and $i > j$ as well as $2\mathcal{C}_k + 2\mathcal{C}_\ell$ for $k \in \{3,4\}$, $\ell \in \{1,2\}$ and $k > \ell + 1$.

With a completely different approach, it was shown by Keszegh, et al. [80] that $\text{sat}^*(n, \mathcal{C}_{i_1} + \cdots + \mathcal{C}_{i_m})$ is bounded whenever $i_1 > i_j$ for all $2 \leq j \leq m$. Moreover, they found a linear lower bound for $2\mathcal{C}_2$ and together with the $2n$ construction consisting of a maximum chain and the set of singletons, it shows $\text{sat}^*(n, 2\mathcal{C}_2) = \Theta(n)$. The lower bound was improved by Martin and Veldt, so current best bounds are $(3n+1)/2 \leq \text{sat}^*(n, 2\mathcal{C}_2) \leq 2n$.

Recently, Ji, Patkós, and Yie [71] showed $\text{sat}^*(n, m\mathcal{C}_k) = O_{m,k}(n)$ for all values of m and k. They also obtained constructions proving $\text{sat}^*(n, 2\mathcal{C}_k + \mathcal{C}_1)$ is constant for any $k \geq 2$ and $\text{sat}^*(n, m\mathcal{C}_2)$ is constant whenever $m = \binom{2t}{t} + 1$. So for an infinite number of odd values, we know that $\text{sat}^*(n, m\mathcal{C}_2)$ is bounded by a constant. In the case of the only even m for which the order of magnitude of $\text{sat}^*(n, m\mathcal{C}_2)$ is known, it is linear in n.

Problem 10 *For each $m \geq 2$, determine $\text{sat}^*(n, m\mathcal{C}_2)$. In particular determine if $\text{sat}^*(n, m\mathcal{C}_2) = O(1)$ for odd $m \geq 3$ and $\text{sat}^*(n, m\mathcal{C}_2) = \Theta(n)$ for even $m \geq 4$.*

4 Ramsey numbers for posets

Ramsey theory originates from a groundbreaking result by Ramsey [111] from 1930, who showed that a sufficiently large uniform hypergraph colored with a fixed number of colors contains a monochromatic sub-hypergraph of any fixed size. Even shortly before that, van der Waerden [120] considered a Ramsey-type question on *arithmetic progressions*. Van der Waerden's work was later generalized by the highly influential Hales-Jewett theorem [63], which laid the foundation for Ramsey theory on various discrete settings such as *combinatorial lines* or *words*. In the last decades, Ramsey theory established itself as an active and major field in discrete mathematics, applied to various discrete settings. To name just a few, there are noteworthy recent advances in arithmetic Ramsey theory [85, 88], Ramsey theory on vector spaces [104], and Euclidean Ramsey theory [30].

4.1 Definition and existence

Consider the following Ramsey question for posets: A *blue/red coloring* of a poset P is a mapping $c\colon P \to \{\text{blue}, \text{red}\}$. This definition extends naturally to colorings with more than two colors. We say that a poset is *monochromatic* if all its vertices have the same color. If every vertex of a subposet is blue, we say that the subposet itself is *blue*. Similarly, if every vertex is red, the subposet itself is *red*.

For posets P and Q, let the *poset Ramsey number* of P and Q be defined as follows:

$$R(P,Q) = \min\big\{N \in \mathbf{N} : \text{every blue/red coloring of } 2^{[N]} \text{ contains either} \\ \text{a blue induced copy of } P \text{ or a red induced copy of } Q\big\}.$$

We shall give an easy argument why this Ramsey number exists, see Axenovich and Walzer [2]. The existence also follows from a classical theorem of Graham and Rothschild [52].

Theorem 4.1 *For any posets P and Q, the poset Ramsey number $R(P,Q)$ is well-defined.*

Proof Let $s = \max\{|P|, |Q|\}$ and $n = 2^s$. We first claim that P and Q can be embedded in $\mathcal{B}_n$. Without loss of generality $s = |P|$ and identify the vertices of $\mathcal{B}_n$ with subsets of P. Consider an embedding $f : P \mapsto 2^{[n]}$ by letting $f(X) = \{Y \in P : Y \leqslant_P X\}$. This is an injection and $X \leqslant_P Y$ iff $f(X) \leqslant_{\mathcal{B}_n} f(Y)$.

Because of this embedding, it is sufficient to show that $R(\mathcal{B}_n, \mathcal{B}_n)$ exists. Consider $2^{[N]}$ with $N = n^2+2n$ and an arbitrary blue/red coloring of its vertices. Let N be the disjoint union $X_0 \sqcup X_1 \sqcup \cdots \sqcup X_{n+1}$, where each X_i has size n. For each nonempty $Y \subseteq X_0$, let $B_Y = \{Y \cup X_1 \cup \cdots \cup X_{|Y|} \cup X : X \subseteq X_{|Y|+1}\}$. Furthermore, let $B_\emptyset = \{X \subseteq X_1\}$. We see that each B_Y is a copy of $\mathcal{B}_n$. If this copy is blue, then B_Y gives a monochromatic copy of $\mathcal{B}_n$. Otherwise, there is a red element Z_Y in each B_Y. For nonempty Y, this element has form $Z_Y = Y \cup X_1 \cup \cdots \cup X_{|Y|} \cup S_Y$, where $S_Y \subseteq X_{|Y|+1}$. The element $Z_\emptyset$ has the form $X \subseteq X_1$. We claim that these elements form a red copy of $\mathcal{B}_n$. Indeed, for each $Y', Y'' \subseteq X_0$ it is the case that $Y' \subseteq Y''$ iff $Z_{Y'} \subseteq Z_{Y''}$. □

Note that this argument gives very weak general upper bound: For $|P| \geq |Q|$, $R(P,Q) \leq 2^{2|P|+1}$. This bound can be significantly improved as we show in the next sections.

4.2 Historical remarks

Nešetřil and Rödl [105] were the first to consider Ramsey-type problems for posets. For posets X, Y, Z, we write $Z \xrightarrow[X]{c} Y$ if in any assignment of colors from $\{1, \dots, c\}$ to the induced copies of poset X in poset Z there is an induced copy of poset Y in Z such that all its subposets isomorphic to X have the same color. We say that for a class of posets $\mathcal{U}$, a poset X is *Ramsey* in $\mathcal{U}$ if for any $Y \in \mathcal{U}$ there is $Z \in \mathcal{U}$ such that $Z \xrightarrow[X]{c} Y$.

Theorem 4.2 (Nesetril-Rödl [105]) *A poset X is Ramsey in the class of all posets if and only if X is complete multipartite.*

We shall consider only the case where X is a singleton poset with exactly one vertex. See also papers by Paoli, Trotter, and Walker [107] and Sokić [115] on this topic. Kierstead and Trotter [81] considered this setting for general posets in a quantitative approach, with the goal of minimizing $p(F)$ for all P with a fixed $p(P)$, where p is a poset parameter such as size, height, or width. If we set the poset parameter p to be the 2-dimension, this leads to the definition of the poset Ramsey number $R(P, Q)$ above, which was introduced by Axenovich and Walzer [2]. Note that the existence result follows from a more general result by Graham and Rothschild [52], who considered so-called *cubes*. Cubes can be interpreted as finite *affine spaces*, see also the *Ramsey Theory* book by Graham, Rothschild, and Spencer [53].

Here, we present an interpretation of the Graham-Rothschild theorem in terms of posets. For any positive integers t and n, the *n-dimensional t-cube* $Q_n^{(t)}$ denotes the poset on vertices $\{1, \dots, t\}^n$ with a partial order $\leq$, such that for vertices X and Y, we have $X \leq Y$ if and only if the ith component of X is at most the ith component of Y for every $i \in \{1, \dots, n\}$. The Graham-Rothschild theorem [52] shows that there exists a sufficiently large N such that any coloring of copies of $Q_q^{(t)}$ in $Q_N^{(t)}$ with r colors contains a monochromatic copy of $Q_n^{(t)}$. In the special case $q = 0$, $r = 2$, and $t = 2$, this shows the existence of the poset Ramsey number $R(\mathcal{B}_n, \mathcal{B}_n)$.

4.3 The diagonal setting: $R(\mathcal{B}_n, \mathcal{B}_n)$

Naturally, one of the central questions in the study of poset Ramsey numbers is to determine the asymptotic behavior of $R(\mathcal{B}_n, \mathcal{B}_n)$ for large n. Since the posets are the same, this is called the *diagonal* setting in Ramsey theory. We immediately see that

$$2n \leq R(\mathcal{B}_n, \mathcal{B}_n) \leq n^2 + 2n. \tag{4.1}$$

The proof of Theorem 4.1 gives the upper bound and the lower bound follows from coloring a vertex $X \in 2^{[2n-1]}$ blue if it has even size and red if it has odd size. Since the height of the resulting red poset is n, as well as the height of the blue poset, we see that there is no monochromatic copy of $\mathcal{B}_n$ (which has height $n+1$).

There were several improvements of these easy bounds. Bohman and Peng [15] improved the lower bound by +1, Lu and Thompson [91] reduced the upper bound by a linear factor. Hence the best-known exact bounds for $n \geq 3$ are:

$$2n + 1 \leq R(\mathcal{B}_n, \mathcal{B}_n) \leq n^2 - n + 2. \tag{4.2}$$

The currently best asymptotic bounds are as follows:

Theorem 4.3 (Winter [124]; Axenovich, Winter [5])

$$\big(2.02 + o(1)\big)n \leq R\big(\mathcal{B}_n, \mathcal{B}_n\big) \leq n^2 - \big(1 - o(1)\big)n \log_2 n.$$

The lower bound from Winter [124] is obtained by an intricate semi-probabilistic construction. See Figure 6. It results in a coloring of $\mathcal{B}_{(2.02+o(1))n}$ in blue and red (black and grey respectively in Figure 6) so that there is no monochromatic copy of $\mathcal{B}_n$ and there is no chain on four vertices with alternating colors.

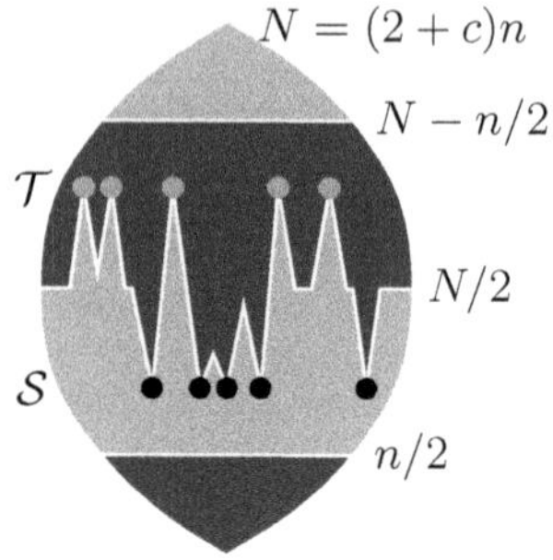

Figure 6: Coloring that illustrates the lower bound in Theorem 4.3

The upper bound from Axenovich and Winter [5] is proved by considering an arbitrary coloring of $2^{[N]}$ for $N = n^2 - (1 - o(1))n \log_2 n$ and iteratively finding monochromatic $\mathcal{B}_n$ by following the *volume* of partially embedded $\mathcal{B}_n$. The volume of a subposet P of $\mathcal{B}_n$ is the size of the union of its vertices, i.e., the total number of ground elements from $[n]$ that are in some vertex of P.

The Ramsey number $R\big(\mathcal{B}_n, \mathcal{B}_n\big)$ is determined exactly for small n. In particular $R\big(\mathcal{B}_2, \mathcal{B}_2\big) = 4$ was verified by Axenovich and Walzer [2] and Lu and Thompson [91]. Falgas-Ravry, Markström, Treglown, and Zhao [43] proved that $R(\mathcal{B}_3, \mathcal{B}_3) = 7$.

Crucially, the asymptotic behavior of $R\big(\mathcal{B}_n, \mathcal{B}_n\big)$ is not known. The known lower bound is linear in n and the upper bound is quadratic in n.

It was conjectured by Lu and Thompson [91] that $R\big(\mathcal{B}_n, \mathcal{B}_n\big) = o(n^2)$. Axenovich and Winter [5] made a stronger conjecture that for any $\varepsilon > 0$ there is $n_0 = n_0(\varepsilon)$ such that $R\big(\mathcal{B}_n, \mathcal{B}_n\big) \leq n^{1+\varepsilon}$ for any $n > n_0$.

Problem 11 *Improve the bounds for* $R\big(\mathcal{B}_n, \mathcal{B}_n\big)$ *– either the exact bounds from* (4.2) *or the asymptotic bounds*

$$\big(2.02 + o(1)\big)n \leq R\big(\mathcal{B}_n, \mathcal{B}_n\big) \leq n^2 - \big(1 - o(1)\big)n \log_2 n.$$

4.4 The off-diagonal setting: $R\big(\mathcal{B}_m, \mathcal{B}_n\big)$

For arbitrary m and n, Axenovich and Walzer [2] showed that

$$m + n - 1 \leq R\big(\mathcal{B}_m, \mathcal{B}_n\big) \leq mn + m + n. \tag{4.3}$$

Both bounds follow from the arguments that give the result in (4.1). Lu and Thompson [91] improved the bound in (4.3) by showing that for a constant $c > 0$,

$$R\big(\mathcal{B}_m, \mathcal{B}_n\big) \leq n\Big(m - 2 + \frac{c}{m}\Big) + m + 3. \tag{4.4}$$

Finally, inequality (4.4) was improved further:

Theorem 4.4 (Axenovich, Winter [5]) *If m and n are integers satisfying the inequalities $2^{25} \le m \le n$, then*

$$R(\mathcal{B}_m, \mathcal{B}_n) \le n\left(m - \left(1 - \frac{2}{\sqrt{\log_2 m}}\right)\log_2 m\right).$$

Moreover, if $m \le n$ and $0 < \varepsilon < 1$, such that $\frac{n+m}{n} \cdot \frac{1}{(1-\varepsilon)\log_2 m} + m^{-\varepsilon} \le \varepsilon$, then

$$R(\mathcal{B}_m, \mathcal{B}_n) \le n\big(m - (1-\varepsilon)^2 \log_2 m\big).$$

It was conjectured by Axenovich and Winter [5] that for any $\epsilon > 0$ there is m_0 such that for any $n > m > m_0$, $R(\mathcal{B}_m, \mathcal{B}_n) \le n \cdot m^{\epsilon}$.

Now we consider the regime for fixed m and growing n. Since $\mathcal{B}_1 = \mathcal{C}_2$, it is easy to derive $R(\mathcal{B}_1, \mathcal{B}_n) = n + 1$. For $\mathcal{B}_2$, Grósz, Methuku, and Tompkins [61] showed that

$$n + 3 \le R(\mathcal{B}_2, \mathcal{B}_n) \le n + \big(2 + o(1)\big)\frac{n}{\log_2 n}.$$

Despite a lot of effort, it remains open whether $R(\mathcal{B}_3, \mathcal{B}_n) = n + o(n)$.

Problem 12 *Determine the asymptotic value of $R(\mathcal{B}_3, \mathcal{B}_n)$.*

4.5 The general diagonal setting: $R(P, P)$

The general diagonal setting $R(P, P)$ was considered for other basic classes of posets P, Walzer [121] and Chen, Chen, Cheng, Li, and Liu [25] determined these numbers for chains, antichains, and special unions of two chains, see also Section 4.8. Recalling the Sperner number from Section 1, $R(\mathcal{A}_n, \mathcal{A}_n) = \mathrm{sp}(2n-1) = \big(1 + o(1)\big)\log_2 n$, and $R(\mathcal{C}_n, \mathcal{C}_n) = 2(n-1)$.

Axenovich and Winter [5] showed that $2\mathrm{sp}(n) \le R(\lozenge_n, \lozenge_n) \le 2\mathrm{sp}(n) + 2$ and $N^*(n) \le R(\mathcal{V}_n, \mathcal{V}_n) \le N^*(n) + 1$, where $N^*(n)$ is expressed as a solution of an equation involving $\mathrm{sp}(n)$. In asymptotic terms, $R(\lozenge_n, \lozenge_n) = \big(2 + o(1)\big)\log_2 n$ and $R(\mathcal{V}_n, \mathcal{V}_n) = \big(d + o(1)\big)\log_2 n$, where $d \approx 1.29382$.

4.6 The general off-diagonal setting: $R(P, \mathcal{B}_n)$, P fixed, n large

This setting is inspired by the so-called Erdős-Hajnal conjecture for graphs. In a graph, a *homogeneous set* is either a clique or an independent set. The Erdős-Hajnal conjecture claims that for any fixed graph H there is $\beta = \beta(H) > 0$ such that if n is sufficiently large, then any graph on n vertices that does not contain H as an induced subgraph has a homogeneous set of size n^{β}. In the context of posets, the parallel claim would be that, for a fixed poset P and any blue/red coloring of the vertices of $2^{[N]}$ (where N is sufficiently large) such that there is no blue induced copy of P, there exists a red induced copy of $\mathcal{B}_n$, where n is some growing function of N.

Note that $R(P, \mathcal{B}_n) \ge n$ for any non-empty poset P as witnessed by a coloring of all vertices of $2^{[n-1]}$ in red. There is a dichotomy in the asymptotic behavior. For a poset that is a parallel composition of chains, Theorem 4.5 shows $R(P, \mathcal{B}_n) \le$

$n+f(P)$. But for all other posets, $R(P,\mathcal{B}_n) \geq n+cn/\log_2 n$. Recall that $w(P)$, $h(P)$, and $\dim_2(P)$ denote width, height, and 2-dimension of P, respectively. Using early work by Axenovich and Walzer [2], the dichotomy was established by Axenovich and Winter.

Theorem 4.5 (Axenovich and Winter [3]) *For every poset P there is an integer n_0 such that for all $n > n_0$ the following holds.*
If P has either an induced copy of $\mathcal{V}$ or an induced copy of Λ, then

$$n + \frac{n}{15\log_2 n} \leq R\big(P,\mathcal{B}_n\big) \leq h(P)n + \dim_2(P);$$

otherwise P is a parallel composition of chains and

$$n + h(P) - 1 \leq R\big(P,\mathcal{B}_n\big) \leq n + h(P) + \mathrm{sp}\big(w(P)\big) - 1.$$

Although the upper bound is linear, it is not clear if the upper bound should be $\big(1+o(1)\big)n$ for all posets P.

Problem 13 *Determine if there exists a poset P and a positive constant c such that $R\big(P,\mathcal{B}_n\big) > (1+c)n$.*

For some posets P that are not a parallel composition of chains, the asymptotic behavior of $R\big(P,\mathcal{B}_n\big)$ is known. In particular, Winter [122] showed that $R\big(P,\mathcal{B}_n\big) = n+\Theta\big(\frac{n}{\log n}\big)$ if P is a complete multipartite poset (see, for example, Figure 1(I)) or a subdivided diamond (see Figure 7(A)). Axenovich and Winter [3] showed the same behavior when P is the N-poset (Figure 1(G)).

(A) subdivided diamond (B) subdivided fork, $\mathcal{V}_{2,4}$

Figure 7: Subdivided posets

There are also sharper bounds for $R\big(P,\mathcal{B}_n\big)$ in case of some posets which are parallel compositions of chains. Winter [123] showed that if $3 \leq t \leq \log_2\log_2 n$, then $R\big(A_t,\mathcal{B}_n\big) = n+3$. For $t_1 \geq t_2 \geq \cdots \geq t_\ell$, Axenovich and Walzer [2] established that

$$n+t_1-1 \leq R\big(\mathcal{C}_{t_1}+\mathcal{C}_{t_2}+\cdots+\mathcal{C}_{t_\ell}\big) \leq n+t_1+\mathrm{sp}(\ell)+1.$$

More precise bounds in some cases were established.

Theorem 4.6 (Axenovich and Walzer [2]; Winter [123]) *If $t_1 \geq t_2$, then*

$$R\big(\mathcal{C}_{t_1},\mathcal{B}_n\big) = n+t_1-1 \qquad \text{and} \qquad R\big(\mathcal{C}_{t_1}+\mathcal{C}_{t_2},\mathcal{B}_n\big) = n+t_1+1.$$

If, in addition, $t_1 \geq t_2+1$, then

$$R\big(\mathcal{C}_{t_1}+\mathcal{C}_{t_1-1}+\mathcal{C}_{t_2},\mathcal{B}_n\big) = n+t_1+2.$$

The tools and methods used to determine $R(P,\mathcal{B}_n)$ include several more precise statements about colored Boolean lattices. For a poset P in a Boolean lattice on a ground set $[N]$ and a subset X of $[N]$, let *projection* of P onto X denoted by $P(X)$ be a set of vertices obtained by deleting the elements not in X from each vertex of P.

We say that a copy Q of $\mathcal{B}_n$ in $2^{[N]}$ is X-*good* for some $X \subseteq [N]$, $|X| = n$ if the projection of $Q(X)$ is equal to 2^X, i.e., it corresponds to the Boolean lattice with ground set X. For example, a copy of $\mathcal{B}_2$ in $\mathcal{B}_5$ with vertices $\{1\}$, $\{1,2,4\}$, $\{1,3,4,5\}$, $\{1,2,3,4,5\}$ is $\{2,3\}$-good since the projection into $\{2,3\}$ is $\{\emptyset, \{2\}, \{3\}, \{2,3\}\} = 2^{\{2,3\}}$. One can show that every copy of $\mathcal{B}_n$ in a Boolean lattice is X-good for some X, $|X| = n$, see Axenovich and Walzer [2]. For a set Y and its ordering $\underline{Y} = (y_1,\dots,y_k)$, we say that a chain C of size $k+1$ in a Boolean lattice with ground set $[N]$ is $\underline{Y}$-*good*, if the projection of C onto Y has elements $\emptyset, \{y_1\}, \{y_1,y_2\},\dots,\{y_1,\dots,y_k\}$.

One of the key tools in this area is the Chain Lemma. It has appeared in different forms in various papers. For example, it was implicit in the paper by Chen, Cheng, Li, and Liu [26] and more explicit in the paper by Gerbner, Methuku and Tompkins [61]. For a short proof of this lemma, see Axenovich and Winter [3].

Lemma 4.7 (Chain Lemma) *For any blue/red coloring of a Boolean lattice on a ground set $[N]$, any partition $[N] = X \sqcup Y$, and any ordering $\underline{Y}$ of Y there is either a red X-good copy of $\mathcal{B}_{|X|}$ or a blue $\underline{Y}$-good chain.*

The Chain Lemma was also generalized from chains to other structures, such as so-called shrubs, etc., such as by Axenovich and Winter [4].

4.7 Erdős-Hajnal number: $\tilde{R}(\dot{P},\mathcal{B}_n)$

A *colored poset* is a pair (P,c_P), where P is a poset and $c_P\colon P \to \{\text{blue},\text{red}\}$ is a blue/red coloring of the vertices of P. If a poset P has a fixed coloring c_P, we write $\dot{P}$ instead of (P,c_P). The poset Erdős-Hajnal number $\tilde{R}(\dot{P},\mathcal{B}_n)$ of a colored poset $\dot{P}$ is the smallest integer N such that every blue/red coloring of $2^{[N]}$ contains a copy of $\dot{P}$, a red copy of $\mathcal{B}_n$, or a blue copy of $\mathcal{B}_n$. Winter [124] introduced this number and gave several general and specific bounds. A colored poset is *diverse* if it has two comparable vertices of different colors, otherwise the colored poset is *non-diverse*.

Theorem 4.8 (Winter [124]) *If $\dot{P}$ is a diverse poset, then*

$$2n \le \tilde{R}(\dot{P},\mathcal{B}_n) \le h(P)n + \dim_2(P).$$

If $\dot{P}$ is a non-diverse poset, i.e., it is a parallel composition of a blue poset P_b and a red poset P_r, then

$$\max\{R(P_b,\mathcal{B}_n), R(P_r,\mathcal{B}_n)\} \le \tilde{R}(\dot{P},\mathcal{B}_n) \le \max\{R(P_b,\mathcal{B}_n), R(P_r,\mathcal{B}_n)\} + 2.$$

More specifically, he showed that if $\dot{\mathcal{B}}_2$ is a colored non-monochromatic copy of $\mathcal{B}_2$, then in three of the cases $\tilde{R}(\dot{\mathcal{B}}_2,\mathcal{B}_n) = 2n$. In the case where $\dot{\mathcal{B}}_2$ is colored such that the minimum or maximum element has a unique color, then we only know that $2n \le \tilde{R}(\dot{\mathcal{B}}_2,\mathcal{B}_n) \le 2n + O(n/\log n)$. For $t \ge 4$, if $\dot{C}_t^a$ is an alternately colored chain on t vertices, then $2.02n \le \tilde{R}(\dot{C}_t^a,\mathcal{B}_n) \le (t-1)n$. If $\dot{A}$ is a non-monochromatic antichain on at least two vertices and n is sufficiently large, then $\tilde{R}(\dot{A},\mathcal{B}_n) \in \{n+2, n+3\}$.

4.8 Multicolor Ramsey numbers

For an integer $k > 0$, define the multicolor poset number $R(P;k)$ to be the smallest integer N such that any coloring of the vertices of $\mathcal{B}_N$ in k colors results in a monochromatic induced copy of P. Of course, $R(P;2) = R(P,P)$.

Recall that $\lambda_N(\mathcal{F})$ denotes the Lubell mass of a family $\mathcal{F} \subseteq 2^{[N]}$. For a poset P, define

$$\ell^*(P) = \limsup_{N\to\infty}\{\lambda_N(\mathcal{F}) : \mathcal{F} \subseteq 2^{[N]}, \mathcal{F} \text{ contains no copy of } P\}.$$

Méroueh [100] (Theorem 2.17) proved that $\ell^*(P)$ exists for any poset P.

Theorem 4.9 (Axenovich, Walzer [2]) *If P is a poset that is not an antichain, then $k \leq R(P;k) \leq k\ell^*(P) + o(k) = O(k)$.*

Proof Consider a coloring $c : 2^{[N]} \to \{1, 2, \ldots, k\}$ with no monochromatic copy of P. Then the Lubell mass of each color class S_i is $\lambda_N(S_i) \leq \ell^*(P) + o(1)$. From the definition we have that $\lambda_N(2^{[N]}) = \lambda_N(S_1)+\cdots+\lambda_N(S_k) \leq k(\ell^*(P)+o(1))$. On the other hand, $\lambda_N(2^{[N]}) = \sum_{i=0}^{N}\binom{N}{i}\binom{N}{i}^{-1} = N+1$. Thus $N+1 \leq k(\ell^*(P)+o(1)) = O(k)$.

By coloring each $X \in 2^{[N]}$ with color $|X|$, we see that $R(P;k) \geq k$. □

Posets for which the multicolored Ramsey numbers are known are chains, antichains, and subdivided forks: Walzer [121] proved that $R(\mathcal{A}_n;k) = \mathrm{sp}(k(n-1)+1)$ and $R(\mathcal{C}_n;k) = k(n-1)$. Chen, Chen, Cheng, Li, and Liu [25] showed that if $m = n = 1$ or $m < n$, then $R(\mathcal{V}_{m,n};k) = nk+1$. Chen, Cheng, Li, and Liu [26] conjectured that $R(\Diamond;k) = 2k(1+o(1))$. See also Katona, Mao, Ozeki, and Wang [79].

4.9 Weak Ramsey number

Although the primary study of Ramsey theory in posets considers the case of induced subposets, there have been results for (weak) subposets. For posets P and Q, let the *weak poset Ramsey number* of P and Q be defined as follows:

$$R_w(P,Q) = \min\{N \in \mathbf{N} : \text{every blue/red coloring of } 2^{[N]} \text{ contains either a blue weak copy of } P \text{ or a red weak copy of } Q\}.$$

In the off-diagonal setting $R_w(P,\mathcal{B}_n)$, where $|P|$ is small and n is large, the asymptotic behavior is clear. Indeed, for for any poset P and $n \gg |P|$, the chain $\mathcal{C}_{|P|}$ is a weak copy of P,

$$n \leq R_w(P,\mathcal{B}_n) \leq R(\mathcal{C}_{|P|},\mathcal{B}_n).$$

From Theorem 4.6, $R(\mathcal{C}_t,\mathcal{B}_n) = n+t-1$, so it follows that $R_w(P,\mathcal{B}_n) = n+\Theta(1)$.

In the diagonal setting, for $P = \mathcal{B}_n$, it was shown by Cox and Stolee [31] that $2n+1 \leq R_w(\mathcal{B}_n,\mathcal{B}_n)$. The upper bound $R_w(\mathcal{B}_n,\mathcal{B}_n) \leq R(\mathcal{B}_n,\mathcal{B}_n) \leq n^2(1-o(1))$ was improved by Axenovich and Winter [5]: If n is sufficiently large n, then

$$R_w(\mathcal{B}_n,\mathcal{B}_n) \leq 0.96n^2.$$

One might wonder if the strong and weak diagonal poset Ramsey numbers grow at the same rate. In the Axenovich-Winter paper [5], Winter made the following conjecture:

Conjecture 4.10 (Winter [5]) $R(\mathcal{B}_n, \mathcal{B}_n) = R_w(\mathcal{B}_n, \mathcal{B}_n) + O(n)$.

Problem 14

- *Determine the asymptotics of* $R_w(\mathcal{B}_n, \mathcal{B}_n)$.
- *Determine the asymptotics of* $R(\mathcal{B}_n, \mathcal{B}_n) - R_w(\mathcal{B}_n, \mathcal{B}_n)$.

4.10 Boolean rainbow Ramsey number $RR(P, Q)$

The *Boolean rainbow Ramsey number* $RR(P, Q)$ for posets P and Q is the smallest integer N such that any coloring of $\mathcal{B}_N$ contains either a monochromatic copy of P or a rainbow copy of Q. Here, a poset is *rainbow* if all its elements have distinct colors. The weak rainbow Ramsey number $RR_w(P, Q)$ is defined analogously by requiring weak copies of P and Q in the above definition. Johnson, Lu, and Milans [73] proved that $RR(P, Q)$ is well-defined by showing that $RR(\mathcal{B}_n, \mathcal{B}_n)$ is finite. Chang, Gerbner, Li, Methuku, Nagy, Patkós, and Vizer [23] determined the weak version $RR_w(P, Q)$ exactly for posets P that are uniformly Lubell bounded, they also gave various bounds on $RR(P, Q)$ using Lubell mass and layered constructions. Chen, Cheng, Li, and Liu [26] determined $RR(P, Q)$ exactly in the case when P is a chain or an antichain and Q is a chain or an antichain. In addition, they determined $RR(P, Q)$ exactly when P is a chain and $Q = \mathcal{B}_n$, for some n. The authors of [26] give the following bounds on the diagonal case:

$$n(2^n - 1) \leq RR(\mathcal{B}_n, \mathcal{B}_n) \leq n^7 2^{6m+9},$$

where the lower bound is given by coloring each layer of $\mathcal{B}_N$, for $N = n(2^n - 1) - 1$ with the same color, so that the total number of colors is $2^n - 1$ (guaranteeing that there is no rainbow $\mathcal{B}_n$) and there are n layers of each color (guaranteeing that there is no monochromatic $\mathcal{B}_n$).

Acknowledgements

The authors would like to thank their co-authors over the years. In addition they would like to thank Gyula O.H. Katona and Jerry Griggs for their pioneering early work in growing the interested in this area. Special thanks to Kirsten Hogenson, Lina Li, and Christian Winter for valuable conversations. The authors greatly appreciate the careful reading and comments from an anonymous reviewer.

References

[1] Maria Axenovich, Jacob Manske, and Ryan Martin, *Q_2-free families in the Boolean lattice*, Order **29** (2012), no. 1, 177–191.

[2] Maria Axenovich and Stefan Walzer, *Boolean lattices: Ramsey properties and embeddings*, Order **34** (2017), no. 2, 287–298.

[3] Maria Axenovich and Christian Winter, *Poset Ramsey numbers: large Boolean lattice versus a fixed poset*, Combin. Probab. Comput. **32** (2023), no. 4, 638–653.

[4] ———, *Poset Ramsey number $R(P, Q_n)$. II. N-shaped poset*, Order **41** (2024), no. 2, 401–418.

[5] ———, *Diagonal poset Ramsey numbers*, Discrete Math. (2025), 18pp, accepted for publication.

[6] József Balogh, Ramon I. Garcia, and Michael C. Wigal, *On the number of P-free set systems for tree posets P*, Order (2025), 14pp.

[7] József Balogh, Ping Hu, Bernard Lidický, and Hong Liu, *Upper bounds on the size of 4- and 6-cycle-free subgraphs of the hypercube*, European J. Combin. **35** (2014), 75–85.

[8] József Balogh, William B. Linz, and Balázs Patkós, *On the sizes of t-intersecting k-chain-free families*, Comb. Theory **3** (2023), no. 2, Paper No. 13, 14pp.

[9] József Balogh, Robert Morris, and Wojciech Samotij, *Independent sets in hypergraphs*, J. Amer. Math. Soc. **28** (2015), no. 3, 669–709.

[10] József Balogh, Richard Mycroft, and Andrew Treglown, *A random version of Sperner's theorem*, J. Combin. Theory Ser. A **128** (2014), 104–110.

[11] Paul Bastide, Carla Groenland, Maria-Romina Ivan, and Tom Johnston, *A polynomial upper bound for poset saturation*, European J. Combin. **129** (2025), Paper No. 103970, 6pp.

[12] Paul Bastide, Carla Groenland, Hugo Jacob, and Tom Johnston, *Exact antichain saturation numbers via a generalisation of a result of Lehman-Ron*, Comb. Theory **4** (2024), no. 1, Paper No. 11, 24pp.

[13] Paul Bastide, Jedrzej Hodor, Hoang La, and William T. Trotter, *Cube height, cube width and related extremal problems for posets*, arXiv:2510.00928, 2025.

[14] Edward Boehnlein and Tao Jiang, *Set families with a forbidden induced subposet*, Combin. Probab. Comput. **21** (2012), no. 4, 496–511.

[15] Tom Bohman and Fei Peng, *A construction for Boolean cube Ramsey numbers*, Order **40** (2023), no. 2, 327–333.

[16] Béla Bollobás, *On generalized graphs*, Acta Math. Acad. Sci. Hungar. **16** (1965), 447–452.

[17] Béla Bollobás, *Sperner systems consisting of pairs of complementary subsets*, J. Combinatorial Theory Ser. A **15** (1973), 363–366.

[18] Béla Bollobás, Imre Leader, and Claudia Malvenuto, *Daisies and other Turán problems*, Combin. Probab. Comput. **20** (2011), no. 5, 743–747.

[19] Béla Bollobás and Andrew Thomason, *Generalized chromatic numbers of random graphs*, Proceedings of the Sixth International Seminar on Random Graphs and Probabilistic Methods in Combinatorics and Computer Science, "Random Graphs '93" (Poznań, 1993), vol. 6, 1995, pp. 353–356.

[20] James Brownlie and Sean Jaffe, *The induced saturation number for $\mathcal{V}_3$ is linear*, arXiv:2507.08353, 2025.

[21] Boris Bukh, *Set families with a forbidden subposet*, Electron. J. Combin. **16** (2009), no. 1, Research Paper 142, 11pp.

[22] Teena Carroll and Gyula O. H. Katona, *Bounds on maximal families of sets not containing three sets with $A \cap B \subset C$, $A \not\subset B$*, Order **25** (2008), no. 3, 229–236.

[23] Fei-Huang Chang, Dániel Gerbner, Wei-Tian Li, Abhishek Methuku, Dániel T. Nagy, Balázs Patkós, and Máté Vizer, *Rainbow Ramsey problems for the Boolean lattice*, Order **39** (2022), no. 3, 453–463.

[24] François Charton, Jordan Ellenberg, Adam Zsolt Wagner, and Williamson Geordie, *PatternBoost: Constructions in mathematics with a little help from AI*, arXiv:23411.00566, 2024.

[25] Hong-Bin Chen, Wei-Han Chen, Yen-Jen Cheng, Wei-Tian Li, and Chia-An Liu, *Ramsey properties for V-shaped posets in the Boolean lattice*, Discrete Math. **347** (2024), no. 7, Paper No. 114045, 11pp.

[26] Hong-Bin Chen, Yen-Jen Cheng, Wei-Tian Li, and Chia-An Liu, *The Boolean rainbow Ramsey number of antichains, Boolean posets and chains*, Electron. J. Combin. **27** (2020), no. 4, Paper No. 4.38, 12pp.

[27] Maria Chudnovsky, *Berge trigraphs*, J. Graph Theory **53** (2006), no. 1, 1–55.

[28] Maurício Collares and Robert Morris, *Maximum-size antichains in random set-systems*, Random Structures Algorithms **49** (2016), no. 2, 308–321.

[29] David Conlon, *An extremal theorem in the hypercube*, Electron. J. Combin. **17** (2010), no. 1, Research Paper 111, 7pp.

[30] David Conlon and Yu-Han Wu, *More on lines in Euclidean Ramsey theory*, C. R. Math. Acad. Sci. Paris **361** (2023), 897–901.

[31] Christopher Cox and Derrick Stolee, *Ramsey numbers for partially-ordered sets*, Order **35** (2018), no. 3, 557–579.

[32] Irina Đanković and Maria-Romina Ivan, *Saturation for small antichains*, Electron. J. Combin. **30** (2023), no. 1, Paper No. 1.3, 9pp.

[33] Annalisa De Bonis and Gyula O. H. Katona, *Largest families without an r-fork*, Order **24** (2007), no. 3, 181–191.

[34] Annalisa De Bonis, Gyula O. H. Katona, and Konrad J. Swanepoel, *Largest family without $A \cup B \subseteq C \cap D$*, J. Combin. Theory Ser. A **111** (2005), no. 2, 331–336.

[35] Robert P. Dilworth, *A decomposition theorem for partially ordered sets*, Ann. of Math. (2) **51** (1950), 161–166.

[36] Andrew Philip Dove, *Generalizations of Sperner's Theorem: Packing posets, families forbidding posets, and supersaturation*, ProQuest LLC, Ann Arbor, MI, 2013, Thesis (Ph.D.)–University of South Carolina.

[37] Ben Dushnik and Edwin W. Miller, *Partially ordered sets*, Amer. J. Math. **63** (1941), 600–610.

[38] David Ellis, Maria-Romina Ivan, and Imre Leader, *Turán densities for daisies and hypercubes*, Bull. Lond. Math. Soc. **56** (2024), no. 12, 3838–3853.

[39] ______, *Turán densities for small hypercubes*, SIAM J. Discrete Math. **39** (2025), no. 2, 1363–1371.

[40] Paul Erdős, *On a lemma of Littlewood and Offord*, Bull. Amer. Math. Soc. **51** (1945), 898–902.

[41] Paul Erdős, Daniel J. Kleitman, and Bruce L. Rothschild, *Asymptotic enumeration of K_n-free graphs*, Colloquio Internazionale sulle Teorie Combinatorie (Roma, 1973), Tomo II, Accad. Naz. Lincei, Rome, 1976, pp. 19–27.

[42] Paul Erdős and Alfréd Rényi, *On the evolution of random graphs*, Magyar Tud. Akad. Mat. Kutató Int. Közl. **5** (1960), 17–61.

[43] Victor Falgas-Ravry, Klas Markström, Andrew Treglown, and Yi Zhao, *Existence thresholds and Ramsey properties of random posets*, Random Structures Algorithms **57** (2020), no. 4, 1097–1133.

[44] Michael Ferrara, Bill Kay, Lucas Kramer, Ryan R. Martin, Benjamin Reiniger, Heather C. Smith, and Eric Sullivan, *The saturation number of induced subposets of the Boolean lattice*, Discrete Math. **340** (2017), no. 10, 2479–2487.

[45] Tomáš Flídr, Maria-Romina Ivan, and Sean Jaffe, *Optimal embeddings of posets in hypercubes*, arXiv:2509.26630, 2025.

[46] Peter Frankl, *Canonical antichains on the circle and applications*, SIAM J. Discrete Math. **3** (1990), no. 3, 355–363.

[47] Andrea Freschi, Simón Piga, Maryam Sharifzadeh, and Andrew Treglown, *The induced saturation problem for posets*, Comb. Theory **3** (2023), no. 3, Paper No. 9, 15pp.

[48] Dániel Gerbner, Balázs Keszegh, Nathan Lemons, Cory Palmer, Dömötör Pálvölgyi, and Patkós Balázs, *Saturating Sperner families*, Graphs Combin. **29** (2013), no. 5, 1355–1364.

[49] Dániel Gerbner, Abhishek Methuku, and Casey Tompkins, *Intersecting P-free families*, J. Combin. Theory Ser. A **151** (2017), 61–83.

[50] Dániel Gerbner, Dániel T. Nagy, Balázs Patkós, and Máté Vizer, *Supersaturation, counting, and randomness in forbidden subposet problems*, Electron. J. Combin. **28** (2021), no. 1, Paper No. 1.40, 20pp.

[51] Dániel Gerbner and Balázs Patkós, *A note on vertex Turán problems in the Kneser cube*, arXiv:2402.02525, 2024.

[52] Ronald L. Graham and Bruce L. Rothschild, *Ramsey's theorem for n-parameter sets*, Trans. Amer. Math. Soc. **159** (1971), 257–292.

[53] Ronald L. Graham, Bruce L. Rothschild, and Joel H. Spencer, *Ramsey theory*, second ed., Wiley-Interscience Series in Discrete Mathematics and Optimization, John Wiley & Sons, Inc., New York, 1990, A Wiley-Interscience Publication.

[54] Ronald L. Graham and Neil J. A. Sloane, *Lower bounds for constant weight codes*, IEEE Trans. Inform. Theory **26** (1980), no. 1, 37–43.

[55] Curtis Greene, Gyula O. H. Katona, and Daniel J. Kleitman, *Extensions of the Erdős-Ko-Rado theorem*, Recent advances in graph theory (Proc. Second Czechoslovak Sympos., Prague, 1974), Academia, Prague, 1975, pp. 223–231.

[56] Jerrold R. Griggs and Gyula O. H. Katona, *No four subsets forming an N*, J. Combin. Theory Ser. A **115** (2008), no. 4, 677–685.

[57] Jerrold R. Griggs and Wei-Tian Li, *The partition method for poset-free families*, J. Comb. Optim. **25** (2013), no. 4, 587–596.

[58] Jerrold R. Griggs, Wei-Tian Li, and Linyuan Lu, *Diamond-free families*, J. Combin. Theory Ser. A **119** (2012), no. 2, 310–322.

[59] Jerrold R. Griggs and Linyuan Lu, *On families of subsets with a forbidden subposet*, Combin. Probab. Comput. **18** (2009), no. 5, 731–748.

[60] Dániel Grósz, Abhishek Methuku, and Casey Tompkins, *An upper bound on the size of diamond-free families of sets*, J. Combin. Theory Ser. A **156** (2018), 164–194.

[61] ———, *Ramsey numbers of Boolean lattices*, Bull. Lond. Math. Soc. **55** (2023), no. 2, 914–932.

[62] Michel Habib, Lhouari Nourine, Olivier Raynaud, and Éric Thierry, *Computational aspects of the 2-dimension of partially ordered sets*, Theoret. Comput. Sci. **312** (2004), no. 2-3, 401–431.

[63] Alfred W. Hales and Robert I. Jewett, *Regularity and positional games*, Trans. Amer. Math. Soc. **106** (1963), 222–229.

[64] Philip Hall, *On Representatives of Subsets*, J. London Math. Soc. **10** (1935), no. 1, 26–30.

[65] Kirsten Ann Hogenson, *Random and deterministic versions of extremal poset problems*, ProQuest LLC, Ann Arbor, MI, 2016, Thesis (Ph.D.)–Iowa State University.

[66] Maria-Romina Ivan, *Saturation for the butterfly poset*, Mathematika **66** (2020), no. 3, 806–817.

[67] ———, *Minimal diamond-saturated families*, Contemp. Math. **3** (2022), no. 2, 81–88.

[68] Maria-Romina Ivan and Sean Jaffe, *Gluing posets and the dichotomy of poset saturation numbers*, arXiv:2503.12223, 2025.

[69] ———, *Saturation for sums of posets and antichains*, arXiv:2509.10294, 2025.

[70] ———, *The saturation number for the diamond is linear*, arXiv:2507.05122, 2025.

[71] Shengjin Ji, Balázs Patkós, and Erfei Yue, *Poset saturation of unions of chains*, arXiv:2505.23128, 2025.

[72] Tao Jiang, Sean Longbrake, Sam Spiro, and Liana Yepremyan, *Tree posets: Supersaturation, enumeration, and randomness*, arXiv:2406.11999, 2024.

[73] Travis Johnston, Linyuan Lu, and Kevin G. Milans, *Boolean algebras and Lubell functions*, J. Combin. Theory Ser. A **136** (2015), 174–183.

[74] Jeff Kahn and Gil Kalai, *Thresholds and expectation thresholds*, Combin. Probab. Comput. **16** (2007), no. 3, 495–502.

[75] Gyula O. H. Katona, *A simple proof of a theorem of Milner*, J. Combin. Theory Ser. A **83** (1998), no. 1, 138–140.

[76] ———, *Excluded subposets in the Boolean lattice*, 2005 European Conference on Combinatorics, Graph Theory and Applications (EuroComb '05) (Berlin, Germany), DMTCS proceedings, 2005, pp. 229–230.

[77] ———, *Forbidden intersection patterns in the families of subsets (introducing a method)*, Horizons of combinatorics, Bolyai Soc. Math. Stud., vol. 17, Springer, Berlin, 2008, pp. 119–140.

[78] Gyula O. H. Katona and Tamás G. Tarján, *Extremal problems with excluded subgraphs in the n-cube*, Graph theory (Łagów, 1981), Lecture Notes in Math., vol. 1018, Springer, Berlin, 1983, pp. 84–93.

[79] Gyula O.H. Katona, Yaping Mao, Kenta Ozeki, and Zhao Wang, private communication, 2025.

[80] Balázs Keszegh, Nathan Lemons, Ryan R. Martin, Dömötör Pálvölgyi, and Balázs Patkós, *Induced and non-induced poset saturation problems*, J. Combin. Theory Ser. A **184** (2021), Paper No. 105497, 20pp.

[81] Henry A. Kierstead and William T. Trotter, *A Ramsey theoretic problem for finite ordered sets*, Discrete Math. **63** (1987), no. 2-3, 217–223, Special issue: ordered sets (Oberwolfach, 1985).

[82] Martin Klazar and Adam Marcus, *Extensions of the linear bound in the Füredi-Hajnal conjecture*, Adv. in Appl. Math. **38** (2007), no. 2, 258–266.

[83] Dénes König, *Gráfok és mátrixok*, Matematikai és Fizikai Lapok **38** (1931), 116–119 (Hungarian).

[84] Lucas Kramer, Ryan R. Martin, and Michael Young, *On diamond-free subposets of the Boolean lattice*, J. Combin. Theory Ser. A **120** (2013), no. 3, 545–560.

[85] Thái Hoàng Lê, *Partition regularity and the primes*, C. R. Math. Acad. Sci. Paris **350** (2012), no. 9-10, 439–441.

[86] Eric Lehman and Dana Ron, *On disjoint chains of subsets*, J. Combin. Theory Ser. A **94** (2001), no. 2, 399–404.

[87] Dingyuan Liu, *Induced saturation for complete bipartite posets*, Discrete Math. **348** (2025), no. 7, Paper No. 114462, 9pp.

[88] Hong Liu, Péter Pál Pach, and Csaba Sándor, *Polynomial Schur's theorem*, Combinatorica **42** (2022), 1357–1384.

[89] Linyuan Lu, *On crown-free families of subsets*, J. Combin. Theory Ser. A **126** (2014), 216–231.

[90] Linyuan Lu and Kevin G. Milans, *Set families with forbidden subposets*, J. Combin. Theory Ser. A **136** (2015), 126–142.

[91] Linyuan Lu and Joshua C. Thompson, *Poset Ramsey numbers for Boolean lattices*, Order **39** (2022), no. 2, 171–185.

[92] David Lubell, *A short proof of Sperner's lemma*, J. Combinatorial Theory **1** (1966), 299.

[93] Jacob Manske and Jian Shen, *Three layer Q_2-free families in the Boolean lattice*, Order **30** (2013), no. 2, 585–592.

[94] Willem Mantel, *Sur un problème concernant les graphes*, Bull. Soc. Math. Belgique **9** (1907), 33–36.

[95] Adam Marcus and Gábor Tardos, *Excluded permutation matrices and the Stanley-Wilf conjecture*, J. Combin. Theory Ser. A **107** (2004), no. 1, 153–160.

[96] Ryan R. Martin, Heather C. Smith, and Shanise Walker, *Improved bounds for induced poset saturation*, Electron. J. Combin. **27** (2020), no. 2, Paper No. 2.31, 9pp.

[97] Ryan R. Martin and Jason J. Smith, *Induced saturation number*, Discrete Math. **312** (2012), no. 21, 3096–3106.

[98] Ryan R. Martin and Nick Veldt, *Induced saturation of the poset $2C_2$*, Ann. Comb. (2025), 14pp.

[99] ______, *Saturation of k-chains in the Boolean lattice*, Electron. J. Combin. **32** (2025), no. 1, Paper No. 1.55, 13pp.

[100] Arès Méroueh, *A LYM inequality for induced posets*, J. Combin. Theory Ser. A **155** (2018), 398–417.

[101] Lev D. Mešalkin, *A generalization of Sperner's theorem on the number of subsets of a finite set*, Teor. Verojatnost. i Primenen. **8** (1963), 219–220.

[102] Abhishek Methuku and Dömötör Pálvölgyi, *Forbidden hypermatrices imply general bounds on induced forbidden subposet problems*, Combin. Probab. Comput. **26** (2017), no. 4, 593–602.

[103] Natasha Morrison, Jonathan A. Noel, and Alex Scott, *On saturated k-Sperner systems*, Electron. J. Combin. **21** (2014), no. 3, Paper 3.22, 17pp.

[104] Peter Nelson and Kazuhiro Nomoto, *The structure of claw-free binary matroids*, J. Combin. Theory Ser. B **150** (2021), 76–118.

[105] Jaroslav Nešetřil and Vojtěch Rödl, *Combinatorial partitions of finite posets and lattices—Ramsey lattices*, Algebra Universalis **19** (1984), no. 1, 106–119.

[106] Deryk Osthus, *Maximum antichains in random subsets of a finite set*, J. Combin. Theory Ser. A **90** (2000), no. 2, 336–346.

[107] Madeleine Paoli, William T. Trotter, and James W. Walker, *Graphs and orders in Ramsey theory and in dimension theory*, Graphs and order (Banff, Alta., 1984), NATO Adv. Sci. Inst. Ser. C: Math. Phys. Sci., vol. 147, Reidel, Dordrecht, 1985, pp. 351–394.

[108] Jinyoung Park and Huy Tuan Pham, *A proof of the Kahn-Kalai conjecture*, J. Amer. Math. Soc. **37** (2024), no. 1, 235–243.

[109] Balázs Patkós, *Induced and non-induced forbidden subposet problems*, Electron. J. Combin. **22** (2015), no. 1, Paper 1.30, 16pp.

[110] Balázs Patkós and Andrew Treglown, *On some extremal and probabilistic questions for tree posets*, Electron. J. Combin. **31** (2024), no. 1, Paper No. 1.19, 23pp.

[111] Frank P. Ramsey, *On a Problem of Formal Logic*, Proc. London Math. Soc. (2) **30** (1929), no. 4, 264–286.

[112] Norbert Sauer, *On the density of families of sets*, J. Combinatorial Theory Ser. A **13** (1972), 145–147.

[113] David Saxton and Andrew Thomason, *Hypergraph containers*, Invent. Math. **201** (2015), no. 3, 925–992.

[114] Saharon Shelah, *A combinatorial problem; stability and order for models and theories in infinitary languages*, Pacific J. Math. **41** (1972), 247–261.

[115] Miodrag Sokić, *Ramsey properties of finite posets*, Order **29** (2012), no. 1, 1–30.

[116] Emanuel Sperner, *Ein Satz über Untermengen einer endlichen Menge*, Math. Z. **27** (1928), no. 1, 544–548.

[117] Hai Tran Thanh, *An extremal problem with excluded subposet in the Boolean lattice*, Order **15** (1998), no. 1, 51–57.

[118] István Tomon, *Forbidden induced subposets of given height*, J. Combin. Theory Ser. A **161** (2019), 537–562.

[119] Paul Turán, *Eine Extremalaufgabe aus der Graphentheorie*, Mat. Fiz. Lapok **48** (1941), 436–452.

[120] Bartel L. van der Waerden, *Beweis einer Baudetschen Vermutung*, Nieuw Archief voor Wiskunde **15** (1928), 212–216.

[121] Stefan Walzer, *Ramsey variant of the 2-dimension of posets*, Master's thesis, Karlsruhe Institute of Technology, Karlsruhe, Germany, 2015.

[122] Christian Winter, *Poset Ramsey number $R(P, Q_n)$. I. Complete multipartite posets*, Order **41** (2024), no. 2, 391–399.

[123] ______, *Poset Ramsey number $R(P, Q_n)$. III. Chain compositions and antichains*, Discrete Math. **347** (2024), no. 7, Paper No. 114031, 14pp.

[124] ______, *Erdős-Hajnal Problems for Posets*, Order **42** (2025), no. 2, 509–527.

[125] Koichi Yamamoto, *Logarithmic order of free distributive lattice*, J. Math. Soc. Japan **6** (1954), 343–353.

Department of Mathematics
Karlsruhe Institute of Technology
Englerstrasse 2, 76131 Karlsruhe
Germany
maria.aksenovich@kit.edu

Department of Mathematics
Iowa State University
428 Carver Hall, 411 Morrill Road, Ames, IA 50011
USA
rymartin@iastate.edu

Department of Combinatorics and its Applications
HUN-REN Alfréd Rényi Institute of Mathematics
Reáltanoda utca 13-15, 1053 Budapest
Hungary
patkos@renyi.hu

A Tables of induced saturation results

Tables 1 and 2 list induced saturation results which are updated from [80]. We will assume that $n \geq 11$ so as to use the simpler $2\sqrt{n}$ bound in Theorem 3.4. For notation, see Section 1 and Section 3.4. We note also that both P and its dual have the same induced saturation numbers and so we don't include both.

poset P	$\mathrm{sat}^*(n,P)$	citation
$\mathcal{C}_k$ $(k \geq 7)$	$\geq \sqrt{k}\,2^{k/2-1.66}$	[99]
$\mathcal{C}_k$ $(k \geq 497)$	$\geq \sqrt{k}\,2^{k/2}$	[99]
$\mathcal{A}_{k+1}$ $(k \geq 4)$	$= kn - \Theta(k \log k)$	[12]

Table 1: Summary of classes of induced saturation results.

P	$\mathrm{sat}^*(n,P)$	citation
$\mathcal{C}_2$	$=1$	
$\mathcal{C}_3$	$=2$	
$\mathcal{C}_4$	$=4$	[48]
$\mathcal{C}_5$	$=8$	[48]+[103]
$\mathcal{C}_6$	$=16$	[48]+[103]
$\mathcal{C}_7$	≤ 30	[48]
$\mathcal{C}_8$	≤ 56	[99]
$\mathcal{C}_9$	≤ 108	[24]
$\mathcal{A}_2$	$=n+1$	
$\mathcal{A}_3$	$=2n$	[44]
$\mathcal{A}_4$	$=3n-1$	[44]
$\mathcal{A}_5$	$=4n-2$	[32]
$\mathcal{A}_6$	$=5n-5$	[32]
$\mathcal{V}$	$=n+1$	[44]
$\mathcal{V}_3$	$\geq n/2$	[20]

P	$\mathrm{sat}^*(n,P)$	citation
$\bowtie$	$\geq n+1$	[66]
	$\leq 6n-10$	[80]
$\Diamond$	$\geq (n+1)/5$	[70]
	$\leq n+1$	[44]
$\mathcal{J}$	$=4$	case anal.
$\mathcal{N}$	$\geq 2\sqrt{n}$	[47]
	$\leq 2n$	[44]
$\mathcal{X}$	$=2n+2$	[47]
$\mathcal{Y}$	$=n+2$	[47]
$\mathcal{V}+\mathcal{C}_1$	$\geq 2\sqrt{n}$	[47]
$2\mathcal{C}_2$	$\geq (3n+1)/2$	[98]
	$\leq 2n$	[80]
$\mathcal{C}_3+\mathcal{C}_1$	≤ 8	[80]
$\mathcal{C}_2+\mathcal{A}_2$	≤ 8	[80]
$3\mathcal{C}_2$	≤ 14	[80]
$5\mathcal{C}_2$	≤ 42	[80]
$7\mathcal{C}_2$	≤ 60	[80]

Table 2: Summary of induced saturation results for selected connected posets.

B Tables of off-diagonal Ramsey results

Table 3 lists off-diagonal Ramsey results which are given by Winter [123]. For notation, see Section 1 and Section 3.4. We note also that both P and its dual have the same Ramsey numbers against the self-dual poset $\mathcal{B}_n$ and so we don't include both.

poset P	$R(P, \mathcal{B}_n)$	citation
$\mathcal{C}_1$	n	
$\mathcal{C}_2$	$n+1$	[123]
$\mathcal{A}_2$	$n+2$	[123]
$\mathcal{C}_3$	$n+2$	[123]
$\mathcal{C}_2+\mathcal{C}_1$	$n+3$	[123]
$\mathcal{A}_3$	$n+3$	[123]
$\mathcal{C}_4$	$n+3$	[123]
$\mathcal{C}_2+\mathcal{C}_2$	$n+3$	[123]
$\mathcal{A}_4$	$n+3$	[123]
$\mathcal{C}_3+\mathcal{C}_1$	$n+4$	[123]
$\mathcal{C}_2+\mathcal{A}_2$	$n+4$	[123]

poset P	$R(P, \mathcal{B}_n) = n + \frac{c(n)n}{\log_2 n}$	citation
$\mathcal{V}$	$c(n) \leq 1 + o(1)$	[3]
$\mathcal{V}+\mathcal{C}_1$	$c(n) \leq 1 + o(1)$	[3]+[121]
$\mathcal{V}_3$	$c(n) \leq 1 + o(1)$	[3]+[122]
$\mathcal{N}$	$c(n) \leq 1 + o(1)$	[3]+[4]
$\Diamond$	$c(n) \leq 2 + o(1)$	[3]+[61]
$\mathcal{Y}$	$c(n) \leq 2 + o(1)$	[3]+[122]
$\mathcal{J}$	$c(n) \leq 2 + o(1)$	[3]+[123]
$\bowtie$	$c(n) \leq 4 + o(1)$	[3]+[122]

Table 3: Summary of off-diagonal Ramsey results. The inequality $c(n) \geq \frac{1}{15}$ was established by Axenovich and Winter [3].

Techniques for Taming Temporal Graphs

Kitty Meeks

Abstract

Temporal graphs, in which the vertex-set remains fixed but the edge-set is subject to discrete changes over time, provide a useful formalism for modelling many real-world datasets, from social networks to transport networks. However, many fundamental concepts and results in graph theory do not transfer in a natural way to the temporal setting, and, in particular, even basic algorithmic questions become much more computationally challenging. In this article, we provide an overview of recent attempts to identify tractable cases for computational problems on temporal graphs, via the framework of parameterised complexity, and reflect on the major challenges in addressing this goal.

1 What are temporal graphs?

When planning a train journey from Glasgow to Cardiff, I can see from looking at a UK train map that I will likely need to change trains at least twice; but, based on the map alone, there are several options, and it is not clear which of these is best. What do I mean by best? There may be a number of considerations. For example, I may want to have the shortest possible journey time. But this might come at the cost of having a very short time to change somewhere along the route, meaning that, even in the event of a short delay to my first train, I would reach Cardiff much later than intended. On the other hand, I am unwilling to break the journey overnight in an intermediate city, and ideally I would like to avoid waiting for several hours to change train. I may also have constraints on what time I will be able to depart Glasgow in the morning, and might have a specific deadline for the time by which I must reach Cardiff. The basic train map, which captures only information about the existence of a train route between pairs of stations, does not provide the information I need to take into account any of these considerations: I also need information about the *times* at which trains run.

This example illustrates the difference between the kind of graphs encountered in an introductory combinatorics course – which, for the rest of this article, will be referred to as *static* graphs – and temporal graphs, in which edges appear and disappear over time. Such graphs are not just useful for planning train journeys: they arise in many other settings, from social networks to communication networks. For example, when considering the spread of disease through a network, the order in which the connections occur is crucial: if I catch a cold on the train while travelling to a conference, I can pass it on to all the other participants, but this is not true if I catch it on the train going home. The same is true for spread of information, where we might be interested either in maximising the efficacy of our advertising campaign or in taking interventions to slow the spread of dangerous misinformation.

Temporal graphs, in which edges are subject to discrete changes over time, therefore offer a useful and flexible formalism for many kinds of real-world networks, and there has been a rapid growth in the volume of literature on the topic in the last decade. However, most intuition from static graph theory fails to transfer to the temporal setting, and natural computational problems – many of which are already known to be intractable in the static setting – typically become much more challenging to solve. In this article, we review some recent successes and failures in the attempt to understand structural restrictions that can be imposed on temporal graphs to rule out some of the more challenging behaviour and hopefully obtain algorithms that are tractable in at least some cases.

We mainly address this question from the perspective of parameterised complexity, a framework that has proved hugely successful in isolating the aspects of computational problems on static graphs that lead to intractability (several textbooks provide extensive background on this field [25, 28, 29, 37]). This involves a multivariate approach to understanding the running time of algorithms, and the goal is to restrict the exponential behaviour to one or more parameters rather than the entire input size. Parameters of interest typically capture some property of the problem (e.g., the number of vertices in a clique we are hoping to find) or of the input (e.g., the maximum degree of the input graph). The goal is to devise a so-called fpt-algorithm, whose running time is bounded by $f(k) \cdot n^c$ where n is the total input size, k is the parameter (or sum of parameters), f is an arbitrary computable function (typically exponential) and c is a constant that does not depend on k; problems admitting such an algorithm are said to be *fixed-parameter tractable* and belong to the class FPT. Crucially, such an algorithm runs in polynomial time if the parameter(s) are bounded by a constant, and, furthermore, because the exponent of the polynomial does not depend on the parameter value. The search for algorithms of this kind that involve a parameter describing the structure of the graph can thus be viewed as a more general approach than that of seeking polynomial-time algorithms for specific graph classes. The theory of parameterised complexity also includes a framework for proving that (up to widely-believed complexity theoretic assumptions) specific parameterised problems do not belong to FPT; this is most commonly done by showing that the problem is hard for some level of the so-called *W-hierarchy*, for example that it is W[1]-hard. In some cases, it is also possible to rule out the existence of an fpt-algorithm using tools from classical complexity, by showing that the problem is already NP-hard for constant values of the parameter.

In the remainder of this section, we introduce our formalism for temporal graphs and some key notions and definitions, then discuss some of the challenges in defining temporal analogues of well-known graph theoretic concepts. The remaining sections discuss two different families of parameters that have been defined for temporal graphs: those that are based in some way on parameters for static graphs, including various measures of "tree-likeness" and other forms of regular structure, and those that are inherently temporal and do not have any analogue in the static setting. We do not attempt to give an exhaustive treatment of all the FPT results in the literature with respect to different families of parameters for temporal graphs; rather, we introduce a handful of well-studied problems on temporal graphs, which serve to illustrate many of the complexity trends observed more generally.

1.1 A mathematical model of temporal graphs

In this article, we will adopt the formalism for temporal graphs which is most widespread in the recent algorithmic literature on the topic; this definition is widely credited to Kempe, Kleinberg and Kumar [50], building on previous work by Berman

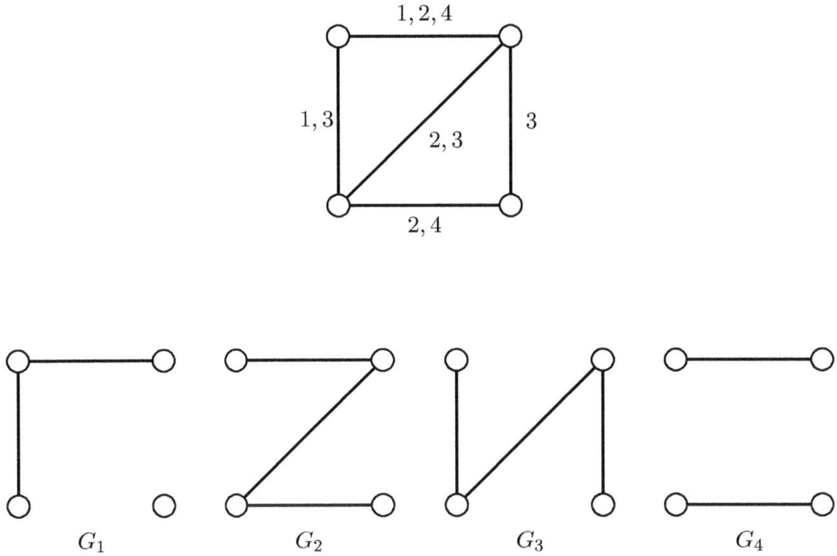

Figure 1: A temporal graph in which edges are labelled with their active times (top). This can alternatively be expressed as a sequence of snapshot graphs, on the same vertex set (below), with the ith graph containing the edges active at time i.

[5]. In this definition, a temporal graph is defined to be a pair $\mathcal{G} = (G, \lambda)$, where $G = (V, E)$ is a graph, and $\lambda : E \to 2^{\mathbb{N}}$ is a function assigning a (non-empty) subset of times to each edge. A graphical illustration of a temporal graph is given in Figure 1. We call the static graph G the *underlying graph* or *footprint*, and, for each edge e, we say that e is *active* or *appears* at the times in $\lambda(e)$. The *lifetime* of $\mathcal{G}$, written $\Lambda(\mathcal{G})$, is the maximum time at which any edge appears, i.e. $\max_{e \in E(G)} \lambda(e)$; we typically assume without loss of generality that at least one edge in the graph is active at time 1. Unless otherwise stated, we assume throughout this article that all temporal graphs have finite lifetime; this is important for algorithmic purposes as it ensures we do not need infinitely large inputs to encode temporal graphs. Temporal graphs with infinite lifetime and finite representation are also possible if we consider *periodic* temporal graphs, in which we specify only the appearances of edges over the first p timesteps, after which they are assumed to repeat infinitely (so an edge e appears at all times $\{ip + t : t \in \lambda(e), i \in \mathbb{N} \cup \{0\}\}$).

Sometimes it is convenient to describe this model in one of two equivalent ways. The first is to describe the temporal graph $\mathcal{G}$ as a triple $(V, \mathcal{E}, \Lambda)$, where V is the vertex-set, $\mathcal{E}$ is a set of *time-edges* (that is, pairs $(\{u, v\}, t)$ where $u, v \in V$ and $t \in \mathbb{N}$, such that the edge uv appears at time t in $\mathcal{G}$), and Λ is the lifetime (we note that the lifetime can be deduced from the set of edges, but, for algorithmic purposes, it is often convenient to assume we can access this value directly without having to examine a large number of time-edges). The second is to represent $\mathcal{G}$ as a sequence

$G_1, G_2, \ldots, G_\Lambda$ of static graphs, where $G_i = (V, E_i = \{e \in E(G) : i \in \lambda(e)\})$. This second representation is often conceptually helpful, but is inefficient of space if many of the edges appear at only a small proportion of the timesteps.

A number of other models, and generalisations of the one described here, have also been considered in the literature; Casteigts, Flocchini, Quattrociocchi, and Santoro [12] discuss the relationships between various models arising in different application areas. Here we comment briefly on two particularly natural generalisations. The first is to associate with each edge a *latency* or travel time in addition to its list of active times. This is a very natural generalisation in the context of transport networks: not all rail journeys take the same time. However, in many cases, we can effectively simulate different (integral) travel times using the simpler model (as described above), in which all edges implicitly have unit travel time, simply by subdividing each edge an appropriate number of times. This strategy can also be used to simulate different travel times at different departure times (such as a high-speed train departing at 9am versus a slower, local train departing at 10am), and only fails in settings where introducing extra vertices causes problems (for example, if we are interested in the number of distinct vertices visited). In the study of certain specific problems, a model has been considered in which different costs are associated with different edges and potentially different departure and arrival times [19]; this could be natural in the context of fuel consumption for a car journey, as it might be possible to take a faster, but geographically longer route, or a slower, but more fuel-efficient option with the same departure time, and moreover, the costs associated with each would vary throughout the day depending on the amount of congestion. However, we mention these models here only for the sake of completeness; in the rest of the article we assume identical travel times and costs for all edges.

We conclude this section by considering what is meant by the size of a temporal graph, as the size of the input is crucial for determining the efficiency or otherwise of algorithms. For a static graph, we would typically consider the size of a graph to be $O(|V|+|E|)$ (since an adjacency-list representation can encode the graph using this number of $\log |V|$-bit words), and describe an algorithm running in time $O(|V|+|E|)$ as linear-time; however, if we are interested only in whether an algorithm runs in time polynomial in the input size, we can just as well take the input size to be $O(|V|^2)$. In the temporal world, however, we also have to take into account the lifetime of the graph, as there is no reason to assume any relationship (or, in particular, a polynomial one) between the number of vertices and the number of timesteps: we might consider a small graph over a very long time period, or a very large one for only a few timesteps. Thus, we consider an algorithm to run in polynomial time on a temporal graph if its running time is bounded by a function that is polynomial in both the number of vertices and the lifetime.

1.2 Turning problems temporal

One of the exciting but also challenging aspects of studying temporal graphs is the fact that there is rarely a single natural way to define the temporal analogue of well-known graph theoretic concepts. This becomes clear even when considering a concept as fundamental as that of walks and paths.

It is easy to argue that a walk or path in a temporal graph should "respect time",

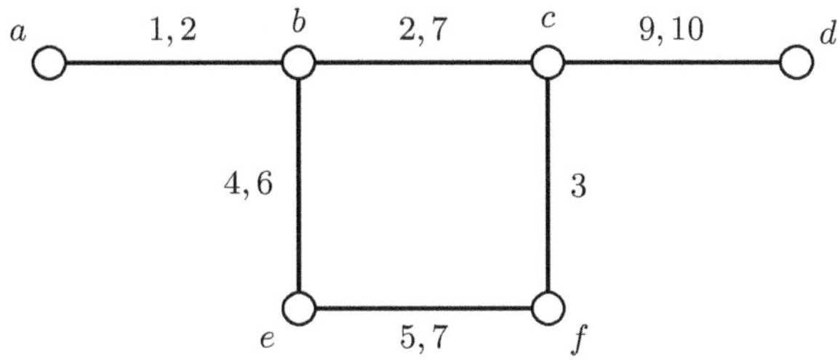

Figure 2: In this temporal graph, there is no 2-restless temporal path from a to d, but there is a 2-restless temporal walk ($abcfebcd$).

meaning that each edge traversed is associated with one of the times at which it is active, and that these times should be increasing, or at least non-decreasing, along the walk or path (if we require that the times are increasing, this gives the notion of a *strict* temporal walk/path, whereas requiring only that they are non-decreasing gives a *non-strict* walk/path). This requirement is natural in the context of most applications of temporal graphs: one cannot depart on a train from a given station before arriving there on the previous leg of the journey, while information or disease can only spread along a sequence of contacts that occur in chronological order. Another reasonable requirement we might want to place on our temporal walks and paths is that the number of timesteps spent waiting at any intermediate vertex does not exceed some given bound; such walks and paths have been termed *restless*. More specifically, a temporal path or walk is said to be Δ-*restless* if the number of timesteps spent at each intermediate vertex is no more than Δ [13]. This restriction is again natural for many applications: we would not normally accept a week-long wait at an intermediate train station on our journey, and, for many diseases, an individual will only remain infectious (and thus able to transmit the disease through a further contact) for a limited time after becoming infected. However, introducing this restriction also has some surprising consequences: as illustrated in Figure 2, it is possible that there is a restless temporal walk but not a restless temporal path from u to v, and moreover, the problem, Restless Temporal Path, of determining the existence or otherwise of a restless temporal path from u to v is NP-complete [13].

There are more choices to make if we want to define an analogue of shortest paths in temporal graphs. Of course, one option is to copy exactly the definition from static graphs, and define the shortest path to be the one containing the fewest edges; this is indeed the standard definition of *shortest temporal path*. However, there are other notions of optimality that might be equally or more relevant in many applications of temporal graphs. Returning to the idea of train journeys, the shortest path corresponds to the journey with fewest changes, but we might prefer the *fastest* temporal path, with the smallest difference between departure and arrival time, or the *foremost* temporal path, which reaches the destination at the earliest time.

Beyond paths, the notion of connectivity poses further challenges when translated to the temporal setting. As for directed paths in digraphs, the existence of a temporal path from u to v does not imply the existence of one from v to u; but

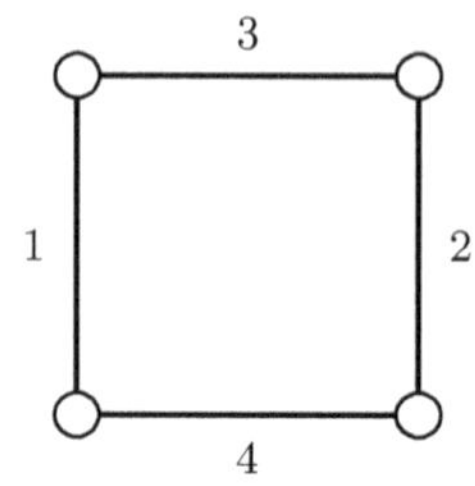

Figure 3: A temporal graph on 4 vertices in which no subtree is a spanner.

here, as well as lacking reflexivity, we also do not have transitivity, as it might be that all temporal paths from y to z depart before any temporal path from x arrives at y. These issues make reasoning about so-called *temporal spanners*, subgraphs containing a temporal path from every vertex to every other (and hence a natural temporal analogue of spanning trees) very challenging. The example in Figure 3 illustrates that temporal graphs do not necessarily admit spanners whose underlying graphs are trees, but it was initially conjectured that any temporal graph on n vertices might still admit a spanner with $O(n)$ edges. This was soon disproved, with a lower bound of $O(n \log n)$ (based on hypercubes [50]), and in fact, it turns out that, in general, it is not possible to guarantee the existence of a temporal spanner with $o(n^2)$ edges [3]. Even in the very special case, in which the underlying graph is complete, the number of edges required for a spanner is not known: in this case, $O(n \log n)$ edges have been shown to suffice [15], but it is open whether linear-sized temporal spanners can be guaranteed here.

Under the broad theme of connectivity, another compelling example of how the choice between different, equally reasonable, temporal interpretations of a question can lead to dramatically different behaviour comes from the search for a temporal analogue of Menger's theorem. Kempe, Kleinberg, and Kumar [50] demonstrated that a natural translation of this classical result to the temporal setting fails, giving an example of a temporal graph in which there are no two disjoint (non-strict) temporal paths between vertices s and z, but there is no single vertex (other than s and z) whose deletion destroys all s-z paths. Building on this, they further demonstrated that it is NP-hard to compute the number of disjoint s-z paths. Later, Mertzios, Michail, and Spirakis [63] showed that a reasonable analogue of Menger's theorem can, however, be obtained by changing the definitions: their requirement for disjointness of temporal paths is that they should be *out-disjoint*, meaning that they never leave from or arrive at the same node at the same time; and, rather than considering the number of vertices that must be deleted to disconnect s and z, they instead consider the number of (vertex,time) pairs (v, t) required so that deleting all edges leaving v at time t destroys all strict temporal paths from s to z.

It is not just defining problems that poses major challenges in the temporal setting, but also solving them algorithmically. We have already seen that deciding whether there exists a specific (and natural) kind of temporal path between two vertices is NP-complete, while the static analogue can easily be solved in linear time. Another example comes from the following problem, a temporal analogue of deciding whether a static graph is Eulerian (also called EULERIAN TRAIL in some literature).

> TEMPEULER
> *Input:* A temporal graph $\mathcal{G}$.
> *Question:* Is there a strict temporal walk in $\mathcal{G}$ that traverses every edge of the underlying graph exactly once?

While the static analogue can be solved in linear time simply by examining the degrees of the vertices, this temporal analogue is NP-complete even under strong restrictions on the input (which will be discussed in more detail later) [10, 61]. An even more extreme example arises in the following problem, whose static analogue is trivial as the answer is always "yes".

> STAREXP(c)
> *Input:* A temporal graph $\mathcal{G}$ whose underlying graph is a star, where the number of active times for each edge is at most c.
> *Question:* Is there a temporal walk, starting and finishing at the centre of the star, that visits every vertex?

This problem is known to be NP-complete for every $c \geq 4$ [1, 10].

2 Temporal analogues of static graph parameters

In this section, we discuss the various ways in which well-known parameters from the static setting can be generalised to temporal graphs – often in several different ways – and how successful these generalisations have been in the quest for tractability.

2.1 Measures of "tree-likeness"

When faced with an NP-hard problem on static graphs, one of the first questions we ask when looking for tractable special cases is whether the problem is polynomial-time solvable on trees (or, more generally, forests); very often, the answer is yes, and, in many cases, the same ideas can be generalised to other classes of graphs that are sufficiently "tree-like" in some suitable sense. There are several ways in which to define how tree-like a graphs is. Perhaps the most obvious option is to consider how much the graph must be modified to obtain a forest: this leads us to two well-studied parameters, the *feedback edge number* and the *feedback vertex number*, which are the smallest number of edges and vertices respectively that must be deleted from the graph in order to obtain a forest. More generally, graphs with small *treewidth* exhibit structure that often makes it possible to generalise dynamic programming approaches used to solve computational problems on trees; formally, treewidth is defined as follows.

Definition 2.1 A pair $(T, \mathcal{B})$ is a *tree decomposition* of G if T is a tree and $\mathcal{B} = \{\mathcal{B}_s : s \in V(T)\}$ is a collection of non-empty subsets of $V(G)$ (or *bags*), indexed by the nodes of T, satisfying:

(1) for all $v \in V(G)$, the set $\{s \in T : v \in \mathcal{B}_s\}$ is nonempty and induces a connected subgraph in T,

(2) for every $e = uv \in E(G)$, there exists $s \in V(T)$ such that $u, v \in \mathcal{B}_s$.

The *width* of the tree decomposition $(T, \mathcal{B})$ is defined to be $\max\{|\mathcal{B}_s| : s \in V(T)\} - 1$, and the *treewidth* of G is the minimum width over all tree decompositions of G.

A celebrated metatheorem of Courcelle [21, 22] tells us that any problem definable in monadic second-order logic[1] can be solved in linear time on graphs whose treewidth is bounded by a constant; this directly implies the existence of an fpt-algorithm for the problem parameterised by treewidth. If either the feedback edge number or the feedback vertex number of a graph is bounded, this implies that the treewidth is bounded, so fixed-parameter tractability for many problems with respect to these parameters also follows from Courcelle's theorem; however, in some cases considering the more restrictive parameter of feedback edge/vertex number allows the design of an fpt-algorithm when no such algorithm is known with respect to treewidth, or allows the design of an fpt-algorithm with a faster running time [16, 4, 55].

An even more restrictive parameter, the *vertex cover number*, is the minimum size of a set X of vertices such that every edge has at least one endpoint in X, so deleting all vertices in X leaves an edgeless graph (which is a very special kind of forest). Bounding this parameter places very strong restrictions on the graph structure (in addition to bounding the feedback edge and vertex numbers), and leads to efficient parameterised algorithms even for problems that are intractable when parameterised by feedback edge or vertex number [30, 36, 62].

So, how can we apply any of these parameters in the temporal setting? A first approach is to ignore the temporal information in the graph, and consider only the structure of the underlying static graph. Several problems related to reachability in temporal graphs become polynomial-time solvable when the underlying graph is a tree [33, 57, 56]. Some path-related problems (counting the number of foremost or fastest temporal paths between a given pair of vertices, or determining the existence of a restless temporal path between two vertices) are in FPT parameterised by the feedback edge number of the underlying graph, although the latter is W[1]-hard when parameterised by the feedback vertex number of the underlying graph.

However, examples such as these seem to be the exception rather than the norm. There are numerous examples of problems that remain hard even when the underlying graph is very highly restricted. We have already seen one example: StarExp is by definition a problem in which the underlying graph is a star, and yet this is NP-hard even if each edge is active at only four times. Another example of a problem that is hard when the underlying graph is a star is the following natural temporal analogue of Vertex Cover [2].

[1] For a formal statement of the theorem, and further background on monadic second-order logic, we refer the reader to [23, 37].

Temporal Vertex Cover
Input: A temporal graph $\mathcal{G} = (G, \lambda)$ with lifetime Λ and a natural number k.
Question: Does there exist a set $X \subseteq V(G) \times \Lambda$, with $|X| \leq k$, such that, for every $uv \in E(G)$, there exists $t \in \lambda(uv)$ such that $\{(u,t),(v,t)\} \cap X \neq \emptyset$?

A third example of a natural temporal problem that remains intractable even on a very restricted class of trees is the following temporal analogue of the maximum matching problem, a classic example of a non-trivial problem that admits a polynomial-time algorithm on static graphs.

Δ-Temporal Matching
Input: A temporal graph $\mathcal{G} = (G, \lambda)$ and a natural number k.
Question: Does there exist a set $Y \subseteq \bigcup_{e \in E(G)} \{(e,t) : t \in \lambda(e)\}$ with $|Y| \geq k$ such that, for all $(e_1, t_1), (e_2, t_2) \in Y$, either $e_1 \cap e_2 = \emptyset$ or $|t_2 - t_1| \geq \Delta$.

In other words, edge appearances in our matching must either have no endpoints in common, or appear sufficiently far apart in time; this is a natural model of a situation in which there is some recovery time Δ which must pass before a vertex can be reused in a new matching edge. This problem is known to be NP-hard even when the underlying graph is a path and $\Delta = 2$ [64].

A second natural way to apply static parameters in the temporal setting would be to restrict the structure of each snapshot. However, this imposes even less structure on the temporal graph than restricting the underlying graph. Indeed, for a large number of problems, it is possible to create an equivalent instance (in the sense that the answer to the problem is guaranteed to be the same), even if the timings of edges are spread out so that very few edges (sometimes just a single edge) appear at each timestep. A simple example of this is given by Bumpus and Meeks [10], who adapt the original NP-hardness proof for the following problem [34] to show that the problem remains NP-hard even when there is at most one active edge at every time. The *(strict) reachability* of a vertex v in a temporal graph is the number of vertices u such that there exists a (strict) temporal path from v to u (with the convention that every vertex reaches itself); this is a crucial concept when considering the spread of information or infection.

Singleton Minimum Temporal Reachability Edge Deletion (SingMinReachDelete)
Input: A temporal graph $\mathcal{G} = (G, \lambda)$, a vertex $v_s \in V(G)$ and positive integer r.
Question: What is the cardinality of the smallest set of time-edges E such that the vertex v_s has temporal reachability at most r after their deletion from $\mathcal{G}$?

More generally, a similar approach is often possible for problems involving paths or reachability (without waiting time constraints).

There are, however, a few examples where very strong restrictions on the individual snapshots lead to efficient algorithms. For example, the problem Temporal

VERTEX COVER introduced above can be solved efficiently if the maximum degree in any snapshot is at most one; but this problem is already intractable when degree two vertices are allowed in each snapshot [49].

These results suggest that we need to take into account the temporal information in the graph in order to obtain more useful measures of how tree-like a temporal graph is. As in the static setting, one very natural way to define such parameters is in terms of the number of deletion operations needed to destroy all cycles. However, in the temporal setting, there are, unsurprisingly, more options for what a deletion operation could involve as well as for what it means to destroy all cycles.

A good illustration of this comes from the attempt to define a temporal analogue of the feedback edge number in static graphs. Haag, Molter, Niedermeier, and Renken [46] introduce four temporal versions of this parameter: the goal is to delete edges (at all times) or edge-appearances (at specific times) so that all temporal cycles are destroyed; there is a further choice as to whether the time labels on consecutive edges in temporal cycles must be strictly or non-strictly increasing. This gives rise to four variants of the problem:

- the *(strict) temporal feedback connection number*, which is the number of edges of the underlying graph that must be deleted to remove all (strict) temporal cycles, and

- the *(strict) temporal feedback edge number*, which is the number of edge-appearances (i.e. (edge,time) pairs) that must be deleted to remove all (strict) temporal cycles.

It is clear that the (strict) temporal feedback connection number is always upper bounded by the feedback edge number of the underlying graph, since destroying all cycles in the underlying graph certainly ensures that there are no temporal cycles remaining. The same is not true, however, for the (strict) temporal feedback edge number, as edges in a minimum feedback edge set for the underlying graph may appear at a very large number of timesteps, making them costly to delete in this version.

In contrast with the static setting, where the feedback edge number can be calculated directly from the numbers of vertices and edges in the graph, it turns out that all four versions of this parameter are NP-hard to compute [46]. The use of these analogues of the feedback edge number in designing fpt-algorithms for temporal graph problems has not yet been thoroughly investigated.

As for temporal versions of the feedback edge number, there are many ways in which the feedback vertex number could be translated to the temporal setting, but only one has seen much attention in the literature thus far, namely, the *timed feedback vertex number*, first introduced by Casteigts, Himmel, Molter, and Zschoche [13].

Definition 2.2 A *timed feedback vertex set* of $\mathcal{G} = (G = (V, E), \lambda)$ is a set $X \subseteq V \times [T]$ of vertex appearances such that the underlying graph of $\mathcal{G}' = ((V, E'), \lambda')$ is a forest, where for each edge $uv \in E$, we have $\lambda'(e) = \lambda(e) \setminus (\{t : (u, t) \in X\} \cup \{t : (v, t) \in X\})$, and E' is defined to be $\{e \in E : \lambda'(e) \neq \emptyset\}$. The *timed feedback vertex number* of $\mathcal{G}$ is the minimum cardinality of a timed feedback vertex set of $\mathcal{G}$.

The timed feedback vertex number tells us the number of vertex appearances that must be deleted so that all cycles are destroyed in the *underlying* graph. This parameter has proved particularly useful in solving intractable problems relating to paths in temporal graphs: both the problem of determining whether there exists a Δ-restless temporal path between two vertices u and v [13], and the problem of counting all optimal foremost or fastest temporal paths between u and v [35] have been shown to be in FPT with respect to this parameter.

Given the huge success of treewidth in the design of efficient parameterised algorithms for static graph problems, a key question in measuring the tree-likeness of temporal graphs is whether there is a useful temporal analogue of treewidth. The answer to this question is both yes and no: there are numerous ways in which we could hope to define and leverage treewidth in temporal graphs, but while each of these approaches has enjoyed some isolated success, none comes close to equalling the broad applicability of static treewidth. One initially appealing approach is to define a relational structure that captures temporal information as well as the structure of the underlying graph, and leverage the generalisation of Courcelle's theorem to relational structures [23, Theorem 9.21]. While there has been some success using this approach, in many cases, an additional parameter – such as the maximum degree of the underlying graph, or the lifetime – also needs to be restricted to recover tractability [34, 56]. Another promising direction is to consider the treewidth of the *static expansion* of a temporal graph, but this has not yet found wide applicability. Given the difficulty of finding a good treewidth analogue for directed graphs [44], it is perhaps not so surprising that this search has had limited success in the temporal setting. For further discussion on the various analogues of treewidth that have been defined in the temporal setting, we refer the reader to an excellent survey article [38].

2.2 Dense but highly structured temporal graphs

Beyond treewidth, numerous other width measures for static graphs have been defined in the literature; see, for example, the recent paper of Brettell, Munaro, Paulusma and Yang [8] for a discussion of some of these and the relationships between them. One of the best known of these other width measures is *cliquewidth*, defined as follows.

Definition 2.3 ([24]) The *cliquewidth* of a static graph $G = (V, E)$ is the number of labels required to construct G using only the following operations:

1. Creating a new vertex with label i.
2. Taking the disjoint union of two labelled graphs.
3. Adding edges to join all vertices labelled i to all vertices labelled j, where $i \neq j$.
4. Renaming label i to label j.

We refer to an algorithm which constructs a graph G using the above operations as a *cliquewidth construction* of G.

The cliquewidth of a graph is bounded whenever the treewidth is bounded, but cliquewidth also takes small values for dense but highly structured graphs (for example, cliques and complete bipartite graphs have cliquewidth one and two respectively). There is also similar algorithmic metatheorem, due to Courcelle, Makowsky and Rotics, which gives fpt-algorithms with respect to cliquewidth for problems definable in a slightly more restrictive fragment of logic [20].

While there is no one natural way to extend the definition of treewidth to the temporal setting, it turns out to be much more obvious how to do this for cliquewidth; this leads to the following definition of temporal cliquewidth, where the only change from the static version is that, when adding all edges between a pair of classes, all such edges must appear at exactly the same timesteps.

Definition 2.4 ([31]) The *temporal cliquewidth* of a temporal graph $\mathcal{G} = (G, \lambda)$ is the number of labels required to construct $\mathcal{G}$ using only the following operations:

1. Creating a new vertex with label i.
2. Taking the disjoint union of two labelled graphs.
3. Adding edges to join all vertices labelled i to all vertices labelled j, where $i \neq j$, such that all the added edges are active at the same set of times T.
4. Renaming label i to label j.

We refer to an algorithm which constructs a temporal graph $\mathcal{G}$ using the above operations as a *temporal cliquewidth construction* of $\mathcal{G}$.

The following NP-hard temporal problem has been shown to admit an fpt-algorithm parameterised by temporal cliquewidth.

> TEMPORAL Δ-CLIQUE
> *Input:* A temporal graph $\mathcal{G} = (G, \lambda)$ and two integers Δ and h.
> *Question:* Is there a set $V' \subseteq V(G)$ such that $|V'| \geq h$ and, for every $u, v \in V'$ and time $t \in [1, \Lambda(G) - \Delta]$, there exists $t' \in [t, t + \Delta]$ such that $t \in \lambda(uv)$?

This problem asks for a set of vertices in which each pair is connected by an edge at least once in every window of Δ consecutive timesteps. Rather than leveraging the definition of temporal cliquewidth directly, the proof of this result relies on the fact that the existence of a temporal Δ-clique is equivalent to the existence of a static clique of the same size in a specific auxiliary graph (which can be constructed efficiently from the temporal graph). The key observation is that the cliquewidth of the input temporal graph can be used to bound the cliquewidth of this auxiliary graph, and the result then follows from the fact that the static CLIQUE problem is in FPT parameterised by cliquewidth.

In order to apply temporal cliquewidth to obtain efficient parameterised algorithms for a wider range of problems, it is natural to ask whether there might exist a Courcelle-style metatheorem for temporal cliquewidth, giving tractability for any temporal graph problem definable in a suitable fragment of logic. Such a metatheorem seems much more feasible for temporal cliquewidth than for any of the temporal analogues of treewidth, but at present this question remains open.

Several parameters for static graphs are known to be more restrictive than cliquewidth, in the sense that boundedness of one of these other parameters implies boundedness of cliquewidth, but the reverse implication does not hold. Such parameters allow us to regain tractability in settings where restricting the cliquewidth is not enough. These specialisations of cliquewidth also seem to lend themselves well to adaptation to the temporal setting.

The first such parameter we consider is modular width. In the static setting, this is defined as follows.

Definition 2.5 (**[42]**) Suppose a static graph G can be constructed by the algebraic expression A which uses the following operations:

1. Creating an isolated vertex.

2. Taking the disjoint union of two graphs.

3. Taking the complete join of two graphs. That is, for graphs $G_1 = (V_1, E_1)$ and $G_2 = (V_2, E_2)$, $V(G_1 \otimes G_2) = V(G_1) \cup V(G_2)$ and $E(G_1 \otimes G_2) = E(G_1) \cup E(G_2) \cup \{(v, w) : v \in V(G_1)$ and $w \in V(G_2)\}$.

4. The substitution of graphs $G_1, \ldots, G_n$ into a graph G' with vertices $v_1, \ldots, v_n$. This gives the graph $G'(G_1, \ldots, G_n)$ with vertex set $\bigcup_{1 \leq i \leq n} V(G_i)$ and edge set $\bigcup_{1 \leq i \leq n} E(G_i) \cup \{(v, w) : v \in V(G_i), w \in V(G_j),$ and $(v_i, v_j) \in E(G')\}$.

The width of an expression A is the maximum number of operands in an occurrence of operation 4 in A. The *modular-width* of G, written $MW(G)$, is this minimum width of an expression A which constructs G.

A key difference between modular width and cliquewidth is that graphs of bounded modular width have bounded diameter, whereas graphs having constant cliquewidth can contain arbitrarily long induced paths. The same is true for the underlying graphs of temporal graphs with bounded cliquewidth and bounded temporal modular width; temporal modular width is formally defined as follows, adapting the static definition in a very similar way as for cliqewidth. It is convenient here to use the alternative formalism for temporal graphs in which we specify the set of time-edges rather than the edge-set and labelling function.

Definition 2.6 (**[31]**) Suppose a temporal graph $\mathcal{G}$ can be constructed by the algebraic expression A which uses the following operations:

1. Creating an isolated vertex.

2. Taking the disjoint union of two temporal graphs.

3. Taking the complete join of two temporal graphs at a set of times T. That is, for graphs $\mathcal{G}_1 = ((V_1, E_1), \lambda_1)$ and $\mathcal{G}_2 = ((V_2, E_2), \lambda_2)$, $V(\mathcal{G}_1 \otimes \mathcal{G}_2) = V(\mathcal{G}_1) \cup V(\mathcal{G}_2)$ and $\mathcal{E}(\mathcal{G}_1 \otimes \mathcal{G}_2) = \mathcal{E}(\mathcal{G}_1) \cup \mathcal{E}(\mathcal{G}_2) \cup \{(vw, t) : v \in V(\mathcal{G}_1), w \in V(\mathcal{G}_2)$ and $t \in T\}$.

4. The substitution of temporal graphs $\mathcal{G}_1, \ldots, \mathcal{G}_n$ into a temporal graph $\mathcal{G}'$ with vertices $v_1, \ldots, v_n$. This gives the graph $\mathcal{G}'(\mathcal{G}_1, \ldots, \mathcal{G}_n)$ with vertex set $\bigcup_{1 \leq i \leq n} V(\mathcal{G}_i)$ and time-edge set $\bigcup_{1 \leq i \leq n} \mathcal{E}(\mathcal{G}_i) \cup \{(vw, t) : v \in V(\mathcal{G}_i), w \in V(\mathcal{G}_j), \text{ and } (v_i v_j, t) \in \mathcal{E}(\mathcal{G}')\}$.

The width of an expression A is the maximum number of operands in an occurrence of the operation 4 in A. The *temporal modular-width* of (G, λ) is this minimum width of an expression A which constructs (G, λ).

Whilst in the static setting, a star has bounded modular width (more generally, any complete bipartite graph can be generated by first constructing two edgeless graphs of the appropriate size, then taking their complete join), this is not true for the temporal analogue if the edges of the star are active at different times. However, any temporal graph whose underlying graph is a star does have bounded temporal cliquewidth; it is immediate from this observation that the problem StarExp, introduced in Section 1.2, is NP-hard on temporal graphs with constant temporal cliquewidth.

It turns out that the additional restrictions imposed by bounding the temporal modular width are enough to make this problem tractable: StarExp(c) is in FPT parameterised by the temporal modular width of the input graph, for any constant c [31]. The crucial observation to prove this result is that, if the temporal modular width of a star is at most k, then the leaves can be partitioned into at most k classes such that, for any two vertices in the same class, the incident edges are active at precisely the same set of times. If there are more than $c/2$ vertices in any one of these classes, we must have a no-instance to the problem, as we cannot even hope to visit all the vertices in this single class on a temporal walk. Thus, it remains only to deal with the case in which there are $O(k)$ vertices in total (recalling that c is constant), in which case we can solve the problem by brute force in time bounded by a function of k.

The argument sketched above is in fact an example of kernelization in the temporal setting. A parameterised problem is said to admit a *kernel* of size $g(k)$ if there exists a polynomial-time algorithm which, given any instance of the problem, outputs an equivalent instance (in the sense that the answer is the same) of size at most $g(k)$. Kernelization is often described as a rigorous theory of efficient preprocessing, and has been a hugely successful approach in the study of parameterised algorithms for static graph problems [25]. In the StarExp example, we obtain a linear kernel by first identifying the classes of equivalent vertices, then for each class of size greater than $c/2$ deleting arbitrarily chosen elements until it contains exactly $\lfloor c/2 \rfloor + 1$ vertices.

Continuing in this vein of defining ever-more restrictive structural parameters which can take small values on highly structured graphs regardless of their density, we now discuss the temporal analogue of *neighbourhood diversity*. This can be thought of as a non-recursive version of modular width, and is defined as follows (first introduced by Lampis [58] and adpated by Ganian [43] to describe uncoloured graphs). In a static graph, we define the *neighbourhood* $N(v)$ of a vertex v as the set of vertices which share an edge with v.

Definition 2.7 Two vertices v, v' have the same *type* if and only if $N(v) \setminus \{v'\} =$

$N(v') \setminus \{v\}$. A graph $G = (V, E)$ has *neighbourhood diversity* at most k if and only if there exists a partition of $V(G)$ into at most w sets where all elements in the same set have the same type. We refer to this partition as a *neighbourhood partition.*

Again, there is a natural generalisation of this parameter to temporal graphs, due to Enright, Hand, Larios-Jones and Meeks [31].

Definition 2.8 The *temporal neighbourhood* of a vertex v in a temporal graph (G, λ) is the set $TN(v)$ of vertex time pairs (w, t), where $(w, t) \in TN(V)$ if and only if $vw \in E(G)$ and $t \in \lambda(vw)$.

Definition 2.9 Two vertices u, v have the same *temporal type* if and only if $\{(w, t) \in TN(u) : w \neq v\} = \{(w, t) \in TN(v) : w \neq u\}$.

Definition 2.10 A graph $\mathcal{G}$ has *temporal neighbourhood diversity* at most k if and only if there exists a partition of $V(\mathcal{G})$ into at most k sets where all elements in the same set have the same temporal type. We refer to this partition as a *temporal neighbourhood partition.*

It is often convenient to refer to the sets of vertices having the same temporal type as the *classes* of the temporal neighbourhood partition.

It is easy to see that the class of graphs having bounded temporal neighbourhood diversity is a subclass of the set of graphs having bounded temporal modular width: we can construct the subgraph induced by each set in the temporal neighbourhood partition using the first three operations repeatedly, and then a single application of operation 4 is needed to add edges between the parts. Again, it turns out that these stronger restrictions lead to tractability for more problems. To illustrate this, we consider the temporal version of the GRAPH BURNING problem. In the static problem, first proposed by Bonato, Janssen and Roshanbin [7], a fire is placed at a new vertex in each time-step, and the fire spreads along incident active edges. The problem asks whether, after h placements of fire, all vertices are burning. GRAPH BURNING is NP-hard on general graphs [6] and was recently shown to be in FPT parameterised by static modular width [51]. To define its temporal analogue, we consider the following process, as described by Enright, Hand, Larios-Jones and Meeks [31].

1. At time $t = 0$ a fire is placed at a chosen vertex. All other vertices are unburnt.

2. At all times $t \geq 1$, the fire spreads, burning all vertices u adjacent to an already burning vertex v where the edge between u and v is active at time t. Then, another fire is placed at a chosen vertex.

3. This process ends once all vertices are burning.

A sequence of vertices at which fires are placed is called a *strategy*, and a strategy is said to be *successful* if every vertex in the graph is burning after a strategy is played. We can now give the definition of the computational problem.

> TEMPORAL GRAPH BURNING
> *Input:* A temporal graph (G, λ) and an integer h.
> *Question:* Does there exist a successful burning strategy for (G, λ) of length less than or equal to h?

In contrast with the static version, TEMPORAL GRAPH BURNING turns out to be NP-hard even when restricted to temporal graphs of constant modular width [31]. However, TEMPORAL GRAPH BURNING is in FPT parameterised by the temporal neighbourhood diversity of the input graph [31]. The key fact that gives rise to an efficient algorithm in this case is that the total number of fires that must be placed in order to obtain a successful strategy depends only on

1. the set of classes in which fires are placed, and
2. the order in which the *first* fire in each such class is placed.

Thus, to search for a successful strategy of the desired length, we can consider in turn each possibility for the set of classes in which fires are placed and the order in which a fire is first placed in these classes, as this number of possibilities is bounded by a function of the temporal neighbourhood diversity. For each possibility, it is then possible (with some care) to determine the minimum length of a successful strategy in polynomial time.

Another problem for which parameterisation by temporal neighbourhood diversity turns out to be somewhat useful is SINGMINREACHDELETE, as introduced in Section 2.1. This problem turns out to be in FPT when parameterised simultaneously by the temporal neighbourhood diversity and the maximum number of times at which any edge appears [31]. The argument uses similar ideas to the algorithm for TEMPORAL GRAPH BURNING – namely that most of the key information about a solution depends on the classes involved, rather than the specific vertices – to show that it suffices to solve a collection of instances INTEGER QUADRATIC PROGRAMMING that meet the conditions of Lokshtanov's fpt-algorithm for this problem [59].

We conclude the discussion of this hierarchy of parameters by discussing briefly the time needed to compute the values of temporal cliquewidth, temporal modular width and temporal neighbourhood diversity on temporal graphs. There exist fairly straightforward polynomial-time algorithms to compute the temporal neighbourhood diversity and temporal modular width of any input temporal graph [31], but NP-hardness of computing temporal cliquewidth follows immediately from the NP-hardness of its static counterpart. It remains open whether there exist efficient approximation algorithms for temporal cliquewidth, as is the case for static cliquewidth.

Finally, in this section, we mention a parameter of a different flavour – not a width measure – that has recently been defined on temporal graphs and is again well-placed to describe graphs that are potentially very dense but have algorithmically useful structure. A static graph is said to be *c-closed* if, for every two distinct vertices u and v which share at least c common neighbours, there is an edge between u and v; this notion is inspired by the triadic closure property often observed in social networks, i.e. that individuals with many common friends are likely to be friends themselves [40, 41]. The *closure number* of a graph, defined as the minimum c for which the graph is c-closed (noting that this is well-defined since every n-vertex graph is necessarily n-closed), has proven to be a useful parameter in the design of fpt-algorithms [51, 52, 53]. A temporal analogue was recently introduced by Davot, Enright, Madathil and Meeks [26], based on the idea that if two individuals have many common contacts during a short time interval, then we should expect them to have contact with each other around the same time, but not necessarily during this same interval. This notion is formalised in the following definition.

Definition 2.11 For integers $\Delta_0, \Delta_1, \Delta_2 \geq 0$ and $c \geq 1$, a temporal graph $\mathcal{G} = (G, \lambda)$ is $(\Delta_0, \Delta_1, \Delta_2, c)$-*closed* if the following condition holds: for any two distinct vertices $u, v \in V(G)$ and any time-interval $[a, b]$ of length at most $\Delta_1 + 1$ (i.e., $b - a \leq \Delta_1$) with $a \geq 1 + \Delta_0$ and $b \leq \Lambda(\mathcal{G}) - \Delta_2$, if u and v have at least c common neighbours during $[a, b]$, then u and v are adjacent to each other during $[a - \Delta_0, b + \Delta_2]$. The $(\Delta_0, \Delta_1, \Delta_2)$-*closure number* of a temporal graph $\mathcal{G}$ is the smallest c for which it is $(\Delta_0, \Delta_1, \Delta_2, c)$-closed.

Thus far, this parameter has only been used to address the problems of enumerating maximal cliques and other dense subgraphs; these problems are shown to be in FPT parameterised simultaneously by the closure number and a second parameter which limits how much the graph can change between consecutive timesteps [26]. The proof mimics the approach used to prove the analogous result for static graphs, providing a rare example of intuition from the static setting translating in a straightforward way to temporal graphs. It will be interesting to see whether other temporal graph problems turn out to be efficiently solvable on graphs whose temporal closure number is small.

3 Inherently temporal parameters

In this section, we discuss several kinds of parameters which have no analogue in the static setting, because they are inherently based on the timing information encoded in the temporal graph. We consider three such families of parameters: "local" restrictions on the assignment of times to edges, restrictions based on the properties of temporal paths and walks in the graph, and finally a hierarchy of parameters capturing an algorithmically useful notion of temporal sparsity.

3.1 Local temporal restrictions

In this section, we discuss some of the simpler ways we can try to restrict the temporal structure of the graph. All of the conditions described here are "local" in one of two senses: they can be verified either by examining independently the sets of times assigned to individual edges (or in some cases pairs of incident edges), or by considering each static snapshot independently.

Three of the most obvious ways to do this are as follows:

- restrict the number of edges appearing at each time;
- restrict the number of times each edges appears; or

- restrict the overall lifetime of the graph.

The first of these can alternatively be viewed as a restriction on the structure of the individual snapshots of the graph (requiring that each snapshot contains only a few edges) and so, as we have seen in Section 2.1, this turns out not to be a particularly effective restriction. We therefore turn our attention to the other two.

The maximum number of times each edge appears is sometimes called the *temporality* of the temporal graph [63]. Many problems remain NP-hard on temporal graphs with temporality one, including SINGMINREACHDELETE [34] and RESTLESS TEMPORAL PATH [13], meaning that this parameter is not often useful on its own; however, it has been used successfully in combination with other structural parameters, as in the example of SINGMINREACHDELETE parameterised simultaneously by temporality and temporal neighbourhood diversity discussed in Section 2.2.

Bounding the lifetime of the temporal graph also turns out to be of limited use: many problems remain NP-hard even when the lifetime is constant. Again, this is true for both SINGMINREACHDELETE (which is NP-hard for lifetime two [34]) and RESTLESS TEMPORAL PATH (which is NP-hard for lifetime three [13]), as well as a number of other reachability problems [56, 57]. Just as for temporality, the lifetime has proved more useful in combination with additional structural parameters (including parameters based on the underlying graph, such as its treewidth) [18, 34, 35, 56].

Some slightly more involved restrictions on the times assigned to edges give rise to the notions of *simple*, *proper* and *happy* temporal graphs [11]. A temporal graph is said to be simple if each edge appears at exactly one timestep (i.e. if the graph has temporality one). It is said to be proper if no vertex is incident with two edges at the same timestep (equivalently, the maximum degree of any snapshot is one). A temporal graph is happy if it is both simple and proper. These restrictions have certain nice properties: for example, in a proper temporal graph there is no distinction between strict and non-strict temporal walks/paths. However, even in happy temporal graphs, it remains NP-hard to compute the size of the largest (open) temporal component [11], and we have already seen that the same is true for SINGMINREACHDELETE [10].

3.2 Path-based temporal restrictions

Several promising parameters have recently been introduced that involve restrictions related to the existence of temporal paths (or related structures such as walks) in the graph.

To motivate the first such collection of parameters, recall that one of the challenges in considering paths and reachability in temporal graphs compared with static (directed or undirected graphs) is that reachability is not transitive: the fact there exist temporal paths from u to v and from v to w does not necessarily imply that there exists a temporal path from u to w. Temporal graphs in which this implication *does* hold are said to be *transitive*. Casteigts, Morawietz and Wolf [14] defined two temporal graph parameters describing the edit distance of a temporal graph from the class of transitive temporal graphs.

To define these parameters, the authors consider the *(strict) reachability graph* of a temporal graph $\mathcal{G} = (G, \lambda)$; this is a directed graph with vertex-set $V(G)$ and arc set $\{(u, v) :$ there exists a (strict) temporal path from u to $v\}$. A temporal graph is then transitive if and only if the corresponding reachability graph is transitive.

Definition 3.1 The *vertex-deletion distance to transitivity* of a temporal graph is the minimum number of vertices whose deletion from its reachability graph results in a transitive directed graph.

Definition 3.2 The *arc-modification distance to transitivity* of a temporal graph is the minimum number of arcs that must be modified (added or removed) in the corresponding reachability graph to obtain a transitive directed graph.

The problem of computing each of these parameters for a given temporal graph belongs to FPT parameterised by the parameter itself.

The two parameters were initially applied to the problem of determining whether there exist large *open connected components* in a temporal graph, where an open connected component is a set C of vertices such that, for every $u, v \in C$, there is a temporal path from u to v and from v to u (but it might be that all such paths use at least one vertex outside C). It was shown that deciding whether there exists an open connected component of size at least k, while NP-hard in general, is in FPT parameterised by either the vertex-deletion distance or the arc-modification distance to transitivity [14]. As these parameters are very new, they have not yet been applied to many other problems, but they seem to offer a promising route to tractable algorithms for problems involving temporal paths and reachability.

The second collection of path-related parameters is based not on the edit distance to a nice class, but rather on the number of simple structures needed to cover the temporal edges of the graph. Deligkas et al. [27] define a number of parameters in this vein, based on the number of (strict or non-strict) temporal walks, paths or trails (which may revisit vertices but not edges) needed to cover each temporal edge exactly once. Computing the number of such structures needed to cover the graph turns out to be NP-hard in nearly all cases, but a promising direction for future research is whether, given a solution to the covering problem (which might credibly be available in some applications, for example networks derived from transport networks where the routes taken by individual vehicles must jointly cover the edges), other NP-hard temporal graph problems become easier to solve.

3.3 Parameters capturing temporal sparsity

In this section, we discuss a hierarchy of parameters that has been developed over the last few years, each of which captures in some sense the idea of a temporal graph being sparse with respect to time.

The first of these parameters to be introduced was the *edge-interval-membership-width* of a temporal graph [9]. Interestingly, this parameter is independent of the structure of the underlying graph (other than the total number of edges), depending only on the times assigned to each edge in the temporal graph. Before defining this parameter, we consider a motivating example.

Suppose that 18th century Königsburg had no bridges, but instead its islands were connected by ferry routes, each of which had its own timetable. An explorer of the city might then hope to use each ferry crossing exactly once before returning to their starting point, giving rise to the following computational problem.

> TEMPEULER
> *Input:* A temporal graph $\mathcal{G} = (G, \lambda)$.
> *Question:* Does there exist a closed (strict) temporal walk using each edge of the underlying graph exactly once?

This problem is also referred to as EULERIAN TRAIL, and is known to be NP-hard even if the input graph has temporality two [61]; it is also NP-hard on graphs with feedback vertex number one and temporality three, and on graphs with vertex cover number two and temporality four [10].

There are some assignments of times to edges which make this problem simple, however. Suppose, in an extreme example, that there is just a single boat operator running all the ferry routes in the city, and that he runs each ferry route for a single, continuous time interval each day. In this case, there would only be a single option for the order in which we could hope to traverse the different crossings (assuming we wish to return to our starting point the same day), and it is easy to determine whether this order gives rise to a closed walk or not.

Now suppose that, due to increasing demand, the boat operator recruits an apprentice on one route, who will continue to run crossings for a short time after the head operator moves onto the next route. In this case, there is a small element of choice in the order we could traverse the crossings: there are two crossings whose order could potentially be swapped. However, this still gives a small number of possible orders to check, meaning that the problem remains tractable. This idea – that the number of edges in the temporal graph which are "relevant" at any point in time is not too large – can be formalised in the definition of edge-interval-membership-width.[2]

Definition 3.3 ([10]) The *edge interval membership sequence* of a temporal graph $\mathcal{G} = (G, \lambda)$ is the sequence $(F_t)_{t \in [\Lambda(\mathcal{G})]}$ of edge-subsets of G, where $F_t := \{e \in E(G) : \min \lambda(e) \leq t \leq \max \lambda(e)\}$. The *edge-interval-membership-width* of $\mathcal{G}$ is the integer $\text{imw}(G, \lambda) := \max_{t \in \Lambda(\mathcal{G})} |F_t|$.

Thus, the edge-interval-membership-width of a temporal graph $\mathcal{G}$ is the maximum, taken over all timesteps t, of the number of edges that are active at time t and/or are active both at some time before t and at some time after t. This is therefore independent of the number of times at which any edge appears, and also of the structure of the underlying graph: two temporal graphs $\mathcal{G}_1 = (G_1, \lambda_1)$ and $\mathcal{G}_2 = (G_2, \lambda_2)$ are certain to have the same edge-interval-membership-width if there exists a bijection $\pi : E(G_1) \to E(G_2)$ such that, for all $e \in E(G_1)$, $\lambda_1(e) = \lambda_2(\pi(e))$, regardless of the (lack of) relationship between G_1 and G_2 – the two underlying graphs might, for example, have very different numbers of vertices.

[2]This parameter was originally referred to simply as "interval-membership-width", but here, as in other recent works, we us the longer name to avoid confusion with the other related width measures discussed in this section.

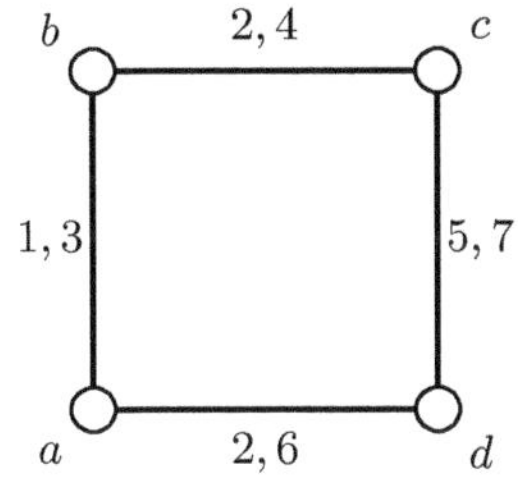

F_1	F_2	F_3	F_4	F_5	F_6	F_7
ab	ab	ab				
	bc	bc	bc			
				cd	cd	cd
	ad	ad	ad	ad	ad	

Figure 4: A temporal graph and its edge interval membership sequence; the edge-interval-membership-width is three.

An example of a temporal graph and its edge interval membership sequence is given in Figure 4. Note that the edge interval membership sequence and edge-interval-membership-width of a temporal graph can both clearly be computed in polynomial time.

As intimated above, TEMPEULER is in FPT parameterised by the edge-interval-membership-width of the input graph; the same is also true for STAREXP(c) for any c [10]. Both results are proved via a straightforward dynamic program over the edge interval membership sequence: at each timestep from 1 to $\Lambda(\mathcal{G})$, we can consider all possibilities for how a partial solution up to this point interacts with the edges in the current set, as this number of possibilities is bounded by a function of the edge-interval-membership-width.

Roughly speaking, a crucial property shared by TEMPEULER and STAREXP that makes edge-interval-membership-width a useful parameter is the fact that the important information about a partial solution at time t can be described in terms of the status of edges (e.g., whether a given edge has been visited), not the vertices. However, for many problems, we would additionally need to keep track of the status of a large number of vertices, and in these cases edge-interval-membership-width does not seem to be so helpful. Indeed, it was shown that SINGMINREACHDELETE is NP-hard on temporal graphs of edge-interval-membership-width one [10]. This motivated the introduction of a vertex-based version of the parameter, which has an analogous definition.

Definition 3.4 The *vertex-interval-membership-sequence* of a temporal graph $\mathcal{G} = (G, \lambda)$ is the sequence $(U_t)_{t \in \Lambda(\mathcal{G})}$ of vertex-subsets of G, where

$$U_t := \{v \in V(G) : \exists i \leq t \leq j \text{ and } u, w \in V(G) \text{ such that } i \in \lambda(uv) \text{ and } j \in \lambda(wv)\}.$$

The *vertex-interval-membership-width* (or VIM-width) of $\mathcal{G}$ is the integer $\text{vimw}(\mathcal{G}) := \max_{t \in \Lambda(\mathcal{G})} |U_t|$.

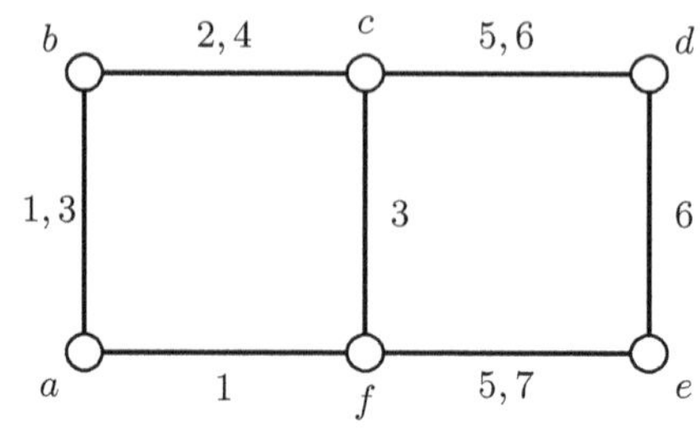

U_1	U_2	U_3	U_4	U_5	U_6	U_7
a	a	a				
b	b	b	b			
	c	c	c	c	c	
				d	d	
				e	e	e
f	f	f	f	f	f	f

Figure 5: A temporal graph and its vertex interval membership sequence; the vertex-interval-membership-width is four.

We note that in the definition above, we could take $u = w$. We can equivalently define VIM-width in terms of the *active window* of each vertex: for any $v \in V(G)$, the active window of v is the interval

$$[\min\{t : \exists\, e \text{ with } t \in \lambda(e) \text{ and } e \text{ incident with } v\},$$
$$\max\{t : \exists\, e \text{ with } t \in \lambda(e) \text{ and } e \text{ incident with } v].$$

The VIM-width is then the maximum, over all times t, of the number of vertices whose active windows contain t.

Once again, it is easy to see that the vertex interval membership sequence and hence vertex-interval-membership-width can be computed in polynomial time. An example of a temporal graph and its vertex interval membership sequence is given in Figure 5.

As for edge-interval-membership-width, the natural way to prove that a problem is in FPT parameterised by VIM-width is via dynamic programming on the vertex interval membership sequence. This approach was used to prove tractability for SingMinReachDelete [10] and Temporal Firefighter [47].[3]

[3] In the static Firefighter problem (where the fire spreads over time, but the graph does not change), the input consists of a graph G together with a root vertex r, which is burning at time $t = 0$. The fire then spreads through the graph as follows: at each time $i \geq 1$, a non-burning vertex is chosen as defended, then the fire spreads to all non-defended vertices that are adjacent to some vertex burning at time $i - 1$. The process continues until the fire can no longer spread, and the computational problem asks whether there exists a sequence of defences that saves at least some number k of the vertices from burning. In Temporal Firefighter, we ask the same question, except that the graph (potentially) changes on each timestep, and at time i the fire will only spread along edges that are active at that time.

Recently, a framework has been defined which abstracts sufficient conditions for a problem to admit a dynamic program of this kind over the vertex interval membership sequence [32]. Partial solutions in this framework must be encodable as so-called (k, X)-states of the static snapshots, where a (k, X)-state consists of a labelling of vertices in the graph with labels from the set X, together with a vector of k integers (where these integers are of size at most polynomial in the size of the graph). For example, to solve TEMPORAL HAMILTON PATH, we could use the labels "visited", "unvisited" and "current" for vertices, together with a single integer recording the number of vertices visited so far. Roughly speaking, a temporal graph problem is then said to be (k, X)-*temporally uniform* if the (k, X) state describing a partial solution up to time t captures all the information needed to determine (in polynomial time) whether this partial solution can be extended to one with a given (k, X)-state at time $t + 1$; it should also be possible to determine whether a candidate solution is indeed a solution to the problem by examining only the corresponding (k, X)-state at the final timestep. To ensure that the problem is in FPT parameterised by VIM-width, we need a further refinement of this notion. A problem is said to be (k, X)-*locally temporally uniform* if it is temporally uniform and moreover respects certain conditions on which vertices' labels are relevant when determining compatibility of two consecutive (k, X)-states and acceptability of a final solution; informally, both of these should depend only on vertices in their active window at the relevant timestep. Such requirements are easily satisfied in the case of TEMPORAL HAMILTON PATH, since the a vertex can only change from "unvisited" to "current" or from "current" to "visited" if it is incident with an edge active at the time under consideration, while we can determine whether we have a valid solution at the final timestep by checking whether the number of vertices visited so far is equal to the total number in the graph. For (k, X)-locally temporally uniform problems, it follows from these definitions that a dynamic program over the vertex interval membership sequence needs only to keep track of the labels of vertices in the current set in the sequence, along with perhaps some summary data (e.g. total number of vertices that have reached some state so far), which can be encoded in the vector of integers. This means that any (k, X)-locally temporally uniform problem admits an fpt-algorithm parameterised by VIM-width, provided that k and $|X|$ are bounded by constants. The meta-algorithm has been used to show that TEMPORAL HAMILTON PATH[4] and Δ-TEMPORAL MATCHING (for Δ a constant) are in FPT parameterised by VIM-width.

While VIM-width seems to be very effective at ensuring tractability for otherwise challenging problems on temporal graphs, this comes at the price of being so restrictive that it is unlikely to take small values on many temporal graphs arising in real-world applications: for the VIM-width to be small (compared with the total number of vertices), we require that, at every timestep, nearly all of the vertices have either never had an active incident edge, or will never again have an active incident edge.

It is natural to ask whether we might overcome this by sacrificing some of the simplicity in the VIM-width definition, and taking into account the component structure at each time step as well as which vertices are isolated or not. The first progress

[4]The problem TEMPORAL HAMILTON PATH asks, given a temporal graph $\mathcal{G}$, whether there exists a temporal path visiting every vertex of $\mathcal{G}$.

in this direction was due to Christodoulou et al. [17], who defined three parameters related to each of edge-interval-membership-width and vertex-interval-membership-width that take into account the connectivity. We begin by defining the edge versions.

The first parameter allows sets in the edge interval membership sequence to be split into subsets, provided that two edges belong to the same subset of F_t whenever they are in the same connected component of the static graph obtained by taking the union of the edges in the first t snapshots. Here, $G_{\leq}(t)$ is used to denote the static graph obtained from the temporal graph $\mathcal{G} = (G, \lambda)$ by taking the vertex-set to be $V(G)$ and the edge-set to be $\{e \in E(G) : [t] \cap \lambda(e) \neq \emptyset\}$. Note that, if u and v belong do different connected components of $G_{\leq}(t)$, then not only do they belong to different connected components of every snapshot up to and including time t, but also there is no path between u and v consisting only of edges that appear at or sometime before time t, and in particular there is no temporal path between the two arriving at or before time t. The intuition behind the following definition is that, when solving many natural problems, any two vertices whose active windows contain the time t but which belong to different connected components of $G_{\leq}(t)$ can, in some sense, be considered independently in partial solutions.

Definition 3.5 ([17]) For each time $t \in [\Lambda]$ and connected component C in $G_{\leq}(t)$, let $\Psi_{\leq}(\mathcal{G}, t, C) = V(C) \cap F_t$ be the *$\leq$-connected bag* at time t of $\mathcal{G}$ and C, where F_t is the bag at time t of the edge interval membership sequence of $\mathcal{G}$. Let $\mathcal{F}_{\leq}(t) = \{\Psi_{\leq}(\mathcal{G}, t, C) : C \text{ is a connected component of } G_{\leq}(t)\}$. The *$\leq$-connected-interval-membership width* is given by $\max_{t \in [\Lambda], \Psi \in \mathcal{F}_{\leq}(t)} |\Psi|$.

Thus, there is a single bag associated with time Λ, and this is identical to the corresponding bag of the edge interval membership sequence, but at earlier times t, the bags of the edge interval membership sequence are split according to their connected components in $G_{\leq}(t)$. The *$\geq$-connected-interval-membership width* is defined analogously, by considering for each time t the set of edges appearing at or later than time t; for this parameter, therefore, we must have a single bag at time 1 and later bags are split by connectivity. These two notions are incomparable, but both are upper-bounded by the edge-interval-membership-width of the temporal graph. These two notions of connected-interval-membership-width are further generalised by the *bidirectional connected-interval-membership-width.*

Definition 3.6 ([17]) Let $\psi_{\leq}(\mathcal{G})$ and $\psi_{\geq}(\mathcal{G})$ be the $\leq$- and $\geq$-connected-interval-membership width of the temporal graph $\mathcal{G}$, respectively. Then, the bidirectional connected-vertex-interval-membership width $\psi_{\sim}(t)$ of a temporal graph $\mathcal{G}$ at time t is

$$\psi_{\sim}(t) = \begin{cases} \max\{\psi_{\leq}(\mathcal{G}_{\leq}(t-1)), \psi_{\geq}(\mathcal{G}_{\geq}(t+1)), |F_t(\mathcal{G})|\} & \text{if } 1 < t < \Lambda, \\ \psi_{\geq}(\mathcal{G}) & \text{if } t = 1, \\ \psi_{\leq}(\mathcal{G}) & \text{if } t = \Lambda, \end{cases}$$

where $F_t(\mathcal{G})$ is the bag at time t of the edge interval membership sequence of $\mathcal{G}$. The *bidirectional connected-interval-membership-width* of $\mathcal{G}$ is $\min_{t \in [\Lambda]} \psi_{\sim}(t)$.

While the definitions of $\leq$-connected-interval-membership width and $\geq$-connected-interval-membership width require that there be a single bag containing all vertices whose active intervals contains Λ or 1 respectively, for bidirectional connected-interval-membership-width, we can choose any one time t such that all vertices whose active intervals contain t belong to a single bag, and split bags of the edge interval membership sequence according to appropriate connectivity criteria at times both before and after t. This is illustrated diagrammatically in Figure 7.

All three of these width parameters can be computed in polynomial time [17]. Moreover, TEMPEULER is also in FPT parameterised by the $\leq$-connected-interval-membership-width, the $\geq$-connected-interval-membership-width or the bidirectional connected-interval-membership-width [17], generalising the positive result for TEMPEULER parameterised by edge-interval-membership-width. Christodoulou et al. [17] also obtain fpt-algorithms with respect to all three parameters for the following temporal version of 2-colouring, which is known to be NP-complete even on temporal graphs whose underlying graphs have bounded treewidth [60].

> TEMPORAL 2-COLOURING
> *Input:* A temporal graph $\mathcal{G} = (G, \lambda)$ and an integer $k \geq 2$.
> *Question:* Does there exist a function $f : V \times [\Lambda(G)] \to [k]$ such that, for every $e = uv \in E(G)$, there exists $t \in \lambda(e)$ with $f(u,t) \neq f(v,t)$.

Christodoulou et al. also defined vertex variants of $\leq$-connected-vertex-interval-membership-width, $\geq$-connected-vertex-interval-membership-width and bidirectional connected-vertex-interval-membership-width in a similar way; these are also computable in polynomial time [17]. TEMPORAL FIREFIGHTER was shown to be in FPT parameterised by $\leq$-connected-vertex-interval-membership-width [17], but turns out to be NP-hard even on temporal graphs having constant bidirectional connected-vertex-interval-membership-width [32].

A further recent generalisation comes in the form of *tree-interval-membership-width (TIM-width)*. The TIM-width of a temporal graph is upper-bounded by the bidirectional-vertex-interval-membership-width, but may be much smaller. Formally, TIM-width is defined as follows.

Definition 3.7 ([32]) We say a triple (T, B, τ) is a *tree-interval-membership decomposition (TIM decomposition)* of a temporal graph $\mathcal{G}$ with lifetime Λ if T is a labelled directed tree, where $B = \{B(s) : s \in V(T)\}$ is a collection of subsets of $V(\mathcal{G})$, called *bags*, and $\tau : B \to [\Lambda]$ is a function which labels each bag with a time, satisfying:

1. For all vertices $v \in V(\mathcal{G})$ and times $t \in [\Lambda]$, there exists a unique node $i \in V(T)$ such that $\tau(B(i)) = t$ and v appears in a unique bag $B(i)$.
2. For all time-edges $(uv, t) \in \mathcal{E}(\mathcal{G})$, there exists a node $i \in V(T)$ such that $\{u, v\} \subseteq B(i)$ and $\tau(B(i)) = t$.
3. The set of arcs of T is given by $\{(i, j) : B(i) \cap B(j) \neq \emptyset \text{ and } \tau(B(j)) = \tau(B(i)) + 1\}$.

The *width* of a TIM decomposition is defined to be $\max\{|B(s)| : s \in V(T)\}$. The *TIM width* of a temporal graph $\mathcal{G}$ is the minimum ϕ such that G has a TIM decomposition of width ϕ.

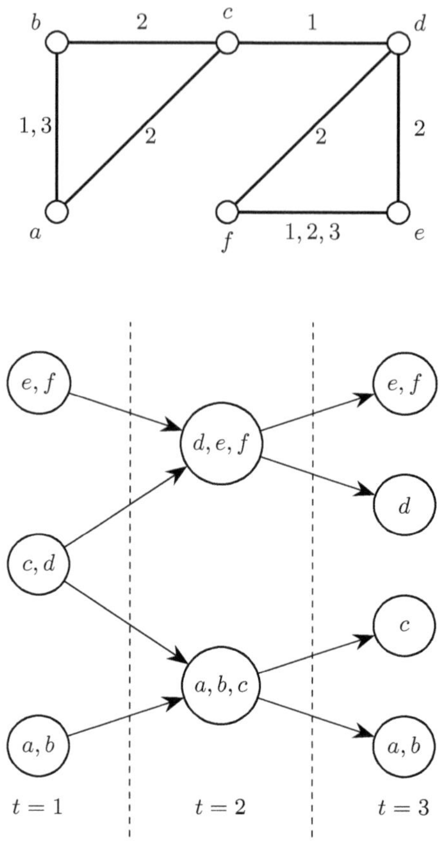

Figure 6: An example of a temporal graph together with a TIM decomposition.

An example of a temporal graph and a TIM decomposition is given in Figure 6. At the time of writing, it is conjectured but not yet proven that the TIM-width of a temporal graph can be computed in polynomial time. Figure 7 illustrates schematically how the sets of vertices in their active window at each time t might potentially be split into bags in each of the three kinds of decompositions.

As for VIM-width, some properties of temporal graph problems that are sufficient to guarantee the existence of an fpt-algorithm via dynamic programming over the TIM decomposition have recently been encapsulated in a meta-algorithm. The requirements are similar to those for VIM-width, but somewhat stronger. Problems satisfying these properties are said to be *component-exchangeable temporally uniform*; roughly speaking, in addition to the requirements of being locally temporally uniform, this enforces that the validity of transitions between consecutive states and acceptability of the final state can be checked by considering different connected components of the relevant snapshot independently [32]. Using this meta-

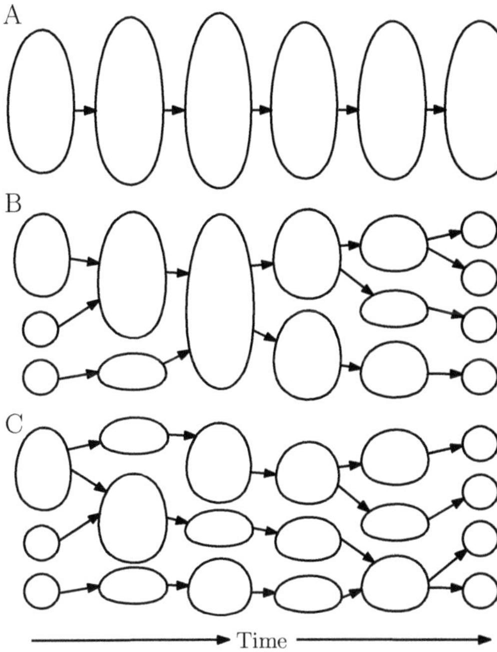

Figure 7: Schematics of possible decompositions of the vertices in their active window at each time t associated with the VIM-width (top), bidirectional vertex-interval-membership-width (middle) and TIM-width (bottom). For VIM-width, all vertices in their active window at time t must belong to the same bag, for all t. For bidirectional vertex-interval-membership-width, there exists *some* t for which all vertices in their active window at time t belong to the same bag (for $\geq$- or $\leq$-connected-interval-membership width this bag must be the first or last respectively). For TIM-width, there need not be any time t at which all vertices in their active window at time t are in the same bag, provided that the bags form a tree structure. Figure reproduced from [32] with the authors' permission.

algorithm, Temporal Hamilton Path, Δ-Temporal Matching and SingMinReachDelete have all been shown to belong to FPT parameterised by TIM-width [32]. However, the authors also show that Temporal Firefighter, which is known to be in FPT parameterised by VIM-width, is NP-hard even on temporal graphs of constant TIM-width.

4 Future challenges

In spite of the large number of temporal graph parameters discussed in this article, as a community, we have only just begun to scratch the surface in understanding what properties of temporal graphs lead to tractability for many natural problems. One of the tools most lacking, when compared with the static setting, is general metatheorems that guarantee problems having particular properties (e.g., being expressible in a suitable fragment of logic) are always tractable with respect to a given parameter. Recent attempts to formalise properties that guarantee the existence of fpt-algorithms with respect to VIM-width and TIM-width make some progress in this direction, but the results are not as easy to apply as the celebrated metatheorems (such as Courcelle's theorem for treewidth) in the static setting. While a widely-applicable logic-based metatheorem for a temporal analogue of treewidth seems a very challenging goal (given that there is not yet a single definition of temporal treewidth that has proved amenable to dynamic programming approaches for a wide range of problems), there is perhaps more hope of obtaining an analogue of the static metatheorem for temporal cliquewidth.

More generally, in the search for restrictions of temporal graphs that give widespread tractability, there are many approaches in the static setting that have not yet been widely explored in the context of temporal graphs. One of these is geometric representations of temporal graphs: in static graphs, many NP-hard problems become polynomial-time solvable in interval graphs [45] or more general geometric models, such as disc graphs. Aside from some preliminary results concerning temporal interval graphs (in which every snapshot must be an interval graph) together with some restrictions on changes permitted between consecutive snapshots [39, 48], there has been very little work on temporal graphs derived from geometric representations in this way. However, the study of such graphs seems well-motivated from an application perspective: interesting and highly temporal networks based on proximity of moving entities (e.g., contact between individuals for epidemiology, or proximity between sensors for communication) are naturally modelled in this way.

Another direction that has proved highly fruitful in the static setting is the study of classes excluding certain (induced) subgraphs or minors. The simplest way to apply this idea in the temporal setting would be to exclude specific substructures from appearing in the underlying graph, or perhaps in any snapshot. However, recall (from Section 2.1) that many temporal problems remain intractable even when the underlying graph (and hence every snapshot) is acyclic, so restricting the underlying graph alone is likely to be of limited use. A more promising approach might be to use a notion of temporal subgraphs and/or minors. Unsurprisingly, there are many reasonable ways in which to define a temporal subgraph; translating the concept of minors to the temporal setting is even less straightforward (as is the case for minors in directed graphs) since it is hard to see how to define edge contractions in a way that does not create new temporal paths. However, in spite of these undeniable challenges, the study of temporal graphs excluding specific temporal substructures seems an important direction for future work.

From a practical perspective, given the relevance of temporal graphs in many applications, one would hope to find algorithmically useful parameters describing properties that are observed in at least some real-world temporal networks. To this end, a much more systematic study of the structural properties of temporal graphs arising in different application domains is needed; ideally, this might inspire new ways to define parameters based on applications under consideration.

Acknowledgements

The author's work in this area has been supported by EPSRC grants EP/T004878/1 and EP/V032305/1. The author is grateful to Laura Larios-Jones and David Kutner for some helpful suggestions while preparing the manuscript.

References

[1] Eleni C. Akrida, George B. Mertzios, Paul G. Spirakis, and Christoforos L. Raptopoulos, *The temporal explorer who returns to the base*, Journal of Computer and System Sciences **120** (2021), 179–193.

[2] Eleni C. Akrida, George B. Mertzios, Paul G. Spirakis, and Viktor Zamaraev, *Temporal vertex cover with a sliding time window*, Journal of Computer and System Sciences **107** (2020), 108–123.

[3] Kyriakos Axiotis and Dimitris Fotakis, *On the size and the approximability of minimum temporally connected subgraphs*, 43rd International Colloquium on Automata, Languages, and Programming, ICALP 2016, July 11-15, 2016, Rome, Italy (Ioannis Chatzigiannakis, Michael Mitzenmacher, Yuval Rabani, and Davide Sangiorgi, eds.), LIPIcs, vol. 55, Schloss Dagstuhl - Leibniz-Zentrum für Informatik, 2016, pp. 149:1–149:14.

[4] Jakub Balabán, Robert Ganian, and Mathis Rocton, *Computing Twin-Width Parameterized by the Feedback Edge Number*, 41st International Symposium on Theoretical Aspects of Computer Science (STACS 2024) (Dagstuhl, Germany) (Olaf Beyersdorff, Mamadou Moustapha Kanté, Orna Kupferman, and Daniel Lokshtanov, eds.), Leibniz International Proceedings in Informatics (LIPIcs), vol. 289, Schloss Dagstuhl – Leibniz-Zentrum für Informatik, 2024, pp. 7:1–7:19.

[5] Kenneth A. Berman, *Vulnerability of scheduled networks and a generalization of Menger's Theorem*, Networks **28** (1996), no. 3, 125–134.

[6] Stéphane Bessy, Anthony Bonato, Jeannette C. M. Janssen, Dieter Rautenbach, and Elham Roshanbin, *Burning a graph is hard*, Discrete Applied Mathematics **232** (2017), 73–87.

[7] Anthony Bonato, Jeannette C. M. Janssen, and Elham Roshanbin, *Burning a graph as a model of social contagion*, Algorithms and Models for the Web Graph - 11th International Workshop, WAW 2014, Beijing, China, December 17-18,

2014, Proceedings (Anthony Bonato, Fan Chung Graham, and Pawel Pralat, eds.), Lecture Notes in Computer Science, vol. 8882, Springer, 2014, pp. 13–22.

[8] Nick Brettell, Andrea Munaro, Daniël Paulusma, and Shizhou Yang, *Comparing width parameters on graph classes*, European Journal of Combinatorics **127** (2025), 104163.

[9] Benjamin Merlin Bumpus and Kitty Meeks, *Edge exploration of temporal graphs*, Combinatorial Algorithms - 32nd International Workshop, IWOCA 2021, Ottawa, ON, Canada, July 5-7, 2021, Proceedings (Paola Flocchini and Lucia Moura, eds.), Lecture Notes in Computer Science, vol. 12757, Springer, 2021, pp. 107–121.

[10] ______, *Edge exploration of temporal graphs*, Algorithmica **85** (2023), no. 3, 688–716.

[11] Arnaud Casteigts, Timothée Corsini, and Writika Sarkar, *Simple, strict, proper, happy: A study of reachability in temporal graphs*, Theoretical Computer Science **991** (2024), 114434.

[12] Arnaud Casteigts, Paola Flocchini, Walter Quattrociocchi, and Nicola Santoro, *Time-varying graphs and dynamic networks*, International Journal of Parallel, Emergent and Distributed Systems **27** (2012), no. 5, 387–408.

[13] Arnaud Casteigts, Anne-Sophie Himmel, Hendrik Molter, and Philipp Zschoche, *Finding temporal paths under waiting time constraints*, Algorithmica **83** (2021), no. 9, 2754–2802.

[14] Arnaud Casteigts, Nils Morawietz, and Petra Wolf, *Distance to transitivity: New parameters for taming reachability in temporal graphs*, 49th International Symposium on Mathematical Foundations of Computer Science, MFCS 2024, August 26-30, 2024, Bratislava, Slovakia (Rastislav Královic and Antonín Kucera, eds.), LIPIcs, vol. 306, Schloss Dagstuhl - Leibniz-Zentrum für Informatik, 2024, pp. 36:1–36:17.

[15] Arnaud Casteigts, Joseph G. Peters, and Jason Schoeters, *Temporal cliques admit sparse spanners*, Journal of Computer and System Sciences **121** (2021), 1–17.

[16] Dipayan Chakraborty, Florent Foucaud, Diptapriyo Majumdar, and Prafullkumar Tale, *Structural parameterization of locating-dominating set and test cover*, Algorithms and Complexity (Cham) (Irene Finocchi and Loukas Georgiadis, eds.), Springer Nature Switzerland, 2025, pp. 187–204.

[17] Filippos Christodoulou, Pierluigi Crescenzi, Andrea Marino, Ana Silva, and Dimitrios M. Thilikos, *Making the Interval Membership Width of Temporal Graphs Connected and Bidirectional*, Combinatorial Algorithms (Cham) (Adele Anna Rescigno and Ugo Vaccaro, eds.), Springer Nature Switzerland, 2024, pp. 247–258.

[18] Lapo Cioni, Riccardo Dondi, Andrea Marino, Jason Schoeters, and Ana Silva, *Matching and edge cover in temporal graphs*, 4th Symposium on Algorithmic Foundations of Dynamic Networks, SAND 2025, June 9-11, 2025, Liverpool, UK (Kitty Meeks and Christian Scheideler, eds.), LIPIcs, vol. 330, Schloss Dagstuhl - Leibniz-Zentrum für Informatik, 2025, pp. 8:1–8:16.

[19] Timothée Corsini, Jessica A. Enright, Laura Larios-Jones, and Kitty Meeks, *Temporal orienteering with changing fuel costs*, CoRR **abs/2507.00728** (2025).

[20] B. Courcelle, J. A. Makowsky, and U. Rotics, *Linear Time Solvable Optimization Problems on Graphs of Bounded Clique-Width*, Theory of Computing Systems **33** (2000), no. 2, 125–150.

[21] Bruno Courcelle, *The monadic second-order logic of graphs I: recognizable sets of finite graphs*, Information and Computation **85** (1990), no. 1, 12–75.

[22] ———, *The monadic second-order logic of graphs III: tree-decompositions, minor and complexity issues*, RAIRO Theoretical Informatics and Applications **26** (1992), 257–286.

[23] Bruno Courcelle and Joost Engelfriet, *Graph structure and monadic second-order logic - A language-theoretic approach*, Encyclopedia of Mathematics and its Applications, vol. 138, Cambridge University Press, 2012.

[24] Bruno Courcelle and Stephan Olariu, *Upper bounds to the clique width of graphs*, Discrete Applied Mathematics **101** (2000), no. 1, 77–114.

[25] Marek Cygan, Fedor V. Fomin, Łukasz Kowalik, Daniel Lokshtanov, Dániel Marx, Marcin Pilipczuk, Michał Pilipczuk, and Saket Saurabh, *Parameterized Algorithms*, Springer International Publishing, Cham, 2015.

[26] Tom Davot, Jessica A. Enright, Jayakrishnan Madathil, and Kitty Meeks, *Temporal triadic closure: Finding dense substructures in social networks that evolve over time*, AAAI-25, Sponsored by the Association for the Advancement of Artificial Intelligence, February 25 - March 4, 2025, Philadelphia, PA, USA (Toby Walsh, Julie Shah, and Zico Kolter, eds.), AAAI Press, 2025, pp. 26939–26946.

[27] Argyrios Deligkas, Michelle Döring, Eduard Eiben, Tiger-Lily Goldsmith, George Skretas, and Georg Tennigkeit, *How many lines to paint the city: Exact edge-cover in temporal graphs*, AAAI-25, Sponsored by the Association for the Advancement of Artificial Intelligence, February 25 - March 4, 2025, Philadelphia, PA, USA (Toby Walsh, Julie Shah, and Zico Kolter, eds.), AAAI Press, 2025, pp. 26498–26506.

[28] Rodney G. Downey and Michael R. Fellows, *Parameterized complexity*, Monographs in Computer Science, Springer, 1999.

[29] ———, *Fundamentals of parameterized complexity*, Texts in Computer Science, Springer, 2013.

[30] Rosa Enciso, Michael R. Fellows, Jiong Guo, Iyad Kanj, Frances Rosamond, and Ondřej Suchý, *What makes equitable connected partition easy*, Parameterized and Exact Computation (Berlin, Heidelberg) (Jianer Chen and Fedor V. Fomin, eds.), Springer Berlin Heidelberg, 2009, pp. 122–133.

[31] Jessica A. Enright, Samuel D. Hand, Laura Larios-Jones, and Kitty Meeks, *Structural parameters for dense temporal graphs*, 49th International Symposium on Mathematical Foundations of Computer Science, MFCS 2024, August 26-30, 2024, Bratislava, Slovakia (Rastislav Královic and Antonín Kucera, eds.), LIPIcs, vol. 306, Schloss Dagstuhl - Leibniz-Zentrum für Informatik, 2024, pp. 52:1–52:15.

[32] ———, *Families of tractable problems with respect to vertex-interval-membership width and its generalisations*, CoRR **abs/2505.15699** (2025).

[33] Jessica A. Enright, Laura Larios-Jones, Kitty Meeks, and William Pettersson, *Reachability in temporal graphs under perturbation*, SOFSEM 2025: Theory and Practice of Computer Science - 50th International Conference on Current Trends in Theory and Practice of Computer Science, SOFSEM 2025, Bratislava, Slovak Republic, January 20-23, 2025, Proceedings, Part I (Rastislav Královic and Vera Kurková, eds.), Lecture Notes in Computer Science, vol. 15538, Springer, 2025, pp. 255–269.

[34] Jessica A. Enright, Kitty Meeks, George B. Mertzios, and Viktor Zamaraev, *Deleting edges to restrict the size of an epidemic in temporal networks*, Journal of Computer and System Sciences **119** (2021), 60–77.

[35] Jessica A. Enright, Kitty Meeks, and Hendrik Molter, *Counting temporal paths*, Algorithmica **87** (2025), no. 5, 736–782.

[36] Jiří Fiala, Petr A. Golovach, and Jan Kratochvíl, *Parameterized complexity of coloring problems: Treewidth versus vertex cover*, Theoretical Computer Science **412** (2011), no. 23, 2513–2523, Theory and Applications of Models of Computation (TAMC 2009).

[37] Jörg Flum and Martin Grohe, *Parameterized complexity theory*, Texts in Theoretical Computer Science, EATCS Series, Springer, 2006.

[38] Till Fluschnik, Hendrik Molter, Rolf Niedermeier, Malte Renken, and Philipp Zschoche, *As time goes by: Reflections on treewidth for temporal graphs*, Treewidth, Kernels, and Algorithms - Essays Dedicated to Hans L. Bodlaender on the Occasion of His 60th Birthday (Fedor V. Fomin, Stefan Kratsch, and Erik Jan van Leeuwen, eds.), Lecture Notes in Computer Science, vol. 12160, Springer, 2020, pp. 49–77.

[39] ———, *Temporal graph classes: A view through temporal separators*, Theoretical Computer Science **806** (2020), 197–218.

[40] Jacob Fox, Tim Roughgarden, C. Seshadhri, Fan Wei, and Nicole Wein, *Finding cliques in social networks: A new distribution-free model*, 45th International Colloquium on Automata, Languages, and Programming, ICALP 2018, July

9-13, 2018, Prague, Czech Republic (Ioannis Chatzigiannakis, Christos Kaklamanis, Dániel Marx, and Donald Sannella, eds.), LIPIcs, vol. 107, Schloss Dagstuhl - Leibniz-Zentrum für Informatik, 2018, pp. 55:1–55:15.

[41] ———, *Finding cliques in social networks: A new distribution-free model*, SIAM Journal on Computing **49** (2020), no. 2, 448–464.

[42] Jakub Gajarský, Michael Lampis, and Sebastian Ordyniak, *Parameterized Algorithms for Modular-Width*, Parameterized and Exact Computation (Cham) (Gregory Gutin and Stefan Szeider, eds.), Lecture Notes in Computer Science, Springer International Publishing, 2013, pp. 163–176.

[43] Robert Ganian, *Using Neighborhood Diversity to Solve Hard Problems*, January 2012, arXiv:1201.3091.

[44] Robert Ganian, Petr Hlinený, Joachim Kneis, Daniel Meister, Jan Obdrzálek, Peter Rossmanith, and Somnath Sikdar, *Are there any good digraph width measures?*, Journal of Combinatorial Theory, Series B **116** (2016), 250–286.

[45] Martin Charles Golumbic, *Algorithmic graph theory and perfect graphs*, Elsevier, 1980.

[46] Roman Haag, Hendrik Molter, Rolf Niedermeier, and Malte Renken, *Feedback edge sets in temporal graphs*, Discrete Applied Mathematics **307** (2022), 65–78.

[47] Samuel D. Hand, Jessica Enright, and Kitty Meeks, *Making Life More Confusing for Firefighters*, 11th International Conference on Fun with Algorithms (FUN 2022) (Dagstuhl, Germany) (Pierre Fraigniaud and Yushi Uno, eds.), Leibniz International Proceedings in Informatics (LIPIcs), vol. 226, Schloss Dagstuhl – Leibniz-Zentrum für Informatik, 2022, pp. 15:1–15:15.

[48] Danny Hermelin, Yuval Itzhaki, Hendrik Molter, and Rolf Niedermeier, *Temporal interval cliques and independent sets*, Theoretical Computer Science **961** (2023), 113885.

[49] Anton Herrmann, Christian Komusiewicz, Nils Morawietz, and Frank Sommer, *Temporal dominating set and temporal vertex cover under the lense of degree restrictions*, 4th Symposium on Algorithmic Foundations of Dynamic Networks, SAND 2025, June 9-11, 2025, Liverpool, UK (Kitty Meeks and Christian Scheideler, eds.), LIPIcs, vol. 330, Schloss Dagstuhl - Leibniz-Zentrum für Informatik, 2025, pp. 16:1–16:18.

[50] David Kempe, Jon M. Kleinberg, and Amit Kumar, *Connectivity and inference problems for temporal networks*, Journal of Computer and System Sciences **64** (2002), no. 4, 820–842.

[51] Tomohiro Koana, Christian Komusiewicz, and Frank Sommer, *Exploiting c-closure in kernelization algorithms for graph problems*, SIAM Journal on Discrete Mathematics **36** (2022), no. 4, 2798–2821.

[52] ———, *Computing dense and sparse subgraphs of weakly closed graphs*, Algorithmica **85** (2023), no. 7, 2156–2187.

[53] ———, *Essentially tight kernels for (weakly) closed graphs*, Algorithmica **85** (2023), no. 6, 1706–1735.

[54] Yasuaki Kobayashi and Yota Otachi, *Parameterized complexity of graph burning*, Algorithmica **84** (2022), no. 8, 2379–2393.

[55] Stefan Kratsch and Pascal Schweitzer, *Isomorphism for graphs of bounded feedback vertex set number*, Algorithm Theory - SWAT 2010 (Berlin, Heidelberg) (Haim Kaplan, ed.), Springer Berlin Heidelberg, 2010, pp. 81–92.

[56] David C. Kutner and Laura Larios-Jones, *Temporal reachability dominating sets: Contagion in temporal graphs*, Algorithmics of Wireless Networks - 19th International Symposium, ALGOWIN 2023, Amsterdam, The Netherlands, September 7-8, 2023, Revised Selected Papers (Konstantinos Georgiou and Evangelos Kranakis, eds.), Lecture Notes in Computer Science, vol. 14061, Springer, 2023, pp. 101–116.

[57] David C. Kutner and Anouk Sommer, *Better late, then? The hardness of choosing delays to meet passenger demands in temporal graphs*, 4th Symposium on Algorithmic Foundations of Dynamic Networks, SAND 2025, June 9-11, 2025, Liverpool, UK (Kitty Meeks and Christian Scheideler, eds.), LIPIcs, vol. 330, Schloss Dagstuhl - Leibniz-Zentrum für Informatik, 2025, pp. 7:1–7:18.

[58] Michael Lampis, *Algorithmic Meta-theorems for Restrictions of Treewidth*, Algorithmica **64** (2012), no. 1, 19–37.

[59] Daniel Lokshtanov, *Parameterized integer quadratic programming: Variables and coefficients*, CoRR **abs/1511.00310** (2015).

[60] Andrea Marino and Ana Silva, *Coloring temporal graphs*, Journal of Computer and System Sciences **123** (2022), 171–185.

[61] ———, *Eulerian walks in temporal graphs*, Algorithmica **85** (2023), no. 3, 805–830.

[62] Kitty Meeks and Fiona Skerman, *The parameterised complexity of computing the maximum modularity of a graph*, Algorithmica **82** (2020), no. 8, 2174–2199.

[63] George B. Mertzios, Othon Michail, and Paul G. Spirakis, *Temporal network optimization subject to connectivity constraints*, Algorithmica **81** (2019), no. 4, 1416–1449.

[64] George B. Mertzios, Hendrik Molter, Rolf Niedermeier, Viktor Zamaraev, and Philipp Zschoche, *Computing maximum matchings in temporal graphs*, Journal of Computer and System Sciences **137** (2023), 1–19.

School of Computing Science
Sir Alwyn Williams Building
University of Glasgow
Glasgow, G12 8RZ
Scotland, UK
kitty.meeks@glasgow.ac.uk

Asymptotic Enumeration via Graph Containers and Entropy

Jinyoung Park

Abstract

The container methods are powerful tools to bound the number of independent sets of graphs and hypergraphs, and they have been extremely influential in the area of extremal and probabilistic combinatorics. We will focus on more specialized graph container methods due to Sapozhenko (1987) that deal with sets in expander graphs. Entropy, first introduced by Shannon (1948) in the area of information theory, is a measure of the expected amount of information contained in a random variable. Entropy has seen lots of fascinating applications in a wide range of enumeration problems. In this survey article, we will discuss recent developments that exploit a combination of the two methods on enumerating graph homomorphisms.

1 Introduction

1.1 Framework

Most of the enumeration problems discussed in this article are within the following framework. Consider a collection $\mathcal{X}$ (actually, a sequence of collections $\mathcal{X}_n$) whose size we want to estimate. Finding the exact size of $\mathcal{X}$ is hard, and in particular, there are no appropriate recurrences that we can leverage to apply a mathematical induction or a traditional generating function approach. Instead, there is a subcollection $\mathcal{X}' \subseteq \mathcal{X}$ that is easily understood. Our goal is to show that (upon finding the "right" $\mathcal{X}'$) $\mathcal{X}'$ accounts for almost all $\mathcal{X}$, thus concluding that the size of $\mathcal{X}'$ in fact provides a good estimate for the size of $\mathcal{X}$.

Of course, the tools to implement the above idea depend on the nature of the enumeration problem under consideration. In this article, we focus on problems related to the "graph container method" and the "entropy method". Before discussing these tools in more detail, we first introduce two classical enumeration problems to illustrate the abstract framework above: Dedekind's problem [19] and calculating the number of independent sets in the hypercube [55], where the corresponding $\mathcal{X}$ are the collections of antichains in the Boolean lattice and independent sets in the hypercube, respectively.

1.2 Dedekind's problem

An *antichain* of a poset $P = (X, \preceq)$ is a subset of X in which any two elements are incomparable in the (partial) order $\preceq$. Dedekind's problem from 1897 asks for the total number $\alpha(n)$ of antichains in the n-dimensional Boolean lattice $\mathcal{B}_n = (2^{[n]}, \subseteq)$, where $[n] := \{1, 2, \ldots, n\}$. (Equivalently, $\alpha(n)$ is the number of monotone Boolean functions $f : \{0,1\}^n \to \{0,1\}$, or the number of elements of the free distributive lattice on n generators.) The Boolean lattice can be viewed as the poset on $\{0,1\}^n$ (the collection of binary strings of length n) with the partial order

$$x \preceq y \ \text{ if and only if } \ x_i \leq y_i \ \text{ for all } i \in [n],$$

where x_i denotes the i-th coordinate of x. In what follows, we will take the latter viewpoint.

It is easy to check that $\alpha(1) = 3$ (the empty collection, the singletons $\{0\}$, and $\{1\}$) and $\alpha(2) = 6$ (the empty collection, the singletons $\{00\}$, $\{01\}, \{10\}, \{11\}$, and $\{01, 10\}$), but there is no closed formula known for $\alpha(n)$ for general n. The sequence $\alpha(n)$ grows very fast and is not easy to compute even for small values of n. Indeed, even with modern computing power, the exact value of $\alpha(n)$ is known only for $n \leq 9$, where the case of $n = 9$ was obtained very recently by two independent groups of researchers [36, 82] more than 30 years after the determination of $\alpha(8)$ [84].

Having the discussion from Section 1.1 in mind, observe that there is a trivial lower bound on $\alpha(n)$:

$$\alpha(n) \geq 2^{\binom{n}{\lfloor n/2 \rfloor}}. \tag{1.1}$$

To see this, denote by $\mathcal{L}_i$ (the "i-th layer" of $\mathcal{B}_n$) the set of elements (binary strings) in $\mathcal{B}_n$ that have precisely i coordinates equal to 1, observing that each $\mathcal{L}_i$ ($i \in [0, n]$) is an antichain. In particular, by taking the subsets of $\mathcal{L}_{\lfloor n/2 \rfloor}$ (which are all antichains), we obtain precisely $2^{\binom{n}{\lfloor n/2 \rfloor}}$ many antichains.

Sperner's theorem [80] – which is usually considered as the first result in extremal set theory – says that a largest antichain in $\mathcal{B}_n$ has size $\binom{n}{\lfloor n/2 \rfloor}$, and the only antichains to achieve this bound are $\mathcal{L}_{\lfloor n/2 \rfloor}$ and $\mathcal{L}_{\lceil n/2 \rceil}$ (the "middle layer(s)" of $\mathcal{B}_n$). So the lower bound in (1.1) is the best one can obtain following the rather naive strategy of taking all subsets of some largest antichain.

Interestingly, this trivial lower bound turned out to be not too far from the truth. A breakthrough result of Kleitman [51] in 1969 pinpointed the asymptotics of the logarithm of $\alpha(n)$, showing that the exponent in the lower bound in (1.1) is actually asymptotically optimal for $\log \alpha(n)$ (in this article, $\log = \log_2$):

Theorem 1.1 (Kleitman [51])

$$\log \alpha(n) < \left(1 + O\left(\frac{\log n}{\sqrt{n}}\right)\right) \binom{n}{\lfloor n/2 \rfloor}.$$

The $O\left(\frac{\log n}{\sqrt{n}}\right)$ term on the right-hand side was subsequently improved to $O\left(\frac{\log n}{n}\right)$ by Kleitman and Markowsky [52]. The problem of finding asymptotics for $\log \alpha(n)$ has been revisited several times: the Kleitman-Markowsky bound was reproved by Kahn [46] using the entropy method, and by Balogh, Treglown, and Wagner [10] using the container method. Pippenger [71] obtained a similar bound with a slightly weaker error term based on the entropy method (whose detailed implementation was different from Kahn's approach).

In 1981, Korshunov [54] determined $\alpha(n)$ itself up to a $(1 + o(1))$ factor, thereby "asymptotically solving" Dedekind's problem. Korshunov's paper was more than 100 pages, and a simpler – though still involved – proof was later given by Sapozhenko [74]. In the statement below, we use $M = \binom{n}{\lfloor n/2 \rfloor}$ for simplicity.

Theorem 1.2 (Korshunov [54], Sapozhenko [74]) *As $n \to \infty$, for n even, we have*

$$\alpha(n) = (1 + o(1)) 2^M \exp\left[\binom{n}{n/2 + 1}\left(2^{-n/2} + \frac{n^2 - 2n - 16}{32} \cdot 2^{-n}\right)\right],$$

while for n odd

$$\alpha(n) = (2+o(1))2^M \exp\left[\binom{n}{\frac{n+1}{2}}\left(2^{-(n+1)/2} + \frac{3n^2-19}{32}\cdot 2^{-(n+1)}\right) + \binom{n}{\frac{n+3}{2}}\left(\frac{1}{2}\cdot 2^{-(n+1)/2} + \frac{n+5}{8}\cdot 2^{-(n+1)}\right)\right].$$

We remark that, while the above formulas might look technical, each of them is actually a natural asymptotic lower bound (for the corresponding parity) on $\alpha(n)$. That is, with writing $\mathcal{A}(\mathcal{B}_n)$ for the collection of antichains of $\mathcal{B}_n$, for each parity of n, there is a (simpler) subfamily $\mathcal{A}'_n \subseteq \mathcal{A}(\mathcal{B}_n)$ with $|\mathcal{A}'_n|$ asymptotically equal to the formula in Theorem 1.2. At the end of Section 1, we will provide a description of $\mathcal{A}'_n$.

So how one proves Theorem 1.2 is as explained in the framework in Section 1.1. Korshunov and Sapozhenko showed that $|\mathcal{A}(\mathcal{B}_n) \setminus \mathcal{A}'_n| = o(|\mathcal{A}'_n|)$, thus concluding that $\mathcal{A}'_n$ accounts for almost all of the antichains in $\mathcal{B}_n$. A main tool in [74] is the graph container method. Note that this proof also provides the structure of "typical" antichains of $\mathcal{B}_n$, as the argument says that almost all antichains in $\mathcal{B}_n$ are the ones in the description of $\mathcal{A}'_n$.

1.3 Independent sets in the hypercube

The n-dimensional (discrete) *hypercube* (or the *Hamming cube*) Q_n is the graph on the vertex set $\{0,1\}^n$ where two vertices are adjacent if and only if they differ in exactly one coordinate. Equivalently, Q_n is the covering graph of the poset $\mathcal{B}_n$. Note that Q_n is an n-regular bipartite graph with the unique bipartition classes $\mathcal{E}$ and $\mathcal{O}$, where $\mathcal{E}$ is the set of vertices (binary strings of length n) which have an even number of 1's in their coordinates. For simplicity, we will write N for $2^n = |V(Q_n)|$. (So $|\mathcal{E}| = |\mathcal{O}| = N/2$.)

An *independent set* in a graph is a set of vertices in which no two vertices are adjacent. Denote by $\mathcal{I}(G)$ the collection of independent sets in a graph G, and $i(G) := |\mathcal{I}(G)|$. We are interested in estimating $i(Q_n)$. This number again grows very fast, and a closed formula is not known.

As in Dedekind's problem, one can easily think of a trivial lower bound on this number:

$$i(Q_n) \geq 2 \cdot 2^{N/2} - 1$$

because each subset of $\mathcal{E}$ or $\mathcal{O}$ is an independent set (and the empty set is counted twice). Korshunov and Sapozhenko [55], using the graph container method, showed that this trivial lower bound is not far from the truth.

Theorem 1.3 (Korshunov, Sapozhenko [55]) *As $n \to \infty$,*

$$i(Q_n) = 2(1+o(1))\sqrt{e}2^{N/2}. \tag{1.2}$$

As in Theorem 1.2, the right-hand side of (1.2) is an easy asymptotic lower bound on $i(Q_n)$. In the next section, we discuss the construction of this lower bound.

1.3.1 Asymptotic lower bound on $i(Q_n)$ Given an independent set I, we say a vertex v is *occupied* if $v \in I$. Let's informally say that an independent set of Q_n is in a "ground state" (or "pure") if it is entirely contained in either $\mathcal{E}$ or $\mathcal{O}$. So there are two ground states ($\mathcal{E}$ and $\mathcal{O}$), and the number of independent sets in each ground state is precisely $2^{N/2}$.

Next, we estimate the number of independent sets with some "flaws" (or "noises") and examine their contribution to $i(Q_n)$. Let's first consider independent sets I that have exactly one vertex (a flaw) from $\mathcal{O}$. The number of such independent sets is

$$N/2 \cdot 2^{N/2-n} = 2^{n-1}2^{N/2-n} = 2^{-1}2^{N/2}$$

($N/2$ is the number of ways to choose a flaw $v \in \mathcal{O}$; once we pick v, then the n vertices in $N(v)$ ($\subseteq \mathcal{E}$) must be unoccupied so we are left with $2^{N/2-n}$ choices for $I \cap \mathcal{E}$). In general, the number of independent sets I that have exactly k "non-nearby" flaws from $\mathcal{O}$ (more precisely: we say v and w are *nearby* if their neighborhoods intersect) is approximately

$$\binom{N/2}{k} \cdot 2^{N/2-kn} \sim \frac{(N/2)^k}{k!} 2^{N/2-kn} = 2^{N/2} \cdot \frac{2^{-k}}{k!} \tag{1.3}$$

as long as k is "not so large" ($\binom{N/2}{k}$ approximates the number of ways to choose k non-nearby flaws from $\mathcal{O}$; those k vertices altogether have a neighborhood of size kn, so we have $2^{N/2-kn}$ choices left for $I \cap \mathcal{E}$).

Now, by summing (1.3) over k from 0 to any (slow) $\omega(1)$ and multiplying it by 2 (since there are two ground states) we obtain the right-hand side of (1.2) as an asymptotic lower bound on $i(Q_n)$.

Remark In connection with the framework in Section 1.1, in the problem of estimating $i(Q_n)$, $\mathcal{X}_n'$ consists of the independents sets described in Section 1.3.1, and Theorem 1.3 asserts that $\mathcal{X}_n'$ accounts for almost all of $\mathcal{I}(Q_n)$. We reiterate that Theorem 1.3 therefore also provides a *structural* result – it says that almost all independent sets in Q_n have the structure described in Section 1.3.1.

1.3.2 Towards the upper bound We first briefly motivate why it is "sound" to expect that the independent sets in Section 1.3.1 should take up almost all of $\mathcal{I}(Q_n)$. In this section, we call an independent set *bad* if I is not of the type described in Section 1.3.1. As a simplest example of bad independent sets, consider independent sets I that have exactly two nearby flaws from $\mathcal{O}$ – i.e., two occupied vertices in $\mathcal{O}$ whose neighborhoods overlap. The number of independent sets of this type is roughly (ignoring some irrelevant constant factors)

$$N/2 \cdot n^2 \cdot 2^{N/2-2n} \approx n^2 \cdot 2^{N/2-n} \ (\ll 2^{N/2}); \tag{1.4}$$

here $N/2$ is the number of ways to choose the first flaw v; now a crucial point is that once we pick v, the second flaw is within the second neighborhood of v (since the two flaws are nearby), so there are at most n^2 choices for it; the two flaws still have the total neighborhood of size roughly $2n$ (because a pair of vertices in Q_n has at most two common neighbors), so we have roughly $2^{N/2-2n}$ choices for $I \cap \mathcal{E}$.

So the contribution from (1.4) is negligible (in fact, exponentially small) and absorbed in the $o(1)$ term in (1.2). Note that the reason why the overall contributions of the independent sets with nearby flaws and those of independent sets with non-nearby flaws are different is because the number of choices for two nearby flaws is significantly smaller than that for two non-nearby flaws, while the sizes of the neighborhoods of the flaws in the two cases remain more or less the same.

While independent sets with a small number of flaws are not so hard to analyze (as we saw in the above computation), the story is completely different when we consider independent sets with a large set of flaws. Notice that the two competing factors in bounding the number of bad independent sets are

the number of choices for flaws vs. the size of the total neighborhood of the flaws.

Informally, the former is the "cost" for specifying flaws, and the latter is the "penalty" from the (given) flaws. The argument surrounding (1.4) illustrates the idea that in order to show the bad independent sets are negligible, we have to show that the penalty from the flaws of the independent sets exceeds their costs. So we need an effective upper bound on the number of subsets of $\mathcal{O}$ (or $\mathcal{E}$) that consist of "nearby vertices" with a prescribed size of their neighborhood. The graph container lemma (Lemma 4.5) due to Sapozhenko [75] serves this purpose, and this is one of the main focuses of this article.

The rest of the article is organized as follows. Section 2 defines the notion of graph homomophism and introduces some well-known examples. In Section 3, we introduce some basics of the entropy function and provide an example to illustrate how the entropy method is used in solving enumeration problems. Section 4 starts from an overview of the container method, and then dives into a more specialized graph container lemma for expanders that can be used to get sharper results when it applies. Section 5 discusses sharp asymptotics for the number of graph homomorphisms, and hints at why a combination of the graph container method and the entropy method can be useful here. In Section 6, we discuss the problem of enumerating antichains of $[t]^n$, a generalized Boolean lattice, in connection with a first successful application of the graph container lemma for expanders in the "irregular" setting. Section 7 discusses counting maximal independent sets in the hypercube, which leads to a development of an interesting isoperimetric inequality on the hypercube.

Notation *Given a graph G, $E(G)$ and $V(G)$ denote the set of edges and vertices of G, respectively. For $u, v \in V(G)$, write $u \sim_G v$ (or $u \sim v$ if the host graph is clear) if $\{u,v\} \in E(G)$. For $U \subseteq V(G)$, $G[U]$ denotes the subgraph of G induced by U. As usual, $N(x)$ denotes the neighborhood of x, and $N(U) := \cup_{x\in U} N(x)$. The codegree of x and y ($x \neq y$) is $|N(x) \cap N(y)|$, and the codegree of G is the maximum of the codgree of all pairs in $V(G)$. For $A, B \subseteq V(G)$, $\nabla(A,B) := \{\{a,b\} \in E(G) : a \in A, b \in B\}$ and $e(A,B) := |\nabla(A,B)|$. $dist_G(u,v)$ (or simply $dist(u,v)$ if the host graph is clear) denotes the length of a shortest path in G between u and v. The diameter of a graph G, denoted by $diam(G)$, is $\max_{u,v\in V(G)} dist_G(u,v)$.*

For an integer k, we write G^k for the k-th power of G, that is, $V(G^k) = V(G)$ and $u, v \in V(G^k)$ are adjacent iff $u \neq v$ and $dist_G(u,v) \leq k$. We say a vertex set

$A \subseteq V(G)$ is k-linked if $G^k[A]$ is connected. The closure of A, denoted by $[A]$, is $[A] := \{v \in V(G) : N(v) \subseteq N(A)\}$.

For integers a and b, $[a] := \{1, 2, \ldots, a\}$ and $[a, b] := \{a, a+1, \ldots, b\}$. We use $\binom{[n]}{k}$ to denote the family of all k-element subsets of $[n]$.

We use standard asymptotic notation, and w.h.p. means "with high probability". Logarithms are always base 2.

We close this section by giving the promised construction of the asymptotic lower bound for Dedekind's problem: we describe the family $\mathcal{A}'_n$ whose size is asymptotically equal to the (asymptotic) values of $\alpha(n)$ in Theorem 1.2. We only describe the construction for even n, for simplicity (the construction for odd n is similar). Recall that $\mathcal{L}_i$ denotes the i-th layer of the Boolean lattice $\mathcal{B}_n$. We define the *comparability graph* of $\mathcal{B}_n$ to be the graph on $V(\mathcal{B}_n)$ where two vertices are adjacent iff they are comparable. Observe that for any $k \in \mathbb{Z}^+$, a vertex set $A \subseteq V(G)$ can be decomposed into its maximal k-linked subsets, and we call these subsets the *k-linked components* of A (with respect to G).

Theorem 1.4 *Let n be even. Let $\mathcal{A}'_n$ be the collection of the antichains I of $\mathcal{B}_n$ that satisfy the following properties:*

(i) $I \subseteq \bigcup_{n-1 \le i \le n+1} \mathcal{L}_i$;

(ii) every 2-linked component of $I \cap (\mathcal{L}_{n-1} \cup \mathcal{L}_{n+1})$ with respect to the comparability graph of $\mathcal{B}_n$ is of size at most 2.

Then, as $n \to \infty$,

$$\alpha(n) = \left(1 + e^{-\Omega(n)}\right) |\mathcal{A}'_n|.$$

We leave it to the readers that $|\mathcal{A}'_n|$ (for even n) is asymptotically equal to the value of $\alpha(n)$ in Theorem 1.2.

2 Graph homomorphisms

Independent sets and many other important combinatorial objects can be incorporated into the notion of graph homomorphisms which we introduce now. Given graphs G and H, a *graph homomorphism* from G to H (or an *H-coloring* of G) is a map $f : V(G) \to V(H)$ such that if $\{v, w\} \in E(G)$, then $\{f(v), f(w)\} \in E(H)$. Write $\mathrm{Hom}(G, H)$ for the collection of graph homomorphisms from G to H. Below are some well-known graph homomorphisms.

Example 2.1 (Independent set) Let H be the graph with $V(H) = \{v_0, v_1\}$ and $E(H) = \{\{v_0, v_0\}, \{v_0, v_1\}\}$.

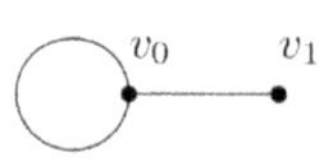

Figure 1: H for independent sets

To see the correspondence between an independent set and an H-coloring, for a given independent set $I \in \mathcal{I}(G)$, define f_I to be

$$f_I(v) = \begin{cases} v_0 & \text{if } v \notin I; \\ v_1 & \text{if } v \in I. \end{cases}$$

Then $f_I \in \mathrm{Hom}(G, H)$, and it is easy to see that the above correspondence between I and f_I provides a bijection between $\mathcal{I}(G)$ and $\mathrm{Hom}(G, H)$. That is, any independent set can be regarded as an H-coloring, and vice versa.

Example 2.2 (Proper coloring) Set $H = K_q$, the clique on q vertices. It is easy to see that there is a one-to-one correspondence between H-colorings of G and (proper) q-colorings of $V(G)$.

We will drop "proper" and just say a q-coloring or a coloring for a proper coloring. In the rest of the paper, we will use $\mathcal{C}_q(G)$ for the collection of q-colorings of $V(G)$, and $c_q(G) := |\mathcal{C}_q(G)|$.

Example 2.3 (Lipschitz function) Given an integer M, say an integer-valued function $g : V(G) \to \mathbb{Z}$ is *M-Lipschitz* if $v \sim w$ implies $|g(v) - g(w)| \leq M$.

Let H be the graph with $V(H) = \mathbb{Z}$ and $E(H) = \{\{m, m\}, \{m, m+1\} : m \in \mathbb{Z}\}$ as in the figure below.

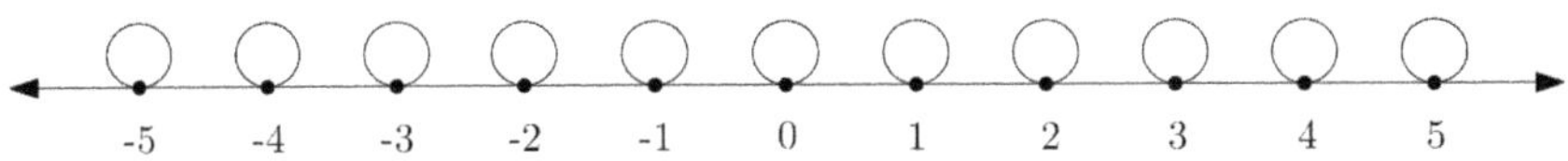

Figure 2: H for 1-Lipschitz functions

Observe that there is a one-to-one correspondence between H-colorings of G and 1-Lipschitz functions from G. Similarly, if we define H_M to be the graph on $\mathbb{Z}$ with $E(H_M) = \{\{m, m\}, \{m, m+1\}, \ldots, \{m, m+M\} : m \in \mathbb{Z}\}$, then there is a bijection between H_M-colorings of G and M-Lipschitz functions from G.

As a last example of this section, an integer-valued function $g : V(G) \to \mathbb{Z}$ is called a *$\mathbb{Z}$-homomorphism* if $v \sim w$ implies $|g(v) - g(w)| = 1$. A $\mathbb{Z}$-homomorphism is an H-coloring where H is the integer line. Note that G must be bipartite to admit a $\mathbb{Z}$-homomorphism.

3 Entropy

The entropy of a random variable $\mathbf{X}$ was first introduced by Shannon in 1948 [79] as a measure of the "information" or "uncertainty" contained in $\mathbf{X}$. Entropy has been used to tackle many enumeration problems, and this article focuses on its usefulness for counting (or bounding the number of) independent sets, colorings, or more generally, graph homomorphisms. In this section, we quickly introduce some very basics of the entropy function, and provide an example to show how entropy is used for enumerating proper colorings of a given graph.

3.1 Basics

This section briefly collects some basics of entropy; for a less hurried introduction, see, e.g., McEliece [59].

For a discrete random variable $\mathbf{X}$, the *entropy* of $\mathbf{X}$ is

$$H(\mathbf{X}) = \sum_x \mathbb{P}(\mathbf{X} = x) \log \frac{1}{\mathbb{P}(\mathbf{X} = x)}$$

where the sum is taken over all x with $\mathbb{P}(\mathbf{X} = x) \neq 0$. (By convention, we take $0 \log \frac{1}{0} = 0$.) The entropy of $\mathbf{X}$ is often understood as the amount of "randomness" of $\mathbf{X}$; for example, one can show that

$$H(\mathbf{X}) \leq \log |\mathrm{Image}(\mathbf{X})| \tag{3.1}$$

($\mathrm{Image}(\mathbf{X})$ is the set of values that $\mathbf{X}$ takes with nonzero probability) by Jensen's inequality using the concavity of the logarithm function. Furthermore, in (3.1) the equality holds iff $\mathbf{X}$ is chosen uniformly at random from $\mathrm{Image}(\mathbf{X})$. So roughly, the "more random" the random variable $\mathbf{X}$ is, the larger the entropy of $\mathbf{X}$ is.

For any event T, we define the entropy of $\mathbf{X}$ given T as

$$H(\mathbf{X}|T) = \sum_x \mathbb{P}(\mathbf{X} = x|T) \log \frac{1}{\mathbb{P}(\mathbf{X} = x|T)}.$$

For another discrete random variable $\mathbf{Y}$, the *conditional entropy* of $\mathbf{X}$ with respect to $\mathbf{Y}$ is

$$\begin{aligned} H(\mathbf{X}|\mathbf{Y}) &= \sum_y \mathbb{P}(\mathbf{Y} = y) H(\mathbf{X}|\mathbf{Y} = y) \\ &= \sum_y \mathbb{P}(\mathbf{Y} = y) \sum_x \mathbb{P}(\mathbf{X} = x|\mathbf{Y} = y) \log \frac{1}{\mathbb{P}(\mathbf{X} = x|\mathbf{Y} = y)}. \end{aligned} \tag{3.2}$$

Below are some standard facts about the entropy; $\mathbf{X}$, $\mathbf{Y}$, and $\mathbf{X}_i$'s are all used for discrete random variables. For any random vector $\mathbf{X} = (\mathbf{X}_1, \ldots, \mathbf{X}_n)$ we have

$$H(\mathbf{X}) \leq H(\mathbf{X}_1) + \cdots + H(\mathbf{X}_n). \tag{3.3}$$

The above inequality can be sharpened using the conditional entropy; for example, for two random variables, we have

$$H(\mathbf{X}, \mathbf{Y}) = H(\mathbf{X}) + H(\mathbf{Y}|\mathbf{X}). \tag{3.4}$$

It also holds that

$$\text{if } \mathbf{Y} \text{ determines } \mathbf{X}, \text{ then } H(\mathbf{Z}|\mathbf{Y}) \leq H(\mathbf{Z}|\mathbf{X}); \tag{3.5}$$

intuitively, the above means that the amount of information in $\mathbf{Z}$ that remains after revealing $\mathbf{X}$ is at least the amount remaining after revealing $\mathbf{Y}$.

The following result is a celebrated inequality due to Shearer, which generalizes the subadditivity property (3.3).

Lemma 3.1 ([15]) *Let $\mathbf{X} = (\mathbf{X}_1, \ldots, \mathbf{X}_k)$ be a random vector, and let $\alpha_A \in \mathbb{R}^+$ be given for every $A \subseteq [k]$. If $\sum_{A \ni i} \alpha_A \geq 1$ for all $i \in [k]$, then*

$$H(\mathbf{X}) \leq \sum_{A \subseteq [k]} \alpha_A H(\mathbf{X}_A),$$

where $\mathbf{X}_A = (\mathbf{X}_i : i \in A)$.

3.2 Enumeration via entropy

In this section, we provide a standard-ish entropy-based argument for bounding the number of colorings of regular bipartite graphs. The proof below, which is a special case of the results in [29], can easily be extended to more general graph homomorphisms. Recall that $\mathcal{C}_q(G)$ is the collection of q-colorings of $V(G)$, and $c_q(G) := |\mathcal{C}_q(G)|$.

Proposition 3.2 *Let G be a d-regular bipartite graph on m vertices. For any q,*

$$\log c_q(G) \le \frac{m}{2} \cdot \left(\log\left(\left\lfloor \frac{q}{2} \right\rfloor \left\lceil \frac{q}{2} \right\rceil\right) + O_q(1/d)\right)$$

where in $O_q(\cdot)$, the implicit constant (only) depends on the value of q.

Observe that, trivially, $\log c_q(G) \ge \frac{m}{2} \cdot \log\left(\lfloor \frac{q}{2} \rfloor \lceil \frac{q}{2} \rceil\right)$ (fix $\lfloor \frac{q}{2} \rfloor$ colors and call this set of colors $\mathcal{C}$; denote the set of the remaining $\lceil \frac{q}{2} \rceil$ colors by $\mathcal{C}'$ and consider every coloring of G (with a bipartition (A, B)) where each vertex in A is given a color from $\mathcal{C}$ and each vertex in B is given a color from $\mathcal{C}'$). So if $q = O(1)$, then Proposition 3.2 provides an asymptotically tight upper bound on $\log c_q(G)$.

We prove the above statement only for $q = 4$ for a simpler exposition, so our target bound is

$$\log c_4(G) \le (1 + O(1/d))m. \tag{3.6}$$

The proof below extends almost identically to any q.

Proof Denote by X and Y a bipartition of G, noting that $|X| = |Y| = m/2$. We write N_x for $N(x)$. For $f \in \mathcal{C}_4(G)$ and $x \in V(G)$, f_x denotes the value of f at x, and for $A \subseteq V(G)$, $f_A = (f_x : x \in A)$ and $f(A) = \{f_x : x \in A\}$ (that is, f_A contains the full information about the color assignment of each $x \in A$, while $f(A)$ (the "palette" of A) only contains the information about what colors are used in A).

Let $\mathbf{f}$ be an element of $\mathcal{C}_4(G)$ chosen uniformly at random. A key point is that, by the equality condition of (3.1),

$$H(\mathbf{f}) = \log c_4(G).$$

Therefore, the conclusion (3.6) will follow if we show that $H(\mathbf{f})$ is bounded above by the right-hand side of (3.6). To this end, first, by (3.4),

$$H(\mathbf{f}) = H(\mathbf{f}_Y) + H(\mathbf{f}_X|\mathbf{f}_Y). \tag{3.7}$$

The second term of the right-hand side of the above is, using a conditional version of (3.3) and then (3.5),

$$H(\mathbf{f}_X|\mathbf{f}_Y) \le \sum_{x \in X} H(\mathbf{f}_x|\mathbf{f}_Y) \le \sum_{x \in X} H(\mathbf{f}_x|\mathbf{f}(N_x));$$

the first term of the right-hand side of (3.7) is (a justification will follow)

$$H(\mathbf{f}_Y) \le \frac{1}{d} \sum_{x \in X} H(\mathbf{f}_{N_x}) = \frac{1}{d} \sum_{x \in X} H(\mathbf{f}(N_x), \mathbf{f}_{N_x}) = \frac{1}{d} \sum_{x \in X} [H(\mathbf{f}(N_x)) + H(\mathbf{f}_{N_x}|\mathbf{f}(N_x))]$$

where for the inequality, we use Lemma 3.1 with

$$\alpha_A = \begin{cases} 1/d & \text{if } A = N_x \text{ for some } x \in X \\ 0 & \text{otherwise} \end{cases}$$

(and the second equality uses (3.4)). Combining the two bounds, (3.7) is

$$H(\mathbf{f}) \leq \sum_{x \in X} \frac{1}{d} [H(\mathbf{f}(N_x)) + H(\mathbf{f}_{N_x}|\mathbf{f}(N_x))] + H(\mathbf{f}_x|\mathbf{f}(N_x)).$$

Let $T(x)$ ($x \in X$) be the summand of the above sum, that is,

$$T(x) = \frac{1}{d} [H(\mathbf{f}(N_x)) + H(\mathbf{f}_{N_x}|\mathbf{f}(N_x))] + H(\mathbf{f}_x|\mathbf{f}(N_x)).$$

We claim that $T(x) \leq 2 + O(1/d)$ from which the conclusion will follow. We will use (a conditional version of) (3.1) repeatedly. The first term of the above expression is easily seen to be $O(1/d)$, since when there are only 4 colors, $H(\mathbf{f}(N_x)) = O(1)$ (the number of possible palettes for N_x cannot exceed 2^4). To bound the last two terms, note that for each possible value (a set of colors) C of $\mathbf{f}(N_x)$, by (3.1),

$$H(\mathbf{f}_x|\mathbf{f}(N_x) = C) \leq \log(4 - |C|); \text{ and}$$

$$H(\mathbf{f}_{N_x}|\mathbf{f}(N_x) = C) \leq \sum_{y \in N_x} H(\mathbf{f}_y|\mathbf{f}(N_x) = C) \leq d \log |C|.$$

Since $\log z + \log(4 - z) \leq 2$ for any z, this bounds $\frac{1}{d}H(\mathbf{f}_{N_x}|\mathbf{f}(N_x)) + H(\mathbf{f}_x|\mathbf{f}(N_x))$ by (using the definition of conditional entropy (3.2))

$$\sum_C \mathbb{P}(\mathbf{f}(N_x) = C) \left[\frac{1}{d}H(\mathbf{f}_{N_x}|\mathbf{f}(N_x) = C) + H(\mathbf{f}_x|\mathbf{f}(N_x) = C)\right] \leq 2,$$

yielding $T(x) \leq 2 + O(1/d)$ for any $x \in X$. □

4 The method of graph containers and sharp asymptotics

The container method has been proven to be an essential tool in the study of independent sets in graphs and hypergraphs. In the graph setting, its roots can be traced back to the work of Kleitman and Winston [53] and Sapozhenko [75]. These tools were extended to the context of hypergraphs and developed into a powerful method in the seminal works of Balogh, Morris and Samotij [7] and Saxton and Thomason [78]. The hypergraph container method has applications in broad areas including extremal graph theory, Ramsey theory, additive combinatorics, and discrete geometry. We refer the readers to [8] for an overview of the hypergraph container method.

A main focus of this article is a more specialized graph container lemma due to Sapozhenko. As a warm-up, in Section 4.1, we introduce some basic concepts of the general graph container method. In Section 4.2, we provide the statement of Sapozhenko's graph container lemma for expanders. This is a powerful tool for the problems of obtaining asymptotically sharp enumeration results and thus gaining an understanding of the structure of typical members of the family under consideration.

4.1 Warm-up

The method of graph containers provides an efficient upper bound on the number of certain vertex subsets (often independent sets) of a given graph.

Definition 4.1 (System of containers) Given a graph G and $\mathcal{F} \subseteq 2^{V(G)}$, suppose a family $\mathcal{C} \subseteq 2^{V(G)}$ satisfies the following property:

$$\text{for any } F \in \mathcal{F}, \text{ there is } C \in \mathcal{C} \text{ such that } F \subseteq C.$$

An element of the family $\mathcal{C}$ is called a *container*, and the family $\mathcal{C}$ is called a *system of containers* for $\mathcal{F}$.

Let some family $\mathcal{F}$ be given, and suppose our goal is to approximate $|\mathcal{F}|$. In the approach using the container method, we are hoping that there is a system of containers $\mathcal{C}$ for $\mathcal{F}$ such that both $|\mathcal{C}|$ and $\max\{|C| : C \in \mathcal{C}\}$ are reasonably small (or close to the truth). Then the naive inequality

$$|\mathcal{F}| \leq \sum_{C \in \mathcal{C}} 2^{|C|} \leq |\mathcal{C}| 2^{\max\{|C|:C\in\mathcal{C}\}} \tag{4.1}$$

will produce a good bound. This approach turned out to work incredibly well (modulo the noble ideas in the works referenced at the beginning of Section 4) for the family of independent sets in graphs (and certain hypergraphs). A very rough intuition behind this approach is that often independent sets in a graph cluster together, so specifying independent sets as subsets of a container should provide an efficient counting.

Below is one of the very first applications of the graph container method due to Sapozhenko [76].

Theorem 4.2 *Let G be a d-regular graph on m vertices. Then*

$$\log i(G) < \left(1 + O\left(\sqrt{\frac{\log d}{d}}\right)\right) \frac{m}{2}.$$

Theorem 4.2 follows from the following lemma that guarantees the existence of a suitable system of containers. (See [2, Theorem 1.6.1] for its proof.)

Lemma 4.3 *Let G be as in Theorem 4.2. For any $\epsilon > 0$, there is a collection $\mathcal{C} \subseteq 2^{V(G)}$ with the following properties:*

(i) For each independent set I in G, there is a member $C \in \mathcal{C}$ such that $I \subseteq C$;

(ii) For every $C \in \mathcal{C}$, $|C| \leq \frac{m}{\epsilon d} + \frac{m}{2-\epsilon}$;

(iii) $|\mathcal{C}| \leq \sum_{i \leq m/(\epsilon d)} \binom{m}{i}$.

To see that Theorem 4.2 follows from the above lemma, using (4.1) (with $\mathcal{F} = \mathcal{I}(G)$),

$$i(G) \leq \sum_{i \leq m/(\epsilon d)} \binom{m}{i} \cdot \exp_2\left[\frac{m}{\epsilon d} + \frac{m}{2-\epsilon}\right].$$

By the well-known estimate that $\log\left[\sum_{i\le m/(\epsilon d)}\binom{m}{i}\right] = O\left(\frac{m\log(\epsilon d)}{\epsilon d}\right)$, we have

$$\log i(G) = O\left(\frac{m\log(\epsilon d)}{\epsilon d} + \frac{m}{\epsilon d} + \frac{m}{2-\epsilon}\right).$$

The right-hand side of the above is minimized when $\epsilon = \Theta(\sqrt{\log d/d})$. By plugging in this value for the above expression, we obtain the conclusion of Theorem 4.2.

4.2 Graph container lemma for expanders

In the previous section, we discussed an upper bound on $\log|\mathcal{F}|$ for $\mathcal{F} = \mathcal{I}(G)$ using the method of graph containers. Notice that the above argument is not enough to give asymptotics for $|\mathcal{F}|$ itself. On the other hand, Sapozhenko showed that, if given graphs satisfy certain regularities in degrees and exhibit certain expansion properties, then one can further refine the container argument and obtain sharp asymptotics (as in Theorems 1.2 and 1.3). In this section, we state a graph container lemma for expanders[1] (Lemma 4.5) from [75], and sketch how to derive the sharp asymptotics in Theorem 1.3 from this lemma. Recall the definition of 2-linkedness from the notation section in Section 1.

Definition 4.4 Let G be a d-regular, bipartite graph with bipartition classes X and Y. For positive integers a and g and $v \in Y$, define

$$\mathcal{G}(a,g,v) = \{A \subseteq X : A \text{ is 2-linked}, |[A]| = a, |N(A)| = g, v \in N(A)\}.$$

(Recall that $[A]$ denotes the closure of A.)

Lemma 4.5 *For each pair of constants $c_1, c_2 > 0$, there is a constant $c = c(c_1, c_2) > 0$ such that the following holds. If G is a d-regular bipartite graph with partition classes X and Y and codegree c_1, then for $g \ge d^4$ and $g - a > \frac{c_2 g \log^3 d}{d^2}$, we have*

$$|\mathcal{G}(a,g,v)| \le 2^{g - c(g-a)/\log d}. \tag{4.2}$$

Sapozhenko's proof [75] for the above lemma uses a container-type argument (but different from that for Lemma 4.3). The original proof is written in Russian, and Galvin [28] provides a more accessible exposition of the proof of Theorem 1.3, including a proof of Lemma 4.5.

Remark 1. Sapozhenko's original formulation of the above lemma does not require strict regularity (as long as certain expansion requirements similar to that in Lemma 4.5 are satisfied). For example, the degree conditions required in [75] are the following: there is a constant C such that

$$\max_{v\in X}\deg(v) \le C\cdot\min_{v\in X}\deg(v); \quad \max_{v\in Y}\deg(v) \le C\cdot\min_{v\in Y}\deg(v),$$

and most crucially,

$$\max_{v\in Y}\deg(v) \le \min_{v\in X}\deg(v). \tag{4.3}$$

[1]This name can be misleading – the expansion that the lemma requires is much smaller than the usual notion of expanders.

A proof of the lemma under the above (more relaxed) degree conditions can be found from [39] in a more general hard-core model setting.

2. The exponential term $-\Omega((g-a)/\log d)$ on the right-hand side of (4.2) is improved to $-\Omega(g-a)$ in [65] (and implicitly in [50]). The weaker bound $-\Omega((g-a)/\log d)$ was already enough for many applications, but the improved bound $-\Omega(g-a)$ was crucial for obtaining sharp estimates for more restricted classes of independent sets including maximal independent sets in Q_n [50] and "balanced independent sets" in Q_n [65].

3. Jenssen and Perkins [42] showed that a combination of the graph container lemma for expanders (more precisely, its variant for the hard-core model by Galvin [27]) and the "cluster expansion method" from statistical physics results in powerful consequences. One consequence is the following vast refinement of the asymptotics in Theorem 1.3:

$$2\sqrt{e}\cdot 2^{N/2}\left(1+\frac{3n^2-3n-2}{8N}+\frac{243n^4-646n^3-33n^2+436n+76}{384N^2}+O\left(\frac{n^6}{N^3}\right)\right).$$

In fact, Jenssen and Perkins provided a formula and an algorithm for computing the asymptotics of $i(Q_n)$ to arbitrary order in N^{-1}. We refer the readers to [42] for the further fruitful applications of this combination of methods. For the cluster expansion method itself, we refer to a survey paper by Jenssen [37]. Jenssen and Perkins' idea of combining the two methods has been further adapted for lots of enumeration problems, including [4, 16, 31, 38, 40, 41, 56, 58]. Jenssen, Malekshahian, and the author [40] used the cluster expansion method to obtain a similar refinement of the asymptotics in Theorem 1.2. For example, for even n,

$$\alpha(n)=\left(1+o\left(2^{-2n}n^{O(1)}\right)\right)2^{\binom{n}{n/2}}\exp\left[\binom{n}{\frac{n}{2}-1}\left(2^{-\frac{n}{2}}+(n^2-2n-16)2^{-n-5}\right.\right.$$
$$\left.\left.+\frac{1}{3}(3n^4-4n^3-60n^2+112n+512)2^{-\frac{3n}{2}-9}\right)\right].$$

Again, as in [42], we also provide a formula and an algorithm for computing the asymptotics of $\alpha(n)$ to arbitrary order in $2^{-n/2}$.

Now we sketch the derivation of Theorem 1.3 from Lemma 4.5. For the sake of clarity, in the example below, we consider only a special type of bad independent set (recall from Sections 1.3.1 and 1.3.2 that bad independent set means it contains at least 2 nearby flaws) in $\mathcal{I}(Q_n)$ and show that the contribution to $i(Q_n)$ is negligible using Lemma 4.5. The (full) derivation of Theorem 1.3 essentially follows by summing up the contributions of all types of bad independent sets. The contribution of each type of bad independent set is exponentially small, so the total contribution of bad independent sets to $i(Q_n)$ is again negligible. A detailed computation can be found in [28, Corollary 4.1].

Example 4.6 Recall that Q_n has the bipartition class $\mathcal{E}$ and $\mathcal{O}$ and $N=2^n=|V(Q_n)|$ (so $|\mathcal{E}|=|\mathcal{O}|=N/2$).

Let $\mathcal{I}_k\subseteq\mathcal{I}(Q_n)$ be the collection of independent sets I such that $I\cap\mathcal{O}$ is 2-linked and $|[I\cap\mathcal{O}]|=k$ where $n^4\le k\le N/4$. (We pose the lower bound on k so that

Lemma 4.5 is applicable; the upper bound $N/4$ is natural since if $I \in \mathcal{I}(Q_n)$, then $\min\{|[I \cap \mathcal{E}]|, |[I \cap \mathcal{O}]|\} \le N/4$ by the expansion property of Q_n that for any $A \subseteq \mathcal{E}$ or $A \subseteq \mathcal{O}$, $|A| \le |N(A)|$.) We claim that $|\mathcal{I}_k|$ is exponentially smaller than $2^{N/2}$ (the main term of the (good) independent set count in Section 1.3.1). To see this, note that

$$\begin{aligned}|\mathcal{I}_k| &= \sum_{\substack{A \subseteq \mathcal{O} \\ A \text{ 2-linked}, |[A]|=k}} 2^{N/2-|N(A)|} \\ &\le 2^{N/2} \sum_{v \in \mathcal{E}} \sum_{g} \sum_{\substack{A \subseteq \mathcal{O}, v \in N(A) \\ A \text{ 2-linked}, |[A]|=k, |N(A)|=g}} 2^{-|N(A)|} \\ &= 2^{N/2} \sum_{v \in \mathcal{E}} \sum_{g} |\mathcal{G}(k,g,v)| \cdot 2^{-g}.\end{aligned}$$

A well-known isoperimetric inequality on the hypercube (see, e.g., [28, Claim 2.5]) says that if $Y \subseteq \mathcal{O}$ satisfies $|Y| \le N/4$, then $|N(Y)| - |Y| = \Omega(|N(Y)|/\sqrt{n})$. Therefore, by applying Lemma 4.5 (and applying the isoperimetric inequality with $Y = [A]$), $|\mathcal{G}(k,g,v)| \cdot 2^{-g}$ is at most $2^{-\Omega(g-k)/\log n} = 2^{-\Omega(g/(\sqrt{n}\log n))}$ if g satisfies $g - k = \Omega(g/\sqrt{n})$; and 0 otherwise. Now, the last line of the above display is at most

$$2^{N/2} \cdot N/2 \cdot \sum_{g \ge k} 2^{-\Omega(g/(\sqrt{n}\log n))} = 2^{N/2} \cdot 2^{-\Omega(k/(\sqrt{n}\log n))}$$

since $k \ge n^4$. This yields the desired conclusion.

4.3 Applications to independent set counts

Lemma 4.5 and its variants have been used for various host graphs to obtain sharp asymptotics for the number of independent sets and/or the structure of typical independent sets. The studied graphs include the discrete torus $\mathbb{Z}_m^n$ for even m by Jenssen and Keevash [38], the middle layers of the hypercube by Balogh, Garcia, and Li [4], and Abelian Cayley graphs by Potukuchi and Yepremyan [73]. Recently, Collares, Erde, Geisler, and Kang [16] extended this approach to a broader class of regular bipartite graphs including the Cartesian product of bipartite, regular base graphs of bounded size. Using the observation that intersecting families are independent sets in a Kneser graph, Balogh, Garcia, Li, and Wagner [5] showed that almost all k-uniform intersecting families in $2^{[n]}$ are "stars" if $n \ge 2k + 100 \ln k$. Their proof uses the graph container method and the Das–Tran removal lemma [18].

The entropy method, graph container method, and/or their combinations also apply to the setting of the "hard-core model", a fundamental model in statistical physics that generalizes the notion of independent sets. The present article will not survey the huge literature on this topic, and just refer to [45, 27, 39] as indicators to the works more closely related to the current discussions.

The independent set enumeration problems have also been considered in sparse random settings. For example, Kronenberg and Spinka [56] obtained asymptotics for the number of independent sets in edge-percolated hypercubes by studying an anti-ferromagnetic Ising model on the hypercube. Geisler, Kang, Sarantis, and Wdowinski [31] extended Kronenberg-Spinka's approach to a broader class of graphs, including even tori of growing side-length. Several cornerstone results in extremal combinatorics such as Erdős-Ko-Rado theorem [22] and Sperner's theorem [80] can also be stated using the language of independent sets. Sparse random versions of those problems were studied in [9, 17, 32, 6], where the approaches based on the graph container methods were again a key ingredient.

5 Sharp asymptotics for the number of graph homomorphisms

In this section, our main concern is getting sharp asymptotics for the number of graph homomorphisms from given graphs. Again, this will also provide us with the typical structure of the graph homomorphisms under consideration.

We start our discussion with graph colorings. (Again, a coloring always means a proper vertex coloring.) Recall that the entropy argument in Proposition 3.2 was not enough to give an asymptotically tight upper bound on $c_q(G)$. On the other hand, a combination of the entropy method and the proof idea for Lemma 4.5 (rather than the statement itself) can produce sharp asymptotics for $c_q(G)$ for certain host graphs. For example, Kahn and the author obtained the asymptotics for the number of 4-colorings of Q_n. Recall that $N = 2^n = |V(Q_n)|$. Throughout the section, we assume n is large, and all the asymptotic notation is as $n \to \infty$.

Theorem 5.1 (Kahn, P. [49])

$$c_4(Q_n) = 6e(1 + o(1))2^N. \tag{5.1}$$

As is usual in the current article, the value on the right-hand side of (5.1) is an asymptotic lower bound on $c_4(Q_n)$: to see this, say the set of given colors is $\{1, 2, 3, 4\}$. Consider a "ground state" (*cf.* Section 1.3.1) that consists of the colorings of $V(Q_n)$ where all the even vertices take colors from $\{1, 2\}$ and all the odd vertices take colors from $\{3, 4\}$. Obviously, this construction produces 2^N distinct (proper) colorings. It is easy to verify that (similarly to the computation in Section 1.3.1) the contribution of the colorings with non-nearby "flaws" – the vertices whose colors do not observe the ground state – together with the colorings in the ground state is asymptotically $e \cdot 2^N$. Now the asymptotic lower bound follows by observing that there are six ground states.

So the main task for the proof of Theorem 5.1 is to show that the number of colorings with nearby (i.e., 2-linked) flaws is negligible (*cf.* Section 1.3.2). The problem of estimating $i(Q_n)$, the number of independent sets in Q_n, was able to be done by invoking Lemma 4.5. However, this is not the case for counting colorings. In the next example, we illustrate why the graph container method alone is not sufficient here.

Example 5.2 Write f for a member of $\mathcal{C}_q(Q_n)$ (that is, f is a q-coloring of $V(Q_n)$) and let $[q]$ be the set of colors (palette). For simplicity, assume q is even, and let $P = [1, q/2]$ and $Q = [q/2 + 1, q]$ (so P and Q equipartition the palette). Similarly to Example 4.6, we consider a special type of bad coloring. Let $\mathcal{C} \subseteq \mathcal{C}_q(Q_n)$ be the collection of f such that

$$f(v) \in \begin{cases} P & \text{if } v \in \mathcal{E}; \\ Q & \text{if } v \in \mathcal{O}, \end{cases} \tag{5.2}$$

(this will be our ground state), noticing that $|\mathcal{C}| = (q/2)^N$. Similarly to $\mathcal{I}_k$ in Example 4.6, define $\mathcal{C}_k \subseteq \mathcal{C}$ to be the collection of q-colorings such that the set A of "flaws" (meaning the vertices that violate (5.2)) in $\mathcal{O}$ is 2-linked and $|[A]| = k$ $(\in [n^4, N/4])$. For the sake of a cleaner exposition, let's further impose the (very strong) assumption that there are no flaws in $\mathcal{E}$.

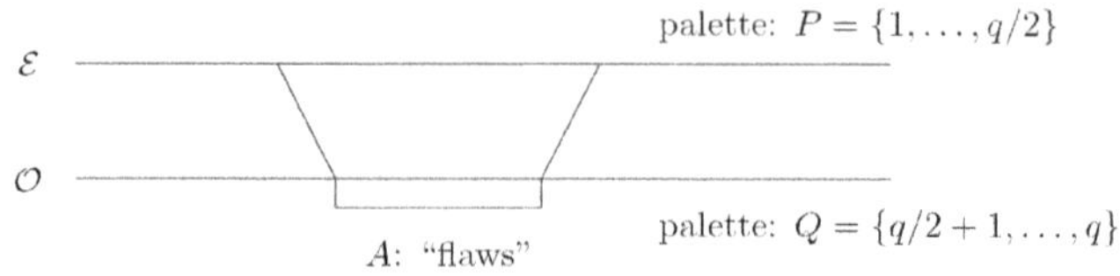

Figure 3: Setting for Example 5.2

Lemma 4.5 bounds the number of choices for A. Given $A \subseteq \mathcal{O}$, let $\mathcal{C}_A$ be the collection of $f \in \mathcal{C}_k$ whose set of flaws is precisely A. Observe that we have a bound

$$|\mathcal{C}_A| \le (q/2)^{N-|N(A)|}(q/2-1)^{|N(A)|}$$

(this is because if $w \in N(A)$, then w is adjacent to a vertex $v \in A$ and the color of v is in P, so $f(w)$ has at most $q/2-1$ choices (rather than $q/2$ choices)), and so

$$|\mathcal{C}_k| \le \sum_{v \in \mathcal{E}} \sum_g \sum_{\substack{A \subseteq \mathcal{O}, v \in N(A) \\ A \text{ 2-linked}, |[A]|=k, |N(A)|=g}} |\mathcal{C}_A|.$$

The inner sum of the above is, by applying Lemma 4.5 (and using the isoperimetric property of Q_n that $|N(A)| - |A| = \Omega(|N(A)|/\sqrt{n})$ as discussed in Example 4.6),

$$\begin{aligned}\sum_{\substack{A \subseteq \mathcal{O}, v \in N(A) \\ A \text{ 2-linked}, |[A]|=k, |N(A)|=g}} |\mathcal{C}_A| &\le |\mathcal{G}(k,g,v)| \cdot (q/2)^N \left(\frac{q/2-1}{q/2}\right)^g \\ &\le (q/2)^N \cdot \left(\frac{2(q/2-1)}{q/2}\right)^g \cdot 2^{-\Omega(g/(\sqrt{n}\log n))}.\end{aligned}$$

Here the point is that the above expression is not necessarily smaller than the "benchmark" $(q/2)^N$ if $q > 4$ (the case of $q = 4$ is also easily seen to be problematic once we consider less specialized types of bad colorings). This issue did not exist in enumerating independent sets, since, crucially, there a vertex has only two possible options for its "color" (either occupied or unoccupied). However, as the number of options for the colors of vertices increases, there tends to be less "penalty" from the specified flaws (in the above computation, the penalty from each vertex in $N(A)$ is $(q/2-1)/(q/2) = 1-2/q$). So in such cases, the penalty is not enough to compensate the cost for flaws given by Lemma 4.5.

Kahn and the author [49] handled the above issue in proving Theorem 5.1 by carefully combining the entropy method and the graph container method. A somewhat vague description of their idea is as the following: Example 5.2 shows that specifying flaws and identifying the colorings observing the given flaws in two separate steps is too expensive. On the other hand, just as independent sets in the given graph cluster together, colorings of the given graph also should "cluster". As in this intuition, Kahn and the author regarded the containers in the proof (not in the statement) of Lemma 4.5 as "containers of colorings", and specified colorings directly from the containers. This way, they could skip the expensive step of specifying flaws. But of course, specifying colorings from containers requires much more involved analysis than specifying colorings given that the flaws are completely specified, and the entropy techniques were crucially used for this treatment.

Kahn and the author's approach for Theorem 5.1 was vastly developed by Jenssen and Keevash [38] and applied to much more general weighted graph homomorphisms from the discrete torus $\mathbb{Z}_m^n$, where m is even, to any fixed graphs (and more). Their proof combines various techniques including entropy, graph containers, the cluster expansion method, and isoperimetric and algebraic properties of the torus.

Balogh, Garcia, and Li [4] suggested the problem of enumerating colorings of the middle layers of the hypercube Q_n for odd n. Note that when n is odd, Q_n has two maximum layers – the $\frac{n\pm1}{2}$-th layers. The subgraph of Q_n induced by those two layers forms a d-regular graph where $d = (n+1)/2$. While this graph is still regular and satisfies some expansion properties (but the expansion is weaker than Q_n), its structure is not as "well-organized" as Q_n itself. In particular, some of the key arguments in [38] based on the symmetric structure of Q_n does not immediately extend to the middle layers of Q_n. Li, McKinley, and the author [58] addressed the question of Balogh, Garcia, and Li when the number of colors is even. They improved Kahn and the author's approach so that, as long as the number of colors is even, the combination of the entropy method and the container approach works without taking advantage of the nice symmetries in the structure of the hypercube. They noted that their proof can be adapted for a broader class of graphs beyond the middle layers of Q_n. The case where the number of colors is odd remains open.

5.1 $\mathbb{Z}$-homomorphisms from the hypercube

Probably somewhat surprisingly, the problem of enumerating colorings of the hypercube was motivated by the study of the typical "range" of $\mathbb{Z}$-homomorphisms from Q_n. The solutions of this problem again involve a combination of entropy and graph containers, but here the use of the entropy method is quite different from that for the proof of Theorem 5.1.

We start with some relevant definitions. For a finite, connected, bipartite graph $G = (V, E)$ with distinguished vertex $v_0 \in V$, set

$$\mathrm{Hom}_{v_0}(G) = \{f : V \to \mathbb{Z} : f(v_0) = 0, \{x, y\} \in E \Rightarrow |f(x) - f(y)| = 1\}.$$

That is, $\mathrm{Hom}_{v_0}(G)$ is the collection of $\mathbb{Z}$-homomorphisms f from G with $f(v_0) = 0$. Note that $\mathrm{Hom}_{v_0}(G)$ is a finite set. For $f \in \mathrm{Hom}_{v_0}(G)$, define the *range* of f by

$$R(f) := |\{f(x) : x \in V\}|.$$

For any set X, we write $x \in_r X$ to denote that x is a uniformly randomly chosen element of X.

Benjamini, Häggström, and Mossel [11] investigated the distribution of $R(f)$ for $f \in_r \mathrm{Hom}_{v_0}(G)$ for various host graphs G, with a motivating question of whether concentration inequalities for *typical* Lipschitz functions are stronger than those which hold for *all* Lipschitz functions. We refer the readers to [11] for more details of this interesting direction of research, and here we just focus on the hypercube.

Observe that, when the host graph is Q_n, $R(f)$ can be as large as $n+1$ (we may assume $f(\bar{0}) = 0$, and then assign $f \equiv i$ to the ith layer for $i \in [1, n]$). However, for $f \in_r \mathrm{Hom}_{v_0}(Q_n)$, it should be highly unlikely that f achieves such a large range. Indeed, it was conjectured in [11] that if $f \in_r \mathrm{Hom}_{v_0}(Q_n)$, then $R(f) = o(n)$ w.h.p. This conjecture was proved in a very strong way by Kahn [47] using a beautiful entropy argument.

Theorem 5.3 (Kahn [47]) *There is a constant b for which the following holds. For $f \in_r Hom_{v_0}(Q_n)$,*

$$\mathbb{P}(R(f) > b) < 2^{-\Omega(n)}.$$

In other words, $R(f) = O(1)$ w.h.p. Kahn further conjectured that, in fact, the constant should be 5, and this conjecture was proved by Galvin [26]. To see why 5 should be the answer: note that a "ground state" for $\mathbb{Z}$-homomorphisms from Q_n is to assign 0 to one part (say, $\mathcal{E}$) and let each of the odd vertices choose a value from $\{\pm 1\}$. This gives $R(f) = 3$. However, similarly to the discussion in Section 1.3.1, it is easy to see that planting some non-nearby flaws in $\mathcal{E}$ and assigning them the values ± 2 produces nontrivial contributions to the homomorphism count. In other words, having $R(f) = 5$ is also not unlikely (in fact, the probabiliy for this event is $\Omega(1)$).

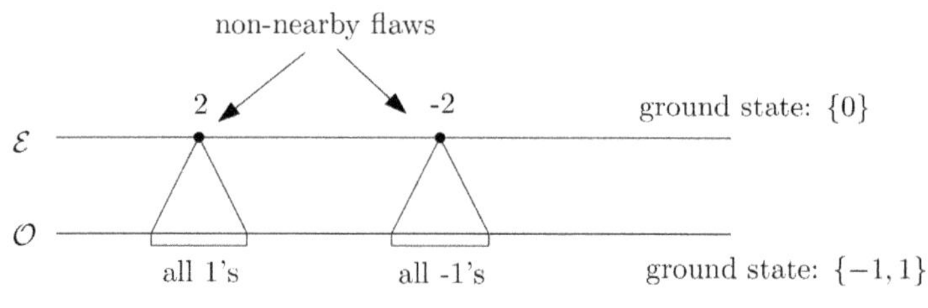

Figure 4: A configuration of f with $R(f) = 5$

On the other hand, having a vertex with the value greater than or equal to ± 3 requires a large set of nearby flaws, which should be highly unlikely.

The starting point of Galvin's proof is Theorem 5.3, and he used the graph container method to find sharp asymptotics for $|\mathrm{Hom}_{v_0}(Q_n)|$ and $|\{f \in \mathrm{Hom}_{v_0}(Q_n) : R(f) \le 5\}|$. By comparing the two asymptotics, the conclusion that the b in Theorem 5.3 should be 5 follows as an immediate corollary. Peled [67] vastly generalized Galvin's result, obtaining results for a general class of tori including $\mathbb{Z}_n^d = (\mathbb{Z}/n\mathbb{Z})^d$ (where n can be large with respect to d) and providing strong bounds on $\mathbb{P}(R(f) \ge k)$ for arbitrary k, for both random $\mathbb{Z}$-homomorphisms and 1-Lipschitz functions.

D. Randall made the interesting observation that there is a bijection from the set $\mathrm{Hom}_{v_0}(Q_n)$ to the set of proper 3-colorings of $V(Q_n)$ with the color of v_0 fixed. The bijection is given by $f \to \chi$ where $\chi(v) = i$ iff $f(v) \equiv i \pmod 3$. As a result, Galvin's enumeration of $\mathbb{Z}$-homomorphisms provides the sharp asymptotics for the number of 3-colorings of Q_n. Later, Engbers and Galvin [21] extended Kahn's entropy approach for Theorem 5.3 to general graph homomorphisms from the discrete torus $\mathbb{Z}_m^n$, where m is even, to any fixed graph. This work was a starting point of Theorem 5.1 and inspired the aforementioned work of Jenssen and Keevash [38].

5.2 Lipschitz functions from expanders

Motivated by the study of the range of random $\mathbb{Z}$-homomorphisms on tree-like graphs ([11]), hypercubes ([47, 26]), and $\mathbb{Z}^d$ for large d ([67]), the typical range of Lipschitz functions on expanders was investigated by Peled, Samotij, and Yehudayoff [69]. We first quickly introduce some relevant definitions.

Recall that, for given $G = (V, E)$, an integer-valued function $f : V(G) \to \mathbb{Z}$ is called an *M-Lipschitz function* from G if $|f(u) - f(v)| \leq M$ for all $\{u, v\} \in E(G)$. For a vertex $v_0 \in V(G)$, let

$$\mathrm{Lip}_{v_0}(G; M) = \{f : V(G) \to \mathbb{Z} \mid f(v_0) = 0 \text{ and } |f(u) - f(v)| \leq M \ \forall \{u, v\} \in E(G)\}.$$

Define the *range* of $f \in \mathrm{Lip}_{v_0}(G; M)$ (*cf.* Section 5.1) by

$$R(f) := \max_{v \in V(G)} f(v) - \min_{v \in V(G)} f(v) + 1.$$

The main focus of this section is the typical range of Lipschitz functions defined on expander graphs. Following [69], say a d-regular N-vertex graph G is a *λ-expander* if it satisfies

$$\left| e(S, T) - \frac{d}{N}|S||T| \right| \leq \lambda \sqrt{|S||T|} \text{ for all } S, T \subseteq V(G).$$

(We always have $(1 - d/n)\sqrt{d} \leq \lambda \leq d$ (see, e.g., [69, Section 1.1]), and the discussion in this section focuses on $\lambda \leq d/5$.) We remark that the above definition of λ-expander (which is motivated by the expander mixing lemma [1]) guarantees some nice expansion properties for G as its name indicates: for example, for any $v \in V(G)$ and $t \in \mathbb{Z}^+$, the volume of the ball in G of radius t centered at v (i.e., the set of vertices w with $\mathrm{dist}_G(v, w) \leq t$) is at least $\min\left\{\frac{N}{2}, \left(\frac{d}{2\lambda}\right)^{2t}\right\}$. So roughly it says that as long as t is "not too large", the volume of the ball grows exponentially. This volume growth poses an upper bound on the range of M-Lipschitz functions from G: it can be derived from the volume growth that any λ-expander G has diameter $\mathrm{diam}(G) \leq \log\left(\frac{d}{2\lambda}\right)^{-1} \log N$, thus trivially, any $f \in \mathrm{Lip}_{v_0}(G; M)$ must satisfy

$$R(f) \leq M \cdot \log\left(\frac{d}{2\lambda}\right)^{-1} \log N + 1. \tag{5.3}$$

On the other hand, it was shown in [69] that if G is a "sufficiently good" expander (roughly, the authors of [69] require

$$|N(A)| \geq \Omega(M \log(dM)|A|) \tag{5.4}$$

for each "not-too-large" set $A \subseteq V(G)$) then for $f \in_r \text{Lip}_{v_0}(G;M)$, w.h.p. $R(f)$ is much smaller than the trivial bound in (5.3); more precisely, they showed that the $\log N$ in (5.3) can be replaced by $\log\log N$.

Consideration of M-Lipschitz functions for $M \to \infty$ is a natural bridge to $\mathbb{R}$-*valued Lipschitz functions*, functions $f : V(G) \to \mathbb{R}$ satisfying $|f(u) - f(v)| \leq 1$ for all $\{u,v\} \in E(G)$. It is not hard to see that (e.g. [68, page 8]) if $f \in_r \text{Lip}_{v_0}(G;M)$, then as $M \to \infty$, f/M converges in distribution to a uniformly random $\mathbb{R}$-valued Lipschitz function on G with $f(v_0) = 0$. With this motivation, it is natural to ask whether one can remove the dependence on M in the expansion requirement (5.4) and still obtain a similar conclusion as in [69]. Indeed, it was shown by Krueger, Li, and the author [57] that requiring an $\Omega(1)$-expansion (i.e., $|N(A)| \geq \Omega(|A|)$) is enough as long as M satisfies the condition in the statement below:

Theorem 5.4 (Krueger-Li-P. [57]) *There exist universal constants $c, C, c', C' > 0$ such that the following holds. Let $M \in \mathbb{Z}^+$, and let G be a connected N-vertex, d-regular λ-expander with $\lambda \leq d/5$, where $M \leq \min\{c\frac{d^{3/2}}{\lambda \log d}, (\log N)^C\}$. Let f be a uniformly randomly chosen M-Lipschitz function on G. Then*

$$\mathbb{P}\left(R(f) \geq C'M\frac{\log\log N}{\log(d/\lambda)} + 2(M+1)\right) \leq N^{-c'}. \tag{5.5}$$

(to read off the $\Omega(1)$-expansion from the theorem above, we note that the expansion for λ-expanders is $\Omega(d/\lambda)$).

It was noted in [69] that the lower bound on $R(f)$ in (5.5) should be tight, and one may apply the techniques of Benjamini, Yadin, and Yehudayoff [12] to prove it. This idea was implemented (with some additional ideas for the case $M \gg 1$) by Işık and the author [35].

The proof of Theorem 5.4 is motivated by the idea of Li, McKinley, and the author [58] in the study of random colorings of the middle layers of the hypercube. The restriction on M in Theorem 5.4 was due to the limit of the techniques used in [57]. As mentioned above, the study of M-Lipschitz functions for $M \to \infty$ provides a transition to $\mathbb{R}$-valued Lipschitz functions, and a natural question is whether $\mathbb{R}$-valued Lipschitz functions still exhibit a small range (much smaller than the diameter of the host graph) on expander graphs (and in what quantitative sense). This question is still wide open, and a possible first step towards this direction seems to be to understand the behavior of M-Lipschitz functions for $M \gg d$.

6 Antichains in $[t]^n$

This section discusses a first successful application of the graph container lemma for expanders in the "irregular" setting. Recall that Dedekind's problem concerns the number of antichains in the Boolean lattice. A natural generalization of the Boolean lattice is the poset $([t]^n, \preceq)$ (we will simply write $[t]^n$) – the ground set consists of all n-tuples $(x_1, \ldots, x_n)$ of integers in $\{0, 1, \ldots, t-1\}$, and the partial order $\preceq$ is defined to be $x \preceq y \Leftrightarrow x_i \leq y_i$ for all $i \in [n]$. We write $\alpha([t]^n)$ for the number of antichains in the poset $[t]^n$.

6.1 Connection to Ramsey-type problems and logarithmic asymptotics

Enumerating antichains in $[t]^n$ is a natural generalization of Dedekind's problem, but it also turns out to have an interesting connection to a Ramsey type question. This connection is not directly related to the focus of the present article so we will try to be brief and refer the readers to [24, 25, 62] for details and omitted definitions. Fox, Pach, Sudakov, and Suk [25] introduced the number $M_k(t,n)$ as the smallest integer M with the property that for any n-colorings of $\binom{[M]}{k}$ one can find a monochromatic "monotone path"[2] of length t. The celebrated results of Erdős and Szekeres [23] can be expressed using $M_2(t,n)$ and $M_3(t,2)$, and Fox et al suggested the problem of estimating $M_k(t,n)$ for general k,t and n. Moshkovitz and Shapira [62] and Milans, Stolee, and West [60] independently observed that

$$\alpha([t]^n) = M_3(t,n) - 1 \quad \text{for all } t, n \geq 2,$$

and using this relationship, Moshkovitz and Shapira obtained the following bounds on $M_3(t,n)$:

Theorem 6.1 (Moshkovitz, Shapira [62]) *For every $t,n \geq 2$,*

$$2^{\frac{2}{3}t^{n-1}/\sqrt{n}} \leq M_3(t,n) \leq 2^{2t^{n-1}}.$$

Moshkovitz and Shapira boldly suggested that, as $n \to \infty$,

$$M_3(t,n)\ (= \alpha([t]^n) + 1)\ = 2^{(1+o_n(1))N(t,n)} \tag{6.1}$$

where $N(t,n)$ is the size of a largest layer of $[t]^n$. Notice the resemblance of the formulation in (6.1) to Theorem 1.1. After partial progress towards (6.1) with some restrictions on the ranges of t and n (e.g., [14, 63, 81, 72, 66]), Falgas-Ravry, Räty and Tomon [24] finally confirmed (6.1) in full, using the container method based on a cleverly constructed normalized matching flow:

Theorem 6.2 (Falgas-Ravry, Räty and Tomon [24]) *There exists a constant $c > 0$ such that for every $n, t \geq 2$, we have*

$$\log_2 \alpha([t]^n) \leq \left(1 + c \cdot \frac{(\log n)^3}{n}\right) \cdot N(t,n). \tag{6.2}$$

6.2 Towards sharp asymptotics

Concerning the sharpness of the error term in (6.2), Falgas-Ravry et al [24] showed that

$$\log_2 \alpha([t]^n) \geq \left(1 + 2^{-\Theta(n)}\right) \cdot N(t,n), \tag{6.3}$$

and said that "it might be reasonable to conjecture that this lower bound is closer to optimal than our upper bound (in Theorem 6.2)". For $t = 2$, an upper bound matching the right-hand side of (6.3) follows from Theorem 1.2, so one might expect that the proof techniques for Dedekind's problem might extend to $\alpha([t]^n)$, at least

[2] Formally, for any sequence of positive integers $x_1 < x_2 < \ldots < x_{l+k-1}$, we say that the k-tuples $(x_i, x_{i+1}, \ldots, x_{i+k-1})$ $(i = 1, 2, \ldots, l)$ form a *monotone path* of length l.

for $t = O(1)$. While Lemma 4.5 works well for Dedekind's problem, somewhat surprisingly, this lemma is not immediately usable for counting antichains in $[t]^n$ for $t \geq 3$. In fact, Sapozhenko [77] himself asked what the number of antichains in $[3]^n$ is. The question of estimating $\alpha([3]^n)$ was also asked by Noel, Scott and Sudakov [63], who proved asymptotics for $\log_2 \alpha([3]^n)$.

We briefly explain why Sapozhenko's approach for Dedekind's problem does not immediately extend to $\alpha([3]^n)$. Recall that, in the graph setting, Sapozhenko's argument crucially assumes the degree condition in (4.3):

$$\max_{v\in Y} \deg(v) \leq \min_{v\in X} \deg(v).$$

Note that the graphs induced by any two consecutive layers of the Boolean lattice $\mathcal{B}_n$ satisfy the above degree condition (by choosing the parts X and Y appropriately), and this was a crucial ingredient for Sapozhenko's (asymptotic) resolution of Dedekind's problem. On the other hand, the graphs induced by two consecutive layers of $[t]^n$ do not necessarily satisfy the degree condition in (4.3) when $t \geq 3$.

Sapozhenko [77] mentioned that the irregular structure of $[3]^n$ is the main obstacle for the application of his method. Recently, Jenssen, Sarantis, and the author [41] found a way to handle this obstacle and obtained the following sharp estimate for $\alpha([3]^n)$. Their approach is still based on the graph container method, but it is the first to extend Lemma 4.5 beyond the degree requirement in (4.3).

Theorem 6.3 (Jenssen, P., Sarantis [41]) *Let $N(3,n)$ be the size of the middle layer of $[3]^n$. Then,*

$$\alpha([3]^n) = 2^{N(3,n)} \exp\left[(1+o(1))\sqrt{\frac{1+2\sqrt{2}}{2\sqrt{2\pi}}}n^{-1/2}\left(\frac{1+2\sqrt{2}}{2}\right)^n\right].$$

As we have discussed several times in the previous sections, the main result of [41] is structural, and the quantitative bound in Theorem 6.3 follows as a corollary from the structural result. The statement about the structure of almost all antichains in $[3]^n$ is very similar to Theorem 1.4, and the estimate in Theorem 6.3 follows by analyzing the contributions of antichains with small flaws.

We note that, for $t \geq 4$, the problem of obtaining a similarly sharp estimate for the number of antichains in $[t]^n$ is still wide open, and we do not even know whether the lower bound in (6.3) is optimal. It might be natural to suspect that for bounded t, a structural result similar to Theorem 1.4 still holds, which would give a much stronger answer to the question of Falgas-Ravry et al following (6.3).

7 Maximal independent sets

Sometimes applications of the graph container method lead to an interesting development of isoperimetric inequalities on the host graphs. One nice example is the problem of counting maximal independent sets in the hypercube.

An independent set $I \in \mathcal{I}(G)$ is called *maximal* if for any $v \in V(G)$, $I \cup \{v\}$ is no longer independent. Write $\mathrm{MIS}(G)$ for the family of maximal independent sets in G, and $\mathrm{mis}(G) := |\mathrm{MIS}(G)|$. In general, finding $\mathrm{mis}(G)$ is #P-hard, but its extremal properties are well-known: for example, Moon and Moser [61] showed that $\text{mis}(G) \leq 3^{N/3}$ for any N-vertex graph G, and the unique extremal graph is a vertex disjoint union of $N/3$ triangles if 3 divides N. Hujter and Tuza [33] showed that if G is triangle-free, then the upper bound is reduced to $\text{mis}(G) \leq 2^{N/2}$, and the unique extremal graph is a matching with $N/2$ edges if 2 divides N.

The problem of enumerating maximal independent sets in Q_n was asked by Duffus, Frankl, and Rödl [20]. Observe that (recalling that $N = |V(Q_n)| = 2^n$)

$$\text{mis}(Q_n) \leq 2^{N/2};$$

in fact, for any bipartite graph G with a bipartition $X \cup Y$, $\text{mis}(G) \leq \max\{2^{|X|}, 2^{|Y|}\}$, since if $I \in \text{MIS}(G)$, then $I \cap Y = Y \setminus N(I \cap X)$ (so $I \cap X$ determines $I \cap Y$). On the other hand, by considering a certain construction of maximal independent sets (*cf.* Section 1.3.1), one can obtain the following asymptotic lower bound.

Observation 7.1

$$\text{mis}(Q_n) > (1 - o(1))2n2^{N/4}.$$

We briefly sketch a justification of the above observation. (For a full justification, see [50, Section 1.1].) An *induced matching* (IM) in a graph is an induced subgraph that is a matching, and a maximum IM in Q_n is constructed as the following. Pick an $i \in [n]$, and consider $Q_0 := \{v \in V(Q_n) : v_i = 0\}$ and $Q_1 = \{v \in V(Q_n) : v_i = 1\}$ where v_i denotes the i-th coordinate of v.

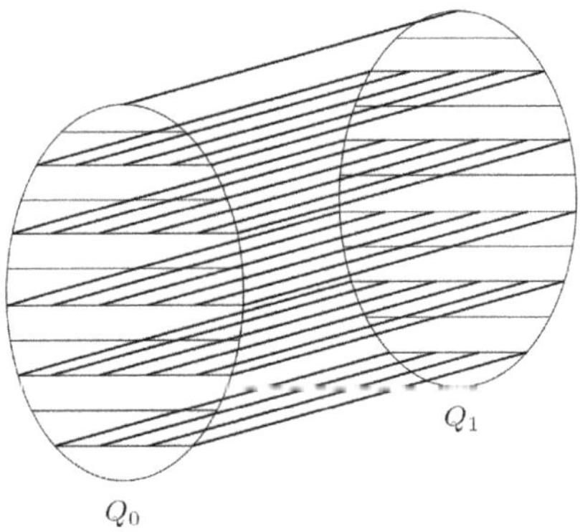

Figure 5: A maximum induced matching between Q_0 and Q_1

Note that $\nabla(Q_0, Q_1)$ is a (perfect) matching. Now, pick the edges of $\nabla(Q_0, Q_1)$ with its end in $V(Q_0)$ having an odd Hamming weight, then these edges form an IM of Q_n with $N/4$ edges. It is easy to see that they are maximum IM's in Q_n, and there are $2n$ of such IM's (n choices for i and 2 choices for the parity of the ends in $V(Q_0)$). We leave it to the readers that those are the only maximum IM's.

Denote by M^* the maximum IM's constructed above. Each M^*, which plays the role of a "ground state" in the MIS count, gives rise to exactly $2^{N/4}$ MIS's, gotten by choosing one vertex from each edge of M^* and extending the resulting independent set to the (unique) MIS containing it. It is also easy to see that the number of MIS's that belong to more than one ground state are negligible. Since there are $2n$ ground states, the asymptotic lower bound in Observation 7.1 follows.

It was conjectured by Ilinca and Kahn [34], who proved that $\log \text{mis}(Q_n) \sim N/4$ as conjectured by Duffus, Frankl, and Rödl [20], that $\text{mis}(Q_n)$ is asymptotically equal to the lower bound in Observation 7.1. This conjecture, which again is from the framework in Section 1.1, was proved by Kahn and the author [50].

Theorem 7.2 (Kahn, P. [50])

$$\text{mis}(Q_n) \sim 2n2^{N/4}.$$

Along with the graph container method, a crucial ingredient for the above result is the isoperimetric inequality in Theorem 7.3 below. The inequality was used to show that (we skip the derivation and refer to [50] for details) almost all MIS's have a large intersection with some maximum induced matching M^*. This step establishes the fact that each M^* indeed forms a ground state, and so it suffices to analyze MIS's with small flaws in each ground state.

In the following statement, μ is the uniform measure on $V = V(Q_n)$. For $x \in V$ and $T \subseteq V$, let $d_T(x)$ be the number of neighbors of x in T and define $h_S : V \to \mathbb{N}$ by

$$h_S(x) = \begin{cases} d_{V\setminus S}(x) & \text{if } x \in S, \\ 0 & \text{if } x \notin S. \end{cases}$$

We define $\beta = \log_2(3/2)(\approx .585)$ and $\int f d\mu = \sum_{x\in V} f(x)\mu(x)$.

Theorem 7.3 *For any $A \subseteq V$,*

$$\int h_A^\beta d\mu \geq 2\mu(A)(1-\mu(A)).$$

We refer to [48] for detailed discussion on the consequences of the above theorem, and here we just state a simple implication to hint at the flavor of Theorem 7.3. Define $\nabla A := \{\{a,b\} \in E(G) : a \in A, b \notin A\}$.

Corollary 7.4 *For $k = 1, 2$, if $|A| = 2^{n-k}$ and $|\nabla A| < (1+\epsilon)|A|\log(2^n/|A|)$, then there is a subcube C with $\mu(C\Delta A) = O(\epsilon)$ (where $C\Delta A$ denotes the symmetric difference between C and A).*

Another question considered in [20] was enumerating maximal independent sets in the graph induced by the middle layers of Q_n for odd n, which we denote by $\mathcal{M}_n$. Again, there is an easy construction that shows that $\log \text{mis}(\mathcal{M}_n) \geq \binom{n-1}{n/2}$, and Ilinca and Kahn showed that this lower bound is asymptotically tight for $\log \text{mis}(\mathcal{M}_n)$, proving a conjecture of Duffus, Frankl, and Rödl [20]. Recently, Balogh, Chen and Garcia [3] obtained sharp asymptotics for $\text{mis}(\mathcal{M}_n)$. The key ingredients of their proof again include the graph container method and isoperimetric inequalities for $\mathcal{M}_n$.

Acknowledgements

The author is grateful to the referee for their careful reading of the manuscript and their constructive comments. We also thank Jozsef Balogh and Matthew Jenssen for their helpful feedback on the draft. The author was supported by NSF Grant DMS-2324978, NSF CAREER Grant DMS-2443706, and a Sloan Fellowship.

References

[1] N. Alon and F. R. K. Chung, Explicit construction of linear sized tolerant networks, *Discrete Mathematics* **72** (1988), 15–19.

[2] N. Alon and J. Spencer, *The probabilistic method*, John Wiley & Sons, 2016.

[3] J. Balogh, C. Chen, and R. I. Garcia, Maximal independent sets in the middle layers of the Boolean lattice, preprint, arXiv:2505.00132, 2025.

[4] J. Balogh, R. I. Garcia, and L. Li, Independent sets in the middle two layers of Boolean lattice, *Journal of Combinatorial Theory, Series A* **178** (2021), 105341.

[5] J. Balogh, R. I. Garcia, L. Li, and A. Zs. Wagner, Intersecting families of sets are typically trivial, *Journal of Combinatorial Theory, Series B* **164** (2024), 44–67.

[6] J. Balogh and R. A. Krueger, A sharp threshold for a random version of Sperner's theorem, *Random Structures & Algorithms* **62(1)** (2023), 3–28.

[7] J. Balogh, R. Morris, and W. Samotij, Independent sets in hypergraphs, *Journal of the American Mathematical Society* **28(3)** (2015), 669–709.

[8] J. Balogh, R. Morris, and W. Samotij, The method of hypergraph containers, in *Proceedings of the International Congress of Mathematicians: Rio de Janeiro 2018* (2018), pp. 3059–3092.

[9] J. Balogh, R. Mycroft, and A. Treglown, A random version of Sperner's theorem, *Journal of Combinatorial Theory, Series A* **128** (2014), 104–110.

[10] J. Balogh, A. Treglown, and A. Zs. Wagner, Applications of graph containers in the Boolean lattice, *Random Structures & Algorithms* **49(4)** (2016), 845–872.

[11] I. Benjamini, O. Häggström, and E. Mossel, On random graph homomorphisms into $\mathbb{Z}$, *Journal of Combinatorial Theory, Series B* **78(1)** (2000), 86–114.

[12] I. Benjamini, A. Yadin, and A. Yehudayoff, Random graph-homomorphisms and logarithmic degree, *Electronic Journal of Probability* **12** (2007), 926–950.

[13] G. Brightwell and P. Winkler, Graph homommorphisms and phase transitions, *Journal of Combinatorial Theory, Series B* **77** (1999), 221–262.

[14] T. Carroll, J. N. Cooper, and P. Tetali, Counting antichains and linear extensions in generalizations of the Boolean lattice, unpublished preprint, 2012.

[15] F. R. K. Chung, P. Frankl, R. Graham, and J. B. Shearer, Some intersection theorems for ordered sets and graphs, *Journal of Combinatorial Theory, Series A* **48** (1986), 23–37.

[16] M. Collares, J. Erde, A. Geisler, and M. Kang, Counting independent sets in expanding bipartite regular graphs, preprint, arXiv:2503.22255, 2025.

[17] M. Collares and R. Morris, Maximum-size antichains in random set-systems, *Random Structures & Algorithms* **49** (2016), 308–321.

[18] S. Das and T. Tran, Removal and stability for Erdős–Ko–Rado, *SIAM Journal on Discrete Mathematics* **30** (2016), 1102–1114.

[19] R. Dedekind, Ueber zerlegungen von zahlen durch ihre grössten gemeinsamen teiler, *Festschrift Hoch. Braunschweig u. ges. Werke* **II** (1897), 103–148.

[20] D. Duffus, P. Frankl, and V. Rödl, Maximal independent sets in the covering graph of the cube, *Discrete Applied Mathematics* **161(9)** (2013), 1203–1208.

[21] J. Engbers and D. Galvin, H-coloring tori, *Journal of Combinatorial Theory, Series B* **102(5)** (2012), 1110–1133.

[22] P. Erdős, C. Ko, and R. Rado, Intersection theorems for systems of finite sets, *The Quarterly Journal of Mathematics, Oxford Second Series* **12** (1961), 313–320.

[23] P. Erdős and G. Szekeres, A combinatorial problem in geometry, *Compositio Math.* **2** (1935), 463–470.

[24] V. Falgas-Ravry, E. Räty, I. Tomon, Dedekind's problem in the hypergrid, preprint, arXiv:2310.12946, 2023.

[25] J. Fox, J. Pach, B. Sudakov, and A. Suk, Erdős–Szekeres-type theorems for monotone paths and convex bodies, *Proceedings of the London Mathematical Society* **105(5)** (2012), 95300982.

[26] D. Galvin, On homomorphisms from the Hamming cube to $\mathbb{Z}$, *Israel Journal of Mathematics* **138** (2003), 189–213.

[27] D. Galvin, A threshold phenomenon for random independent sets in the discrete hypercube, *Combinatorics, Probability and Computing* **20(1)** (2011), 27–51.

[28] D. Galvin, Independent sets in the discrete hypercube, preprint, arXiv: 1901.01991, 2019.

[29] D. Galvin and P. Tetali, On weighted graph homomorphisms, *DIMACS Series in Discrete Mathematics and Theoretical Computer Science* **63** (2004), 97–104.

[30] D. Galvin and P. Tetali, Slow mixing of Glauber dynamics for the hard-core model on regular bipartite graphs, *Random Structures & Algorithms* **28(4)** (2006), 427–443.

[31] A. Geisler, M. Kang, M. Sarantis, and R. Wdowinski, Counting independent sets in percolated graphs via the Ising model, preprint, arXiv:2504.08715, 2025.

[32] A. Hamm and J. Kahn, On Erdős–Ko–Rado for random hypergraphs II, *Combinatorics, Probability and Computing* **28** (2019), 61–80.

[33] M. Hujter and Z. Tuza, The number of maximal independent sets in triangle-free graphs, *SIAM Journal on Discrete Mathematics* **6** (1993), 284–288.

[34] L. Ilinca and J. Kahn, Counting maximal antichains and independent sets, *Order* **30** (2013), 427–435.

[35] S. Işık and J. Park, Random Lipschitz functions on graphs with weak expansion, preprint, arXiv:2411.09640, 2024.

[36] C. Jäkel, A computation of the ninth Dedekind Number, preprint, arXiv: 2304.00895, 2023.

[37] M. Jenssen, The cluster expansion in combinatorics, *Surveys in combinatorics* **493** (2024), 55–88.

[38] M. Jenssen and P. Keevash, Homomorphisms from the torus, *Advances in Mathematics* **430** (2023), 109212.

[39] M. Jenssen, A. Malekshahian, and J. Park, A refined graph container lemma and applications to the hard-core model on bipartite expanders, preprint, arXiv:2411.03393, 2024.

[40] M. Jenssen, A. Malekshahian, and J. Park, On Dedekind's problem, a sparse version of Sperner's theorem, and antichains of a given size in the Boolean lattice, preprint, arXiv:2411.03400, 2024.

[41] M. Jenssen, J. Park, and M. Sarantis, On the number of antichains in $\{0, 1, 2\}^n$, in preparation.

[42] M. Jenssen and W. Perkins, Independent sets in the hypercube revisited, *Journal of the London Mathematical Society* **102** (2020), 645–669.

[43] M. Jenssen, W. Perkins, and A. Potukuchi, Approximately counting independent sets in bipartite graphs via graph containers, *Random Structures & Algorithms* **63(1)** (2023), 215–241.

[44] M. Jenssen, W. Perkins, and A. Potukuchi, Independent sets of a given size and structure in the hypercube, *Combinatorics, Probability and Computing* **31(4)** (2022), 215–241.

[45] J. Kahn, An entropy approach to the hard-core model on bipartite graphs, *Combinatorics, Probability and Computing* **10(3)** (2001), 219–237.

[46] J. Kahn, Entropy, independent sets and antichains: a new approach to Dedekind's problem, *Proceedings of the American Mathematical Society* **130** (2001), 371–378.

[47] J. Kahn, Range of cube-indexed random walk, *Israel Journal of Mathematics* **124** (2001), 189-201.

[48] J. Kahn and J. Park, An isoperimetric inequality for the Hamming cube and some consequences, *Proceedings of the American Mathematical Society* **149** (2020), 4213–4224.

[49] J. Kahn and J. Park, The number of 4-colorings of the Hamming cube, *Israel Journal of Mathematics* **236** (2020), 629–649.

[50] J. Kahn and J. Park, The number of maximal independent sets in the Hamming cube, *Combinatorica* **42(6)** (2022), 853–880.

[51] D. Kleitman, On Dedekind's problem: The number of monotone Boolean functions, *Proceedings of the American Mathematical Society* **21** (1969), 677–682.

[52] D. Kleitman and G. Markowsky, On Dedekind's problem: the number of isotone Boolean functions. II, *Transactions of the American Mathematical Society* **213** (1975), 373–390.

[53] D. Kleitman and K. Winston, On the number of graphs without 4-cycles, *Discrete Mathematics* **41(2)** (1982), 167–172.

[54] A. D. Korshunov, The number of monotone Boolean functions, *Problemy Kibernetiki* **31** (1981), 5–108.

[55] A. D. Korshunov and A. A. Sapozhenko, The number of binary codes with distance 2, *Problemy Kibernet* **40** (1983), 111–130.

[56] G. Kronenberg and Y. Spinka, Independent sets in random subgraphs of the hypercube, preprint, arXiv:2201.06127, 2022.

[57] R. A. Krueger, L. Li and J. Park, Lipschitz functions on weak expanders, preprint, arXiv:2408.14702, 2024.

[58] L. Li, G. McKinley, and J. Park, The number of colorings of the middle layers of the Hamming cube, *Combinatorica* **45** (2025), 1–47.

[59] R. J. McEliece, *The Theory of Information and Coding*, Addison-Wesley, London (1977).

[60] K. Milans, D. Stolee, and D. West, Ordered Ramsey theory and track representations of graphs, *Journal of Combinatorics* **6** (2015), 445–456.

[61] J. W. Moon and L. Moser, On cliques in graphs, *Israel Journal of Mathematics* **3** (1965), 23–28.

[62] G. Moshkovitz and A. Shapira, Ramsey theory, integer partitions and a new proof of the Erdős–Szekeres theorem, *Advances in Mathematics* **262** (2012), 1107–1129.

[63] J. Noel, A. Scott, and B. Sudakov, Supersaturation in posets and applications involving the container method, *Journal of Combinatorial Theory, Series A* **154** (2018), 247–284.

[64] D. Osthus, Maximum antichains in random subsets of a finite set, *Journal of Combinatorial Theory, Series A* **90** (2000), 336–346.

[65] J. Park, Note on the number of balanced independent sets in the Hamming cube, *The Electronic Journal of Combinatorics* **29** (2022), 2–34.

[66] J. Park, M. Sarantis, and P. Tetali, Note on the number of antichains in generalizations of the Boolean lattice, *Combinatorial Theory* (to appear).

[67] R. Peled, High-dimensional Lipschitz functions are typically flat, *The Annals of Probability* **45(3)** (2017), 1351–1447.

[68] R. Peled, W. Samotij, and A. Yehudayoff, Grounded Lipschitz functions on trees are typically flat, *Electronic Communications in Probability* **18** (2013), 1–9.

[69] R. Peled, W. Samotij, and A. Yehudayoff, Lipschitz functions on expanders are typically flat, *Combinatorics, Probability and Computing* **22(4)** (2013), 566–591.

[70] R. Peled and Y. Spinka, Long-range order in discrete spin systems, preprint, arXiv:2010.03177, 2024.

[71] N. Pippenger, Entropy and enumeration of Boolean functions, *IEEE Transactions on Information Theory* **45** (1999), 2096–2100.

[72] C. Pohoata and D. Zakharov, On the number of high-dimensional partitions, *Proceedings of the London Mathematical Society* **128(2)** (2024), e12586.

[73] A. Potukuchi and L. Yepremyan, Enumerating independent sets in Abelian Cayley graphs, preprint, arXiv:2109.06152, 2021.

[74] A. A. Sapozhenko, On the number of antichains in multilevelled ranked posets, *Discrete Mathematics and Applications* **1(2)** (1991), 149–169.

[75] A. A. Sapozhenko, On the number of connected subsets with given cardinality of the boundary in bipartite graphs, *Metody Diskretnogo Analiza* **45** (1987), 42–70.

[76] A. A. Sapozhenko, On the number of independent sets in expanders, *Discrete Mathematics* **13(1)** (2001), 56–62.

[77] A. A. Sapozhenko, Systems of containers and enumeration problems, in *Stochastic Algorithms: Foundations and Applications* (ed. O. B. Lupanov, O. M. Kasim-Zade, A. V. Chaskin, and K. Steinhöfel), Springer Berlin Heidelberg, Berlin, Heidelberg (2005), pp. 1–13.

[78] D. Saxton and A. Thomason, Hypergraph containers, *Inventiones mathematicae* **201(3)** (2015), 925–992.

[79] C. Shannon, A mathematical theory of communication, *Bell System Technical Journal* **27(3)** (1948), 379–423.

[80] E. Sperner, Ein Satz über Untermengen einer endlichen Menge, *Mathematische Zeitschrift* **27(1)** (1928), 544–548.

[81] S. F. Tsai, A simple upper bound on the number of antichains in $[t]^n$, *Order* **36** (2019), 507–510.

[82] L. Van Hirtum, P. De Causmaecker, J. Goemaere, T. Kenter, H. Riebler, M. Lass, and C. Plessl, A computation of $D(9)$ using FPGA Supercomputing, preprint, arXiv:2304.03039, 2023.

[83] D. Weitz, Counting independent sets up to the tree threshold, *Symposium on the Theory of Computing* (2006), 140–149.

[84] D. Wiedemann, A computation of the eighth Dedekind number, *Order* **8** (1991), 5–6.

Department of Mathematics
Courant Institute of Mathematical Sciences, NYU
251 Mercer Street, New York, NY 10012
USA
jinyoungpark@nyu.edu

Graph Bootstrap Percolation – A Discovery of Slowness

David Fabian, Patrick Morris, Tibor Szabó

Abstract

Graph bootstrap percolation is a discrete-time process capturing the spread of a virus on the edges of K_n. Given an initial set $G \subseteq K_n$ of infected edges, the transmission of the virus is governed by a fixed graph H: in each round of the process, any edge e of K_n that is the last uninfected edge in a copy of H in K_n gets infected as well. Once infected, edges remain infected forever. The process was introduced by Bollobás in 1968 in the context of weak saturation and has since inspired a vast array of beautiful mathematics. The main focus of this survey is the extremal question of *how long* the infection process can last before stabilising. We give an exposition of our recent systematic study of this *maximum running time* and the influence of the infection rule H. The topic turns out to possess a wide variety of interesting behaviour, with connections to additive, extremal, and probabilistic combinatorics. Along the way, we encounter a number of surprises and attractive open problems.

1 Introduction

The central notion of this survey is *graph bootstrap percolation*, a term coined by Balogh, Bollobás, and Morris [13], although the process itself appeared much earlier, in a different context, in work of Bollobás [24]. Given graphs H and G, let $n_H(G)$ denote the number of copies of H in G. The *H-bootstrap percolation process* (*H-process* for short) on a graph G is the sequence $(G_i)_{i\geq 0}$ of graphs defined by $G_0 := G$ and for $i \geq 1$, $V(G_i) := V(G)$ and

$$E(G_i) := E(G_{i-1}) \cup \left\{ e \in \binom{V(G)}{2} \ : \ n_H(G_{i-1}+e) > n_H(G_{i-1}) \right\}.$$

Informally, one can imagine a virus spreading in discrete time steps on the edges of the complete graph on $V(G)$. Initially just the edges of G are infected. The graph H governs the spread of the virus: any edge e of K_n which is the last uninfected edge in a copy of H in K_n gets infected as well. Once infected, edges remain infected forever. We call G the *starting graph* and H the *infection rule* of the process.

Note that for finite G, the H-process on G stabilises and we refer to $\tau_H(G) := \min\{t \in \mathbf{N} : G_t = G_{t+1}\}$, i.e., the number of rounds it takes for that to happen, as the *running time* of the H-process on G. We define G_τ with $\tau = \tau_H(G)$ to be the *final graph* of the process, and denote it $\langle G \rangle_H := G_\tau$. If eventually all edges get infected, that is, the final graph $\langle G \rangle_H$ is the clique $K_{v(G)}$, we say that the starting graph G is *H-percolating*.

Example 1.1 (The triangle) Let $H = K_3$ and G be an arbitrary n-vertex graph. Note first that no two components of G become connected during the K_3-process as two non-adjacent vertices become adjacent during a round of the K_3-process if and only if they have a common neighbour already. Now if x and y are in the same component in G_{i-1} then in the next round their distance along a shortest path is reduced to roughly half: $\text{dist}_{G_i}(x,y) = \lceil \frac{1}{2} \text{dist}_{G_{i-1}}(x,y) \rceil$. In particular, after $\lceil \log_2(\text{dist}_G(x,y)) \rceil$ rounds their distance is one and they are adjacent. Thus we have that $\tau_H(G) = \max_{x,y} \lceil \log_2(\text{dist}_G(x,y)) \rceil$ with the maximum taken over pairs of vertices x, y in the same connected component of G, and $\langle G \rangle_H$ is a union of disjoint cliques on the vertex sets of the connected components.

1.1 Weak saturation

The graph percolation process was introduced by Bollobás [24] in 1968 for cliques, under the name *weak saturation*, and has since inspired a vast array of beautiful mathematics. In our terminology, Bollobás was interested in the smallest possible number of infected edges that percolate in the K_k-process. Subsequently this extremal question turned out to be greatly influential, as it inspired one of the first applications of algebra in extremal combinatorics. For an arbitrary graph H, $\text{wsat}(n,H)$ denotes the minimum number of edges in an H-percolating graph on n vertices. The concept is motivated by work of Erdős, Hajnal and Moon [41], who determined the smallest number of edges on n vertices which K_k-percolate in a *single round*. Graphs that are H-free and H-percolate in one time step are called *H-saturated*. The minimum number of edges in an H-saturated graph is denoted by $\text{sat}(n,H)$, whilst the maximum number is the *Turán number* $\text{ex}(n,H)$ of H. Clearly $\text{wsat}(n,H) \le \text{sat}(n,H)$. Considering Example 1.1 we see that $\text{wsat}(n,K_3) = \text{sat}(n,K_3) = n-1$ as any tree on n vertices percolates with a star doing so in one round, whilst any graph with less than $n-1$ edges will not percolate.

1.2 Percolation thresholds

Whilst weak saturation looks at the *smallest* possible starting graphs that H-percolate, the statistical physics origins of bootstrap percolation motivates the investigation of *random* initial infection graphs. Balogh, Bollobás, and Morris [13] initiated the study of the threshold probability for when the random graph $G(n,p)$ becomes H-percolating. Here $G(n,p)$ denotes the Erdős-Rényi binomial random graph which is obtained by taking every edge on n vertices independently with probability $p = p(n)$. We say an event holds asymptotically almost surely (a.a.s. for short) in $G(n,p)$ if the probability it holds tends to 1 as n tends to infinity. As the property of being H-percolating is monotone increasing, the classical result of Bollobás and Thomason [29] gives that there is some threshold $p_c = p_c(n,H)$, such that if $p = \omega(p_c)$ then a.a.s. $G(n,p)$ is H-percolating whilst if $p = o(p_c)$ then a.a.s. $G(n,p)$ is not H-percolating. Considering Example 1.1, we see that $p_c(n,K_3)$ is precisely the threshold for connectivity in $G(n,p)$, namely $\log n/n$ [42].

1.3 Maximum running times

Another relevant question concerning percolation processes is *how long* the virus could be spreading for. The study of the corresponding extremal parameter, concerning the worst case behaviour, was suggested by Bollobás (cf. [27]). We define

$$M_H(n) := \max\{\tau_H(G) : G \text{ is an } n\text{-vertex graph}\},$$

to be the *maximum running time* for the H-process on n vertices. Returning to our Example 1.1, we see that determining $M_{K_3}(n)$ is equivalent to maximising the distance between two vertices in the same connected component of an n-vertex graph G. This is achieved when G is a path, giving $M_{K_3}(n) = \lceil \log_2(n-1) \rceil$.

1.4 The content and the goal of this survey

In all three settings outlined above, investigations focus on understanding how the infection rule H influences the extremal functions of interest: the weak saturation number $\mathrm{wsat}(n, H)$, the percolation threshold $p_c(n, H)$, or the maximum running time $M_H(n)$. The purpose of this survey is to explore exactly this, presenting the current progress and highlighting the gaps in our knowledge. Indeed, as in all good mathematical problems, there are many mysteries and a full understanding of H-processes presents a formidable challenge.

We will look at both weak saturation and percolation thresholds in this context and report the foundational work as well as recent exciting breakthroughs and key unanswered questions. However, this survey is mostly concerned with maximum running times. In particular, our primary goal is to give an exposition of our recent systematic exploration [45, 46, 47] of the parameter $M_H(n)$. We present general proof methods and discover a rich landscape of theorems with several surprises therein. Along the way we build connections to other topics in extremal graph theory as well as probabilistic and additive combinatorics. We intend this survey to act as an invitation to the topic, collecting and organising the state of the art and highlighting plenty of tantalising open problems and conjectures!

1.5 Organisation and notation

Maximum running times were first studied for cliques by Bollobás, Przykucki, Riordan, and Sahasrabudhe [27] and Matzke [66], and subsequently by Balogh, Kronenberg, Pokrovskiy, and the third author [14]. It is these works that we treat in Section 2, describing the development of insights that lay the foundation for our results and highlighting the motivations and connections to key concepts already present in works on weak saturation and percolation thresholds. In Section 3 we include a novel concise proof of a lower bound of Balogh et al. [14] on the running time of the K_5-process. This introduces central facts and definitions and explains the connection between our problem and additive combinatorics, a link that was first utilised in [14]. In Section 4 we touch on a question that merges the perspective of maximum running time and percolating graphs. Namely, we prove that for a broad family of infection rules H, the slowest (up to a constant factor) initial infection configuration can also be chosen to be H-percolating. In Section 5 we start presenting our results, putting them in context and introducing key graph properties that contribute to extremely slow running times. Sections 6 and 7 investigate what property of an infection rule causes the process to stabilise quickly for every starting graph. Finally, Sections 8 and 9 are devoted to bipartite and disconnected infection rules, respectively.

Given two graphs H and H', the graph $H \cap H'$ denotes the graph on vertex set $V(H) \cap V(H')$ with edge set $E(H) \cap E(H')$. The graph $H \cup H'$ is the graph with vertex set $V(H) \cup V(H')$ and whose edge set is $E(H) \cup E(H')$. The notation $H \sqcup H'$

is used to emphasise when the union is a *disjoint union*, i.e. when $V(H) \cap V(H') = \emptyset$. Throughout the survey we will often identify graphs with their edge sets.

1.6 What is not included

To close this introduction, we briefly mention some highly relevant directions that are omitted in order to keep this survey (relatively) concise. Indeed, there are vast bodies of work that provide context or study variants to the topics covered here. Luckily, there are already fantastic surveys available, by Currie, Faudree, Faudree and Schmitt [37] for (weak) saturation, and by Morris for bootstrap percolation [67] and monotone cellular automata in general [68].

Neighbour bootstrap percolation. Besides its intrinsic mathematical interest, graph bootstrap percolation [13] was inspired by the classical notion of r-neighbourhood bootstrap percolation. There, one initially infects some *vertices* of a specific (possibly infinite) graph G and at each time step, a vertex becomes infected if at least r of its neighbours are infected. This is a monotone cellular automaton (à la von Neumann [83] and Ulam [82]). The process was introduced by Chalupa, Leath and Reich [34] on the m-regular infinite tree, as a simplified model of ferromagnets. Subsequently an immense amount of literature developed around this concept in combinatorics and probability theory [32, 53, 54] as well as statistical physics [1, 2, 3]. Analogously to the graph bootstrap model, there is great interest in when the model *percolates* with both random starting configurations [10, 11, 12, 23, 33] and extremal starting configurations [9, 12, 38, 69] being studied. Running times of neighbour bootstrap percolation processes have recently been addressed from an extremal perspective by Przykucki and Shelton [75], Przykucki [74] and Benevides and Przykucki [20, 21] and when the starting configuration is random by Bollobás, Holmgren, Smith, and Uzzell [26], Bollobás, Smith, and Uzzell [28] and Balister, Bollobás, and Smith [8].

Hypergraphs. The study of weak saturation with respect to hypergraphs has been an important topic [40, 72, 73, 81]. For the case of threshold probabilities, Noel and Morrison [70] considered a variant of the H-process for hypergraphs H where only certain copies of H are activated (with some probability), as part of a more general theorem generalising work of Korándi, Peled and Sudakov [61]. For maximum running times, hypergraph cliques H have also been considered by Noel and Ranganathan [71], Espuny Díaz, Janzer, Kronenberg and Lada [44] and Hartarsky and Lichev [52]. Whilst we do not dwell on the topic of hypergraphs for maximum running times here, we believe that it would be interesting to explore it further.

2 Cliques and chains

We begin by exploring the K_k-process for all integers $k \geq 3$.

2.1 Weak saturation for cliques

Erdős, Hajnal and Moon [41] proved that for cliques K_k with $k \geq 3$, one has $\mathrm{sat}(n, K_k) = (k-2)(n-k+2) + \binom{k-2}{2} = \binom{n}{2} - \binom{n-k+2}{2}$ and the unique graph that achieves this minimum is $K_{k-2} \vee \bar{K}_{n-k+2}$, the disjoint union of a $(k-2)$-clique and an independent set of order $(n-k+2)$ with a complete bipartite graph in between.

For $k = 3$, this matches what we observed in the introduction, and $\text{wsat}(n, K_3) = \text{sat}(n, K_3) = n - 1$. Bollobás [24] proved that the weak saturation number and saturation number coincide for cliques K_k with $3 \leq k \leq 7$ and later also conjectured [25] that there is equality for all values of k. The conjecture was proved by Kalai who gave two proofs, one appealing to the geometric notion of rigidity [56] and one using exterior algebras [57]. An elegant proof inferring it from the Skew Set-Pairs Inequality appears in a paper of Alon [5]. The inequality can be derived from the pioneering work of Lovász [65], which introduced the use of exterior algebra into combinatorics, but was first stated and proved explicitly by Frankl [49] (see also [55, 5]). Despite the keen interest and multiple proofs, the determination of $\text{wsat}(n, K_k)$ remains one of the intriguing combinatorial problems that to this date does not have a purely combinatorial solution.

One indication for why this is the case is the large number of non-isomorphic extremal examples. Indeed, unlike the saturation problem, where $K_{k-2} \vee \bar{K}_{n-k+2}$ is the unique graph achieving $\text{sat}(n, K_k) = \text{wsat}(n, K_k)$, when considering K_k-percolating graphs, as observed already by Bollobás [24], one can build many more. Indeed, to obtain further constructions one can arbitrarily add a new vertex of degree $k - 2$ to a K_k-percolating graph. The following provides a K_4-percolating example with the minimum number of edges, that looks very different to $K_2 \vee \bar{K}_{n-2}$.

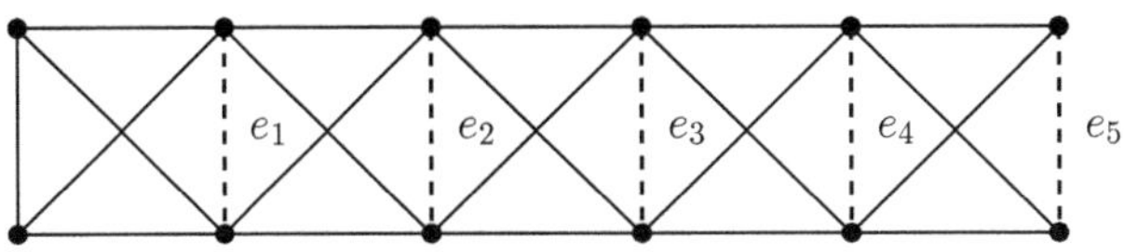

Figure 1: A simple K_4-chain.

Definition 2.1 (A simple K_k-chain) Let $d \in \mathbf{N}$ and $n := (k-2)d + 2$. A *simple K_k-chain* $(H_i, e_i)_{i=1}^d$ *of length* d is defined as follows. We take $H_1, \dots, H_d$ to be a sequence of copies of K_k such that for $1 \leq i < j \leq d$ we have $V(H_i) \cap V(H_j)$ empty unless $j = i + 1$, in which case $V(H_i) \cap V(H_{i+1})$ has size two and the edge at the intersection is e_i. We let e_d be an edge of H_d disjoint from e_{d-1} and we define F_d to be the graph obtained by taking the union $\cup_{i=1}^d H_i$ after removing all the edges e_i.

Note that for any even n, the graph F_d with $d = n/2 - 1$ indeed provides an example of a K_4-percolating graph G with $\text{wsat}(n, K_4) = 2n - 3$ edges.

2.2 Percolation thresholds for cliques

In their paper introducing the percolation threshold $p_c(n, H)$, Balogh, Bollobás and Morris [13] showed that for cliques of order $k \geq 3$ we have

$$p_c(n, K_k) = n^{-1/\lambda(K_k)+o(1)}, \tag{2.1}$$

where $\lambda(H) := \frac{e_H - 2}{v_H - 2}$ and the $o(1)$ in the exponent is hiding explicit poly-logarithmic factors in both the upper bound and lower bounds. For a sanity check we note that $\lambda(K_3) = 1$, as it should be to obtain the connectivity threshold in (2.1). In order to see where the exponent is coming from, consider the simple K_k-chains from

Definition 2.1. Note that if an edge $e \in E(K_n)$ plays the role of e_d for some copy of F_d in $G(n,p)$ then certainly e will be infected as one can see that the K_k-process on F_d infects edge e_i at time i for $1 \leq i \leq d$. Now $n^{-1/\lambda(K_k)}$ is the threshold for when a fixed edge $e \in E(K_n)$ plays the role of a non-edge in a copy of K_k^{--}, the graph obtained from K_k by removing a matching of size 2. Indeed, the expected number of copies of K_k^{--} that have e playing the role of a non-edge is $n^{k-2}p^{e(K_k^{--})}$ which is constant at $p = n^{-1/\lambda(H)}$. Therefore, if p is significantly above $n^{-1/\lambda(H)}$ (by some poly-logarithmic factors, say), then *any* $e \in E(K_n)$ plays the role of a non-edge in K_k^{--}, and we can string together these copies to start to build some simple K_k-chain (or rather the resulting graph F_d) in reverse, that has e as its final edge e_d. There are in fact many ways to do this and extending these to length $d \geq \log n$ say, we find one in which the final copy of K_k^{--} added (which will be the first copy in the chain graph F_d) is in fact a copy of K_k^{-} (that is, there is an extra edge) and so then we have a chain that causes the edge e to be infected. With some more care one can apply the second moment method to show that *all* edges of K_n play the role of e_d in some copy of F_d in $G(n,p)$ and will hence get infected.

There is a poly-logarithmic gap between the upper and lower bounds in (2.1). For K_4 however, Balogh, Bollobás and Morris [13] could determine $p_c(K_4)$ up to a constant. Subsequent work of Angel and Kolesnik [6] and Kolesnik [60] established matching upper and lower bounds respectively, giving a sharp threshold at $p_c(K_4) = 1/\sqrt{3n \log n}$. For larger cliques, Bartha and Kolesnik [15] removed the poly-logarithmic factor from the upper bound in (2.1). In a very recent breakthrough, Bartha, Kolesnik, Kronenberg and Peled [17] announced the resolution of a sharp threshold for $k \geq 5$ at $p_c(K_k) = (\gamma_k n)^{-1/\lambda(K_k)}$ for some explicit constants γ_k that depend on *Fuss-Catalan* numbers. Interestingly there is a change in behaviour for $k \geq 5$, evidenced by the lack of poly-logarithmic factors in the threshold.

2.3 Maximum running times for cliques

As we saw in the introduction, we have that $M_{K_3}(n)$ is logarithmic in n. The situation changes for K_4. Indeed, it is not difficult to see that the starting graph F_d corresponding to a simple K_4-chain in Definition 2.1 has running time $d = \frac{n}{2} - 1$ and so $M_{K_4}(n)$ is at least linear in n. This construction is in fact a bit wasteful. Adding a new vertex to a percolating starting graph and connecting it to the endpoints of an edge that only gets infected in the last round, increases the running time by one. (This is in contrast with F_d, where the second endpoint of e_i gets the same neighbours as the first, hence its addition does not increase the running time.) Starting with K_4^-, that is K_4 minus an edge, and iterating leads to a construction with running time $n-3$. Bollobás, Przykucki, Riordan and Sahasrabudhe [27] and, independently, Matzke [66] showed that this is best possible and $M_{K_4}(n) = n-3$, for all $n \geq 3$. The upper bound here adopts the idea of inductively "growing" a clique, showing that if the process does not stabilise by step i then there must be a clique of size $i+3$ in G_i.

For larger cliques K_k, simple chains again give a linear lower bound but there is a key difference with the case of the K_4-chain. Indeed, for $k \geq 5$ the *only* edges added in the K_5-process on F_d are the edges $e_1, \dots, e_d$ with e_i added in the ith time step. This was not the case for K_4, where the process on F_d percolates. The sparse final graph of the process on a simple K_k-chain gives hope that a long chain of copies of K_k can somehow be made to wrap back onto itself whilst maintaining that each edge e_i gets infected at time i. Bollobás et al. [27] and Matzke [66] both managed to do this and achieve super-linear running times. This birthed the concept of a "chain construction" which has since been the principal idea for deriving lower bounds for running times of H-processes. We therefore introduce the relevant definitions in generality allowing us to formalise and extend the ideas above. Moreover, *chains* as defined below also play a crucial role in deriving upper bounds (see Section 5.3.2).

Definition 2.2 (H-chains) Let H be a connected graph, and let $\tau \geq 1$. An *H-chain* of *length* τ is a sequence $(H_i, e_i)_{i\in[\tau]}$ of copies H_i of H together with edges $e_i \in E(H_i)$ such that $e_i \in E(H_i) \cap E(H_{i+1})$ for $i \in [\tau - 1]$. We call $G := \cup_{i=1}^{\tau} H_i$ the *underlying graph* of the H-chain and $G' := G - \{e_1, \dots, e_\tau\}$ the *starting graph* of the H-chain.

Note that simple K_k-chains are certainly examples of K_k-chains but Definition 2.2 is much more general as there is no restriction on how the copies of H can intersect (other than the edges e_i necessarily belonging to H_i and H_{i+1}).

Our aim is to derive lower bounds on running times by showing that the edge e_i is infected at time i in the K_k-process on the starting graph of some K_k-chain. What extra conditions do we need for this? For the edge e_i to get infected at all, we do need some restriction on the intersections of edge sets of the graphs H_i. Also in order that edges e_i are not infected too early in the K_k-process, we want that no other edge is infected. Therefore we restrict the copies of K_k^-, the graph obtained by removing an edge from K_k. This leads to the following key lemma.

Lemma 2.3 (A lower bound on running times for chains) *Let $k \geq 5$ and $\tau \geq 1$ and suppose that $(H_i, e_i)_{i\in[\tau]}$ is a K_k-chain such that:*

(†) *For $1 \leq i < j \leq \tau$, we have $E(H_i) \cap E(H_j)$ empty unless $j = i + 1$, in which case the intersection is just e_i.*

(*) *For any copy F of K_k^- in $G := \cup_{i=1}^{\tau} H_i$, there is an $i \in [\tau]$ such that $F \subseteq H_i$.*

Then, taking $G' := G - \{e_1, \dots, e_\tau\}$ to be the starting graph of the chain, we have that $\tau_H(G') \geq \tau$.

Proof Setting $G_0 = G'$ to be the starting graph, we show by induction that $G_i = G_0 + \{e_1, \dots, e_i\} = G_{i-1} + e_i$. The base case $i = 0$ follows from the definition. For $i \geq 1$, supposing that the claim holds for previous i, we have that $H_i - e_i \subseteq G_{i-1}$ and so certainly e_i is added to G_i. To see that no other edge is added, suppose an edge $f \in G_i \setminus G_{i-1}$ completes a copy $F \subseteq G_{i-1}$ of K_k^- to a copy of K_k at time i. As $G_{i-1} \subseteq G$ by the induction hypothesis, property **(*)** gives that $F \subseteq H_j$ for some $j \in [\tau]$. As $f \notin G_{i-1}$, we have $j \geq i$. If $j > i$, then $H_j \cap G_{i-1} = H_j - \{e_{j-1}, e_j\}$ by the induction hypothesis and condition **(†)**. Therefore there are not enough edges in $H_j \cap G_{i-1}$ to contain F and so we must have $j = i$, $F \subseteq H_i$ and $f = e_i$. □

Simple K_k-chains, by construction, satisfy condition **(†)** for every $k \geq 3$. It is also easy to see that **(*)** holds when $k \geq 5$.

Observation 2.4 *Suppose that $d \in \mathbf{N}$ and $(H_i, e_i)_{i\in[d]}$ is a simple K_k-chain, for some $k \geq 5$. If F is some copy of K_k^- in $G := \cup_{i\in[d]} H_i$, then there is some $i \in [d]$ such that $F \subseteq H_i$.*

Indeed, if F was not contained in any H_i then there would be a choice of e_j such that removing the vertices of e_j disconnects $F \cong K_k^-$, a contradiction. Note that this does *not* hold in the case of K_4.

Chains satisfying the hypothesis of Lemma 2.3 are what Bollobás, Przykucki, Riordan and Sahasrabudhe [27] call *good chains* and they used them to give lower bounds on $M_{K_k}(n)$ for $k \geq 5$. Matzke [66] also uses chains and gave an explicit construction of a good chain showing that $M_{K_k}(n) \geq \Omega(n^{3/2})$ for all $k \geq 5$. Bollobás et al. [27] went further, showing that

$$M_{K_k}(n) \geq n^{2-\frac{1}{\lambda(K_k)}-o(1)}, \tag{2.2}$$

with a random construction. They build their chain by choosing one H_i at a time, taking a uniformly random choice for the next $k-2$ vertices to complete the vertices of H_i and then taking $e_i \in H_i - e_{i-1}$ also as a random choice. The exponent in (2.2) matches that in (2.1) and this is no coincidence. Indeed, the authors of [27] show that in terms of unwanted (that is, not contained in some H_i) copies of graphs $F \subseteq K_k$ in $G = \cup_i H_i$, the graph G resembles $G(n,p)$ of the same edge density. Therefore if a typical edge $e \in E(K_n)$ lies in some (unwanted) copy of K_k^{--} in G, we can no longer use e as it would create an unwanted copy of K_k^-. Thus the process for building the random chain lasts until a typical edge is no longer usable.

The exponents in (2.2) tend to 2 as k tends to infinity, and as sketched above, are optimal with respect to a random construction of a good chain. Bollobás, Przykucki, Riordan and Sahasrabudhe [27] contemplated that while it is "tempting" to think that their construction is also optimal in general, they had no real reason to believe this. Instead, motivated by the lack of non-trivial upper bounds in the problem, they opted for conjecturing only $M_{K_k}(n) = o(n^2)$, that is, that no K_k-process should be able to add a constant fraction of the edges of K_n in distinct rounds. Surprisingly, even this more modest prediction turned out not to hold for $k \geq 6$. Balogh, Kronenberg, Pokrovskiy and the third author [14] gave an explicit construction of a starting graph with quadratic running time, hence proving that $M_{K_k}(n) = \Theta(n^2)$ for all $k \geq 6$. Theirs was also a chain construction but with a different flavour. In order to get longer running times, intuitively one needs the chains to intersect more significantly. One key observation of [14] is that the condition (†) in Lemma 2.3 can be substantially weakened. Indeed, we do not really care if distinct H_i and H_j with $j > i+1$ intersect, as long as the edges e_{j-1}, e_j are not added before they should be. Thus (†) can be replaced by

(†′) For $1 \leq i < j \leq \tau$, we have $e_j \notin E(H_i)$.

One can check that the proof of Lemma 2.3 goes through with this weaker condition.

In [14], this observation was exploited with the construction of a *ladder chain* which we now sketch for the case of $H = K_6$. In fact, the construction we see here is a slight variation of the one from [14] and we instead follow [47] which generalised the construction to other H (and introduced the terminology used here). We will return to highlight the difference at the end of the following section after seeing chain constructions in detail.

The idea is to superimpose linearly many simple chains on the same vertex set. Each of these simple chains $(H_i^a, e_i^a)_{i\in[\tau_0]}$ will have some linear length, say $\tau_0 = \lfloor n/8 \rfloor$. These can then be linked together via disjoint simple chains of constant length to form one long continuous chain which will have quadratic length (we will see in the next section the details of such a linking procedure).

To construct the linearly long simple chains, we start by considering a bipartition of the vertices into two *sides* which we think of as the *left* side and a *right* side. On each side we place a sequence of $2\tau_0$ copies of K_3, with consecutive copies intersecting in a singular vertex. Using these triangles, for any *slope* $a = 0, 1, \ldots, \tau_0$, we can consider a chain $(H_i^a, e_i^a)_{i\in[\tau_0]}$ which matches the triangle in position j on the left to the triangle in position $j + a$ on the right and places a complete bipartite graph between them (see Figure 2). We will take an appropriate collection A of slopes that will define our family of simple chains.

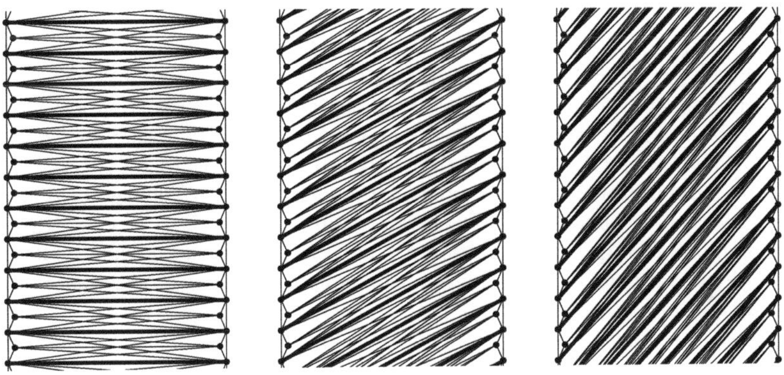

Figure 2: Sections of simple K_6-chains with slopes 0 (left), 4 (middle) and 8 (right). The edges e_i of the simple chains are bold.

Crucial to the analysis of this chain and establishing the key condition (*) of Lemma 2.3 is the fact that the graph $G := \cup_{a\in A, i\in[\tau]} H_i^a$ induces a sequence of triangles on both the left and right side. This in turn forces any copy F of K_6^- in G to use three vertices on the left and three vertices on the right. Using this, one can show that if F contains edges of the underlying graph $\cup_{i\in[\tau_0]} H_i^a$ for different slopes $a \in A$, then two of these must have difference at most three. We can avoid this by choosing only every fourth integer into A.

Interestingly, the construction does not work for K_5. Indeed, the naive generalisation would place a sequence of triangles on the left side as before, and a path on the right side. For any pair of simple K_5-chains with slopes $a \neq a'$ using this bipartition, one can find copies of K_5^- using one triangle on the left and edges corresponding to the different slopes a, a' for the remaining vertices v, v' of the copy. If we try to include these two simple chains as segments of a long chain, this will lead to unwanted copies of K_5^-, contradicting (*). However with an alternative, more subtle approach the same authors [14] achieved almost quadratic time:

$$M_{K_5}(n) \geq n^{1-O(1/\sqrt{\log n})} = n^{2-o(1)}.$$

In fact, their construction gives that $M_{K_5}(n) \geq n \cdot r_3(n)$ where $r_3(n)$ is the size of a largest subset of $[n]$ free of three-term arithmetic progressions. The famous construction of Behrend [19] gives that $r_3(n) \geq n^{1-O(1/\sqrt{\log n})}$, whilst it is well-known that $r_3(n) = o(n)$, as was originally shown by Roth [76] in 1953.

We close this section with what is probably the most pertinent outstanding question in the pursuit of maximum running times, namely determining the asymptotics of $M_{K_5}(n)$.

Conjecture 2.5 (Bollobás-Przykucki-Riordan-Sahasrabudhe [27]) *The maximum running time for infection rule K_5 is $M_{K_5}(n) = o(n^2)$.*

3 Proof highlight: $M_{K_5}(n)$ lower bound

In this section, we prove a lower bound for $M_{K_5}(n)$ of the form $n^{2-o(1)}$. This was originally proved by Balogh et al. [14], using a construction similar to the *ladder chains* discussed in the previous section, although it uses a tripartition and slopes defined by Behrend's set free of arithmetic progressions. Our proof here uses *dilation chains* [47] and is quite distinct, even though it also relies on Behrend's construction. In [47], we developed a theory of chain constructions (and in particular the dilation chains which we see here) to give general results that apply to a wide array of infection rules, see for example Section 5.4. In this section, we introduce the key ideas of chain constructions and dilation chains in the context of K_5, where many technicalities can be avoided.

As with the ladder chains of the previous section, we will first superimpose linearly many simple chains on the same vertex set, each of linear length. Note that each of the chains $(H_i^a, e_i^a)_{i\in[\tau_0]}$ there can be obtained from the chain $(H_i^0, e_i^0)_{i\in[\tau_0]}$ with slope 0 by leaving the left side fixed and translating the vertices on the right side by a triangles. The main difference for dilation chain constructions will be the way we choose the permutation that "mixes up" the vertices of a simple chain to obtain other simple chains in our family. The vertex set will be labelled by the elements of $\mathbf{Z}_p \setminus \{0\}$, which is equipped with nicely interacting additive and multiplicative operations. In the basic simple chain $(H_i^1, e_i^1)_{i\in[\tau_0]}$ the copies of K_5s will be laid out one after the other, following the additive structure of $\mathbf{Z}_p \setminus \{0\}$. The family of simple chains $(H_i^a, e_i^a)_{i\in[\tau_0]}$ will be created using the multiplicative operation: the vertices of the basic simple chain are multiplied by a constant $a \in \mathbf{Z}_p \setminus \{0\}$. We will again find that if we choose the set A of dilation constants a appropriately, then there will be no unwanted copy of K_5^- when we superimpose the corresponding simple chains. Let us see this idea formally.

3.1 The construction

Let $p \in \mathbf{N}$ be a prime and define $\tau_0 := \lfloor \frac{p-3}{3} \rfloor$. Let $A := \{a_1, \ldots, a_q\} \subseteq \mathbf{Z}_p \setminus \{0\}$ be a subset of q dilations.

Dilation chains. For an element $a \in \mathbf{Z}_p \setminus \{0\}$, the *dilation K_5-chain* $(H_i^a, f_i^a)_{i\in[\tau_0]}$ is defined on the vertex set $W = \{w_i : i \in \mathbf{Z}_p \setminus \{0\}\}$ such that

$$V(H_i^a) = \{w_{a(3i-2)}, w_{a(3i-1)}, w_{a(3i)}, w_{a(3i+1)}, w_{a(3i+2)}\} \tag{3.1}$$

and $f_i^a = w_{a(3i+1)}w_{a(3i+2)}$ for $i = [\tau_0]$. We also define $f_0^a = w_a w_{2a}$.

The chain $(H_i^1, f_i^1)_{i\in[\tau_0]}$ is clearly simple and the others are obtained from it by multiplying the indices of the vertices by a. Since p is chosen to be a prime, the map $x \mapsto ax$ is a bijection of $\mathbf{Z}_p \setminus \{0\}$ to itself so $(H_i^a, f_i^a)_{i\in[\tau_0]}$ is also a simple K_5-chain.

Linking chains. For each element $a_j \in A$, the dilation chain $(H_i^{a_j}, f_i^{a_j})_{i\in[\tau_0]}$ will feature as subchains of our final chain $(H_i, e_i)_{i\in[\tau]}$. For ease of notation we use $(H_i^j, f_i^j)_{i\in[\tau_0]}$ instead. In order to link these simple chains up, we introduce short simple K_5-chains $(L_i^j, g_i^j)_{i\in[3]}$ for $j \in [q-1]$, each of length 3. The jth linking chain will link the last K_5 of the a_j-dilation chain to the first K_5 of the a_{j+1}-dilation chain. That is, two vertices of L_1^j (disjoint from g_1^j) will be the vertices of the edge $f_{\tau_0}^j$ and g_3^j will coincide with f_0^{j+1}, see Figure 3. All other 7 vertices of $U_j := \cup_{i\in[3]} V(L_i^j)$ will be disjoint from W and also disjoint from $U_{j'} \setminus (W \cap U_{j'})$ for choices of $j' \neq j$.

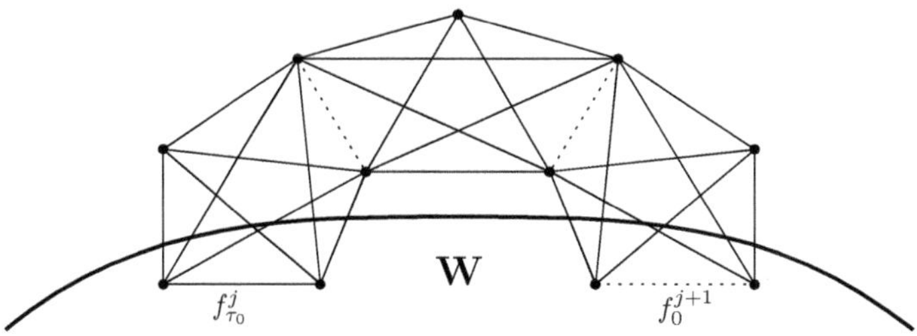

Figure 3: A linking K_5-chain $(L_i^j, g_i^j)_{i\in[3]}$ with the edges g_i^j as dashed lines.

Note that in order to have that $(L_i^j, g_i^j)_{i\in[3]}$ is a well-defined simple K_5-chain, we need that $f_{\tau_0}^j \in E(L_1^j)$ and $g_3^j = f_0^{j+1} \in E(L_3^j)$ are vertex-disjoint in W. We will ensure this when we select the set A of dilations.

Bringing everything together. Finally, we define our chain $(H_i, e_i)_{i\in[\tau]}$ which is simply obtained by concatenating the sequences, starting with $(H_i^1, f_i^1)_{i\in[\tau_0]}$ followed by $(L_i^1, g_i^1)_{i\in[3]}$ and then $(H_i^2, f_i^2)_{i\in[\tau_0]}$ and so on, alternating between dilation chains and linking chains, until we finish with the sequence $(H_i^q, f_i^q)_{i\in[\tau_0]}$. This defines a K_5-chain which has length $\tau := \tau_0 q + 3(q-1)$ and lies on $V = W \cup \left(\cup_{j\in[q-1]} U_j\right)$ which has size $|V| = p - 1 + 7(q-1)$.

3.2 The set of dilations.

In this section, we collect properties of the set A that we will need for our dilation chain construction above to work.

First of all, as we promised above, we make sure that $f_{\tau_0}^j \cap f_0^{j+1} = \emptyset$ for every $j \in [q-1]$, so the jth linking chain is simple. Furthermore, in order to use Lemma 2.3 for the running time of $(H_i, e_i)_{i\in[\tau]}$, we need to establish properties **(*)** and **(†)** for it. Both of these depend on certain restrictions about how the edges of different dilation chains interact with each other. At the heart of the proof of all of these lies a general observation, relating the edges of the graph $G^j := \cup_{i\in[\tau_0]} H_i^j$ of the jth dilation chain to the existence of a congruence modulo p.

Observation 3.1 *If $w_x w_y \in E(G^j)$ for $x, y \in [p-1]$ and $j \in [q]$, then there is some $\alpha \in \mathbf{Z} \setminus \{0\}$ with $|\alpha| \le 4$ such that*

$$x - y = \alpha a_j \pmod{p}.$$

Proof As $w_x w_y \in E(G^j)$, there is some index $i \in [\tau_0]$ such that $w_x, w_y \in V(H_i^j) = \{w_{a_j(3i-2)}, \dots, w_{a_j(3i+2)}\}$. Therefore, there are some distinct $\ell_x, \ell_y \in \{-2, -1, 0, 1, 2\}$ such that $x = a_j(3i + \ell_x)$ and $y = a_j(3i + \ell_y)$ modulo p. Then $x - y = \alpha a_j$ modulo p, where $\alpha = \ell_x - \ell_y \neq 0$. □

The properties we need for our main proof forbid that edges from two or three different dilation chains form certain small subgraphs in G. The above observation translates these subgraphs to certain linear equations modulo p, in at most three variables, having a solution from A. This leads us to the next theorem, based on the famous construction of Behrend avoiding 3-term arithmetic progressions. It gives us large subsets A of $\mathbf{Z}_p$ which do not have non-trivial solutions to *any* of the linear equations we will care about.

Theorem 3.2 (Behrend [19]) *There exists $C' > 0$ such that for any prime $p \in \mathbf{N}$ there is some set $A \subset \mathbf{Z}_p \setminus \{0\}$ such that $|A| \ge p^{1-C'/\sqrt{\log p}}$ and if*

$$\alpha_1 a_1 + \alpha_2 a_2 + \alpha_3 a_3 = 0 \pmod{p}$$

for some $a_i \in A$ and $\alpha_i \in \mathbf{Z}$ with $|\alpha_i| \le 4$ for $i \in [3]$ then $\sum_{i=1}^{3} \alpha_i = 0$ and $a_i = a_j$ for all $i, j \in [3]$ with $\alpha_i, \alpha_j \neq 0$.

Note that in particular, by fixing $\alpha_1 = \alpha_2 = 1$ and $\alpha_3 = -2$, the set A from Theorem 3.2 avoids three-term arithmetic progressions $x, x+d, x+2d$ with $x, d \in \mathbf{Z}_p$ and $d \neq 0$. The original proof of Behrend [19] simply gave sets avoiding these arithmetic progressions but it is well-known [77, 79] that his construction can be adapted to avoid non-trivial solutions to any constant number of linear equations in at most three variables.

Distinct dilation chains are edge-disjoint. To guarantee (†) we need to make sure that G^j and $G^{j'}$ do not share an edge if $j \neq j'$. If there was some edge $w_x w_y \in G_j \cap G_{j'}$ then by Observation 3.1 we have that there are some $\alpha, \alpha' \in \mathbf{Z} \setminus \{0\}$ with $|\alpha|, |\alpha'| \le 4$ such that $\alpha a_j - \alpha' a_{j'} = 0$ modulo p. This contradicts our choice of A from Theorem 3.2, since $a_j, a_{j'} \in A$ and $a_j \neq a_{j'}$.

Corollary 3.3 *The graphs G^j with $j \in [q]$ are pairwise edge-disjoint.*

Preventing "unwanted" copies of K_5^-. Condition (*) forbids that a copy F of K_5^- in G shares an edge with more than one H_i. The most problematic part of showing this turns out to be if F shares an edge with more than one dilation chain. Here is where we use the full power of the properties of our set A from Theorem 3.2.

Lemma 3.4 *For any copy F of K_5^- in $\cup_{j\in[q]} G^j$, there is a $j \in [q]$ with $F \subseteq G^j$.*

The proof of this relies on a simple combinatorial observation implying that in order to avoid copies of K_5^- that have edge from different dilation chains, it is enough to avoid triangles with that property.

Fact 3.5 *Any non-monochromatic edge-colouring of K_5^- contains a copy of K_3, that is non-monochromatic.*

Proof of Lemma 3.4 Consider a colouring of $\cup_j G^j$ that colours an edge j if it belongs to G^j. If F is not a subgraph of any of the G^j then it is not monochromatic and hence there is a triangle F' which is also not. Let $x, y, z \in \mathbf{Z}_p$ such that the vertices of F' are $w_x, w_y, w_z \in W$ and let $j_1, j_2, j_3 \in [q]$ such that $w_x w_y \in G^{j_1}$, $w_x w_z \in G^{j_2}$ and $w_y w_z \in G^{j_3}$. By Observation 3.1, we have that for $i = 1, 2, 3$ there are $\alpha_i \in \mathbf{Z} \setminus \{0\}$ such that $|\alpha_i| \leq 4$ and

$$x - y = \alpha_1 a_{j_1}, \quad z - x = \alpha_2 a_{j_2} \quad \text{and} \quad y - z = \alpha_3 a_{j_3},$$

modulo p. Summing these equations gives that $\sum_{i=1}^{3} \alpha_i a_i = 0$ modulo p and thus, as all the α_i are non-zero and A satisfies the conclusion of Theorem 3.2, we have that $a_{j_1} = a_{j_2} = a_{j_3}$, contradicting that F' is non-monochromatic. □

Linking chains are simple. Finally, we also rectify what we left open at the end of the definition of linking chains: to ensure that they are simple. Namely, we need to see that for every $j \in [q-1]$ the edge $f^j_{\tau_0}$ and the edge f^{j+1}_0 are disjoint. The vertices of f^{j+1}_0 are $w_{a_{j+1}}$ and $w_{2a_{j+1}}$ and the vertices of $f^j_{\tau_0}$ are among $w_{a_j(p-\alpha)}$, where $\alpha \in [4]$. If two of these were to coincide then there was $\alpha' \in [2]$ and $\alpha \in [4]$ such that $\alpha' a_{j+1} - \alpha a_j = 0$ modulo p, which contradicts the property of our set A.

Corollary 3.6 *Every linking chain is simple.*

3.3 Finishing the proof

Finally we show that for appropriate choice of p and A, the K_5-chain $(H_i, e_i)_{i \in [\tau]}$ satisfies the conditions of Lemma 2.3, hence the starting graph $G' := (\cup_{i \in [\tau]} H_i) - \{e_1, \ldots, e_\tau\}$ has running time $\tau = n^{2-o(1)}$. To this end, for $n \in \mathbf{N}$ sufficiently large, we let $p \in \mathbf{N}$ be a prime such that $n/4 \leq p \leq n/2$, which exists by Bertrand's postulate/Chebyshev's theorem, and fix $\tau_0 := \lfloor \frac{p-3}{3} \rfloor$ as before. Let $A := \{a_1, \ldots, a_q\} \subseteq \mathbf{Z}_p \setminus \{0\}$ with $q \geq p^{1 - C'/\sqrt{\log p}}$ for some $C' > 0$ be the set from Theorem 3.2. Then for an appropriate $C > 0$ we have

$$\tau := \tau_0 q + 3(q-1) \geq pq/4 \geq n^{2 - C/\sqrt{\log n}} = n^{2-o(1)}.$$

The chains all lie on $V = W \cup \left(\cup_{j \in [q-1]} U_j\right)$ which has size $|V| = p - 1 + 7(q-1) \leq n$.

We are left to show conditions (*) and (†) of Lemma 2.3. For (†) note that by construction we do have that $E(H_i) \cap E(H_{i+1}) = e_i$ for every $i \in [\tau]$. For $2 \leq i + 1 < i' \leq \tau$ we must show that $E(H_i) \cap E(H_{i'}) = \varnothing$. Corollary 3.3 applies here unless H_i and $H_{i'}$ belong to the same dilation chain or same linking chain. But these are simple (cf. Corollary 3.6) and the property holds by construction.

To establish property (*) we fix F to be some copy of K_5^- in $G := \cup_{i \in [\tau]} H_i$ and show that there is some $i \in [\tau]$ such that $F \subseteq H_i$. Let us first assume that F has a vertex u outside of W, say $u \in U_j \setminus W$ for some $j \in [q-1]$. In case $V(F) \subseteq U_j$ then we are done by Observation 2.4 as $G[U_j]$ only contains the edges of the jth linking chain which is simple by Corollary 3.6. Otherwise there is some $w \in V(F) \setminus U_j$, which we show is impossible. On the one hand u and w should have at least three common neighbours as they are the non-adjacent vertices of a copy of K_5^- in G. This is because all neighbours of u are in U_j (and w is not). On the other hand, we claim that their common neighbourhood is contained in one of the border edges f_τ^j and f_0^{j+1} of the jth linking chain, providing a contradiction. Indeed, all U_j-neighbours of w are in the union $f_{\tau_0}^j \cup f_0^{j+1}$, yet u, whose neighbourhood is contained in U_j, cannot have neighbours in both $f_{\tau_0}^j$ and f_0^{j+1} (since these edges are disjoint by Corollary 3.6 and the linking chain has three copies of K_5).

Thus we have $V(F) \subseteq W$ and hence $F \subseteq \cup_{j\in[q]} G^j$. In this case Lemma 3.4 implies that $F \subseteq G^j$ for some $j \in [q]$. As each dilation chain is simple by construction, Observation 2.4 can then be used to deduce that there is some H_i containing F. □

The idea of the linking chains which we used for our lower bound on $M_{K_5}(n)$ originated in the proof of Balogh et al. [14] for K_5. In our recent work [47], we axiomatised the procedure and give general conditions for when chains can be linked to give longer chains. This reduces the problem of finding one long chain to finding many chain segments, which is often conceptually much easier. Indeed, for example for the ladder chains discussed in Section 2.3, by using linking chains one can give the simple chains with different slopes separately and does not need to worry about one leading into another. In fact, in their original proof showing that $M_{K_k}(n)$ is quadratic for $k \geq 6$, Balogh et al. [14] did not use linking chains. Instead, they used unequal sides of the partitions and wrapped the triangles around so that each time the chain visits a certain triangle, it does so with respect to a different slope.

4 H-percolating graphs

Having explored K_k-processes in detail, the rest of the survey will be concerned with the influence of the infection rule H on the H-process, our main focus being on the running time $M_H(n)$. Before this, let us briefly take stock and recap what we have seen in the clique case. In particular, it is interesting to consider the interplay between the running time of the bootstrap process and whether or not it percolates. In Section 2, we saw that $\text{wsat}(n, K_k) = \text{sat}(n, K_k)$ and so there are edge-minimal K_k-percolating starting graphs G that percolate in just one step. The set of extremal constructions for weak saturation is larger though and there are edge-minimal K_k-percolating graphs which are much slower to percolate, for example the simple K_4-chains of Definition 2.1 which take linear time to percolate, asymptotically meeting the slowest possible running time $M_{K_4}(n)$. For larger $k \geq 5$, as noted by Matzke [66], the wide class of tight examples for $\text{wsat}(n, K_k)$ given by Bollobás [24] using an inductive argument, all percolate in linear time and so fall far short of the maximum running time $M_{K_k}(n)$ which is quadratic (or at least almost quadratic for K_5).

We also looked at the threshold $p_c(n, K_k)$ for when a random graph $G(n,p)$ is K_k-percolating. Above this threshold, we saw that typical edges of K_n play the role of the final edge e_d of some F_d generated by a K_k-chain after removing the edges e_i. Moreover, we can take that this chain is logarithmic in size and so random graphs above the threshold will also percolate fast. In fact, this was studied in detail by Gunderson, Koch and Przykucki [51] who showed that just above the threshold $p_c(n, K_k)$ the random graph will a.a.s. K_k-percolate in $O(\log\log n)$ steps and this time decreases as p increases.

On the other hand, the chain constructions that we saw to lower bound $M_{K_k}(n)$ for $k \geq 5$ are far from being K_k-percolating. Indeed, it was crucial to our analysis via Lemma 2.3 that only the edges e_i of the chain will be added during the K_k-process. This begs the question as to whether it is possible to have a starting graph G such that the K_k-process percolates but does so slowly. The main topic of this section is a new result which indicates that it is possible for H-percolating graphs to be slow with respect to the H-process, even asymptotically achieving the optimal maximal running time $M_H(n)$. Theorem 4.2 below was noted, but not proven, by Bollobás, Przykucki, Riordan and Sahasrabudhe [27] for the case $H = K_k$. Here we show the result for the large class of $(2,1)$-*inseparable* infection rules H.

Definition 4.1 A graph H with an edge is called $(2,1)$-*separable* if it can be disconnected by deleting at most two vertices (and all incident edges) and one further edge. Otherwise H is called $(2,1)$-*inseparable.*

Note that every $(2,1)$-inseparable graph is 3-connected as it has at least five vertices and cannot be disconnected by deleting two vertices. Furthermore every 4-connected graph is $(2,1)$-inseparable. Indeed, if the removal of an edge e and two vertices z_1, z_2 disconnected a 4-connected graph H then so would the removal of z_1, z_2 and an appropriate endpoint of e (as H has at least five vertices). The notion of $(2,1)$-inseparability plays a crucial role with respect to chain constructions. Indeed, it was precisely this property of K_5 that was used to prove Observation 2.4 showing that simple chains do not induce unwanted copies of K_5^-.

Theorem 4.2 *For any $(2,1)$-inseparable H, there is some $C = C(H)$ such that for all $n \in \mathbf{N}$, there is an H-percolating graph $\tilde{G}$ with $v(\tilde{G}) \leq Cn$ and $\tau_H(\tilde{G}) \geq M_H(n)$.*

Following [27], we define

$$M'_H(n) := \max\{\tau_H(G) : G \text{ is an } n\text{-vertex } H\text{-percolating graph}\}.$$

Then clearly one has that $M_H(n) \geq M'_H(n)$. Whilst Theorem 4.2 does not quite give that $M_H(n) = O(M'_H(n))$ due to our lack of understanding (see Conjecture 8.8) of the functions $M_H(n)$ and $M'_H(n)$, it does give a strong indication that this is the case. The $(2,1)$-inseparability is crucial in our proof of Theorem 4.2 as we adopt ideas from chain constructions. As we will see later, there are graphs that are $(2,1)$-separable for which we can still use chain constructions, for example the wheel graph or bipartite graphs which satisfy an analogue of inseparability [47]. For these, it is likely that the proof ideas for Theorem 4.2 can be extended. For other graphs H for which we know $M_H(n)$, the constructions for the lower bounds are often H-percolating already. We therefore conjecture the following.

Conjecture 4.3 *For all infection rules H, we have $M_H(n) = \Theta(M'_H(n))$.*

Before proving Theorem 4.2, we briefly survey what is known about general infection rules H in the context of weak saturation and percolation thresholds.

4.1 Minimal H-percolating graphs

Other than $H = K_k$, there are several results that determine $\mathrm{wsat}(n, H)$ exactly. Restricting here to connected H, Borowiecki and Sidorowicz [31] noted that $\mathrm{wsat}(n, C_k) = n - 1$ if k is odd and n when k is even. Indeed, any spanning tree will percolate if k is odd and if k is even an extra edge is needed, otherwise the final graph will be complete bipartite. For trees T, Faudree, Gould and Jacobson [48] showed that $\mathrm{wsat}(n, T)$ does not depend on n and for any t-vertex tree, one has $t - 2 \leq \mathrm{wsat}(n, T) \leq \binom{t-1}{2}$ with both bounds being tight for certain trees. Further weak saturation numbers for trees were determined by Pu and Cui [36] and recently by Chen, Liu and Yang [35] who show that for any rational $\alpha \in [1, 2]$ there is a family of t-vertex trees whose weak saturation numbers are of order t^{α}. Also recently, Kronenberg, Martins and Morrison [62] determined $\mathrm{wsat}(n, K_{k,k})$ and $\mathrm{wsat}(n, K_{k,k+1})$ and complete bipartite graphs whose sizes grow with n were considered in [4].

Moving away from exact results to asymptotics, Alon [5] proved that for every H, there is some constant $w_H \geq 0$ such that $\mathrm{wsat}(n, H) = (w_H + o(1))n$. What controls the constant w_H? Faudree, Gould and Jacobson [48] established that it is closely related to the minimum degree $\delta(H)$ of H, showing that

$$\frac{\delta(H)}{2} - \frac{1}{\delta(H)+1} \leq w_H \leq \delta(H) - 1. \tag{4.1}$$

In fact, Terekhov and Zhukovskii [80] found an error in the lower bound which they then fixed, as well as showing that every multiple of $1/(\delta(H) + 1)$ in the interval is realisable as the limit $w_H := \lim \mathrm{wsat}(H)/n$ for some graph H. Very recently Ascoli and He [7] went further, characterising all the rationals that are realisable as w_H for some H. Interestingly, they showed that all possible rationals at least $3/2$ are realisable whereas below $3/2$ the realisable values are far more sparse with just one accumulation point at 1. They conjecture that all possible values of w_H are rational.

We conclude this section by remarking that there are still many open questions regarding the behaviour of $\mathrm{wsat}(n, H)$ and the associated constants w_H. In particular, it would be interesting to determine what properties of H control the position of w_H in the interval (4.1). What about for a typical H?

Problem 4.4 *What is $w_H = \lim \frac{\mathrm{wsat}(n,H)}{n}$ when $H = G(k, p)$?*

4.2 H-percolating random graphs

With the same proof as for cliques K_k, Balogh, Bollobás, and Morris [13] showed an upper bound on $p_c(n, H)$ of the form $n^{-1/\lambda(H)+o(1)}$ for *balanced* infection rules H, where we recall that $\lambda(H) := \frac{e_H - 2}{v_H - 2}$ (2.1) and the $o(1)$-term hides logarithmic factors. Here *balanced* means that $\frac{e_F - 1}{v_F - 2} \leq \lambda(H)$ for every proper subgraph $F \subset H$ with $v_F \geq 3$. The balanced condition is necessary due to their use of the second moment method to show that typical edges lie at the end of simple H-chains. The authors also show that all cycles C_k act like K_3 and have $p_c(n, C_k)$ coinciding with the connectivity threshold $\log n/n$. Moreover, they show that the threshold for $K_{2,3}$ is also $p_c(n, K_{2,3}) = \log n/n$ which gives an example H where the exponent is not equal to $-1/\lambda(H)$. In fact, the paper also provides indication that the general picture is much more complex than just what $\lambda(H)$ offers. It is not difficult to see [13] that the existence threshold for a copy of H minus an edge in $G(n,p)$, namely $\Omega(n^{-1/\lambda'})$ with $\lambda' = \lambda'(H) = \min_{e\in H}\max_{F\subseteq H-e} e_F/v_F$, provides a lower bound on the H-percolation threshold for all infection rules H, and it was shown in [13] that this lower bound is tight when the infection rule H has a pendant edge. They also provide further examples H where the exponent of the threshold is $-\frac{1}{\lambda'(H)}$ and where it is strictly in between $-\frac{1}{\lambda'(H)}$ and $-\frac{1}{\lambda(H)}$.

Bayraktar and Chakraborty [18] establish that for the complete bipartite graph $K_{r,s}$ the critical exponent coincides with the upper bound $-1/\lambda(K_{r,s})$ of Balogh et al. [13] when $3 \le r \le s \le (s-2)^2 + s$, i.e. when $K_{r,s}$ is balanced.

For general H with $\delta(H) \ge 2$ and $v(H) \ge 4$, Bartha and Kolesnik [15] proved a lower bound $n^{-1/\lambda_*(H)+o(1)}$ which is governed by a more subtle function $\lambda_*(H) = \min \frac{e_H - e_F - 1}{v_H - v_F}$, where the minimum is taken over all subgraphs $F \subset H$ with $2 \le v_F < v_H$. They show that for balanced graphs $\lambda^* = \lambda$, so together with the upper bound of [13] the asymptotics of the exponent of the threshold is determined for all balanced graphs. Recently Bartha, Kronenberg, and Kolesnik [16] proved $H = G(k, 1/2)$ is balanced a.a.s. and hence the critical exponent for the H-process percolating is equal to $-1/\lambda(H)$ for "most" infection rules H.

For non-balanced H, one has that λ_* is strictly smaller than λ and may even be smaller than λ'. Therefore, for non-balanced infection rules H, the problem of determining the percolation threshold $p_c(n,H)$ is still an outstanding mystery. The graphs $K_{2,t}$ seem like an interesting example that is yet to be understood. Indeed, Bidgoli, Mohammadian and Tayfeh-Rezaie [22] offer both upper bounds and lower bounds on the exponent for $p_c(n, K_{2,t})$. For $K_{2,3}$, their upper bound exponent had already been observed to be tight in [13]. In [22] the tightness of their upper bound is also established for $K_{2,4}$. There the answer is $-10/13$, which is interestingly different from $-1/\lambda, -1/\lambda'$ and $-1/\lambda_*$.

4.3 Slow H-percolating graphs

In this section, we prove Theorem 4.2 giving H-percolating graphs that percolate slowly. This proof has not appeared elsewhere and we believe it is of considerable interest to the community. Nonetheless, it is somewhat delicate and a bit of a detour from the main theme of the survey. Therefore the reader may want to skip ahead to Section 5 on first read.

In order to prove Theorem 4.2, we follow a natural strategy. Given $n \in \mathbf{N}$, fix a starting graph G that achieves running time $\tau := M_H(n)$ and an edge e_τ that is added at time τ in the process on G, the aim is to create a new starting graph $\tilde{G}$ on $O(n)$ vertices that contains G and does not interfere with the H-process on G for the first τ steps of the process. The addition of e_τ should then trigger the rest of the graph to percolate. In particular, we want that for every edge f on $V(G)$ that does *not* lie in $\langle G\rangle_H$, there is some gadget that is triggered by e_τ and forces f to be infected. A natural candidate for such a gadget is a *simple H-chain*. Here, as with our simple K_k-chains, a simple H-chain is just a sequence $(H_i, e_i)_{i\in\ell}$ of copies of H with consecutive copies H_i and H_{i+1} intersecting in a single edge $e_i \in H_i \cap H_{i+1}$ (and no other intersections between the H_i). As we saw for $H = K_5$ in Observation 2.4, we have the following.

Observation 4.5 *If H is $(2,1)$-inseparable and $(H_i, e_i)_{i\in\ell}$ is a simple H-chain then for any $e \in E(H)$ and any copy F of $H - e$ in $\cup_{i\in[\ell]} H_i$, there is some $i \in [\ell]$ such that $F \subseteq H_i$.*

As a consequence of this observation, note that if $e_0 \in H_1 - e_1$ the graph $G' = \cup_{i=1}^{\ell} H_i - \{e_0, \dots, e_\ell\}$ is H*-stable*, that is, no edge is added in the H-process on G'. Indeed, for each $i \in [\ell]$, we have that $V(H_i)$ hosts $e(H) - 2$ edges in G' and so we cannot embed some copy of $H - e$ into H_i.

Now armed with these simple H-chains, in a similar fashion to the linking chains used to lower bound $M_{K_5}(n)$, for each $f \in \binom{V(G)}{2} - \langle G \rangle_H$, we can place a short simple chain $(H_i, e_i)_{i\in[\ell]}$ such that e_τ is some edge of $H_1 - e_1$, $f = e_\ell$ and all other vertices of the chain are disjoint from $V(G)$, with the new vertices introduced by these chains being distinct for different choices of f. One immediate problem with this strategy is that we might be adding too many vertices. Indeed, it could be that there are quadratically many edges $f \in \binom{V(G)}{2} - \langle G \rangle_H$ and each one requires adding constantly many new vertices. For this, there is an easy solution. Indeed, we do not have to infect *all* of the edges f, just enough of them so as to trigger all edges to be added eventually. Indeed, as we saw in Section 4.1, there is some graph G' on n vertices with $(w_H + o(1))n \leq v(H)n$ edges that is H-percolating. Placing an auxiliary copy of G' on $V(G)$ and infecting all edges f which lie in G', will be enough to trigger the process to infect all $\binom{V(G)}{2}$ edges.

There is a second problem which is more serious. Indeed, the above sketch shows how to infect all of the edges on $V(G)$ but in doing so, we added new vertices for our small chains $(H_i, e_i)_{i\in[\ell]}$ used to infect the edges f. There is no guarantee that the edges incident to these new vertices will be infected by this process. Indeed, the small chains were chosen to be vertex-disjoint from $V(G)$ *precisely because* we do not want the process on the small chains $(H_i, e_i)_{i\in[\ell]}$ and the process on G to interact. One way to guarantee that vertices gain many neighbours in the H-process is if they have $\delta(H)-1$ neighbours in a set that will eventually become a large clique, as $V(G)$ will become after we ensure that all edges on $V(G)$ become infected.

Lemma 4.6 *Let G be a graph, and $U, W \subseteq V(G)$. If W is a clique of size at least $v(H) - 1$ in $\langle G \rangle_H$, and every vertex in U has at least $\delta(H) - 1$ G-neighbours in W, then $U \cup W$ is also a clique in $\langle G \rangle_H$.*

Proof Let $u \in U$, and let $w_1, \dots, w_{\delta(H)-1} \in W$ be pairwise distinct G-neighbours of u. For $w \in W \setminus \{w_1, \dots, w_{\delta(H)-1}\}$, any map $\varphi : V(H) \to V(G)$ that sends a vertex of minimum degree to u, its neighbourhood to $\{w, w_1, \dots, w_{\delta(H)-1}\}$, and the remaining vertices of H to the rest of W is an embedding of H minus an edge into $\langle G \rangle_H$. Thus uw completes a copy of H and there must be a complete bipartite graph between U and W in $\langle G \rangle_H$. All edges in $\binom{U}{2}$ then complete some copy of H with other vertices chosen from W and so indeed $U \cup W$ is a clique in $\langle G \rangle_H$. □

We will use this to trigger new vertices to join a clique with $V(G)$. However, we cannot add $\delta(H) - 1$ edges from every new vertex to $V(G)$ without interfering with the H-process on G. We will be able to do this for some subset U of the new vertices though. This leads us to consider a new gadget which is similar to a simple H-chain, but has the added property that if we trigger the infection of a clique on some vertices U in the middle of the chain then the infection will cause the whole gadget to percolate. The following lemma details what we need from this gadget.

Lemma 4.7 (The gadget graph) *For any $(2,1)$-inseparable H on k vertices, there exists a graph Γ on $\ell \leq k^{4k}$ vertices together with an edge $e \in E(\Gamma)$ and a non-edge $f \in \binom{V(\Gamma)}{2} \setminus E(\Gamma)$ such that*

1. *$\Gamma - e$ is H-stable (no edge is added in the H-process on $\Gamma - e$) and $f \in E(\langle \Gamma \rangle_H)$,*
2. *$\mathrm{dist}_\Gamma(e,f) \geq k$,*
3. *there is $U \subseteq V(\Gamma)$ such that $\Gamma \cup \binom{U}{2}$ is H-percolating and $\mathrm{dist}_\Gamma(U, e \cup f) \geq k$.*

Proof Let $\ell := \lceil \frac{2k^{3k+3}-2}{k-2} \rceil$, and let $(H_i, e_i)_{i \in [\ell]}$ be a simple H-chain of length ℓ. Label the vertices of the chain by $v_1, \ldots, v_r$, where $r := 2 + (k-2) \cdot \ell \geq 2k^{3k+3}$, such that the ith copy of H has vertex set $\{v_j : j \in [(i-1) \cdot (k-2) + 1, i \cdot (k-2) + 2]\}$. Write $\delta := \delta(H)$. Define Γ by

$$V(\Gamma) = \{v_1, \ldots, v_r\} \quad \text{and} \quad E(\Gamma) = \cup_{i=1}^{\ell} E(H_i) \setminus \{e_1, \ldots, e_\ell\} \cup E' \cup E'',$$

where

$$E' = \Big\{ v_i v_j : i \in [k^{3k}], j \in k^2 \cdot [(i-1)(\delta-1)+1, i(\delta-1)] \Big\}$$
$$E'' = \Big\{ v_{r+1-i} v_{r+1-j} : i \in [k^{3k}], j \in k^2 \cdot [(i-1)(\delta-1)+1, i(\delta-1)] \Big\},$$

and let $e := v_1 v_2$, $f := e_\ell$, and $U := \{v_i : i \in [k^{3k}+1, r-k^{3k}]\}$.

To show (1), note that e_i appears by step i in the H-process on $\cup_{i=1}^{\ell} H_i - \{e_1, \ldots, e_\ell\}$ and in particular f appears by step ℓ. Since $\cup_{i=1}^{\ell} H_i - \{e_1, \ldots, e_\ell\} \subseteq \Gamma$ we obtain that $f \in E(\langle \Gamma \rangle_H)$. To see that $\Gamma - e$ is H-stable we show that there are no copies of H minus an edge in $\Gamma - e$. Indeed, if F is such a copy then F is not contained in $\cup_{i=1}^{\ell} H_i - \{e, e_1, \ldots, e_\ell\}$ because $(H_i, e_i)_{i \in [\ell]}$ is a simple H-chain (as noted after Observation 4.5). For this reason at least one edge of F lies in $E' \cup E''$. Without loss of generality we can assume that $E(F) \cap E' \neq \varnothing$ and $E(F) \cap E'' = \varnothing$ (using here that the largest vertices in an edge of E' are of distance greater than k in Γ from the smallest vertices in an edge of E''). Pick the largest $j \in [r]$ such that v_j is an endpoint of an edge e' in $E(F) \cap E'$. Note that $N^{k-1}_{F-e'}(v_j) \subseteq \{v_i : |i-j| \leq (k-1)^2\}$ where $N^{k-1}_{F-e'}(v_j)$ denotes all the vertices that can be reached from v_j by a path in $F - e' \subseteq \Gamma - e'$ with at most $k-1$ edges. We used here that any edge $e'' \in E'$ with endpoints $v_{i''}, v_{j''}$ and $i'' < j'' < j$ must have $j'' \leq j - k^2$. Hence there can be no path between v_j and the other endpoint of e' in $F - e'$, which contradicts the assumption that H is $(2,1)$-inseparable.

As to (3), by construction every vertex v_i with $i \leq k^{3k}$ has at least $\delta - 1$ neighbours in $\{v_{i+1}, \ldots, v_{\lfloor r/2 \rfloor}\}$, and for $i \geq r - k^{3k} + 1$, v_i is adjacent to at least $\delta - 1$ vertices in $\{v_{\lceil r/2 \rceil}, \ldots, v_{i-1}\}$. By repeated applications of Lemma 4.6, we see that $\Gamma \cup \binom{U}{2}$ indeed H-percolates. Let $s := \mathrm{dist}_\Gamma(e, U)$ and choose a shortest path $v_{i_1} \ldots v_{i_s}$ path from e to U. For any $i \in [k^{3k}]$ and $j \in [r]$ with $v_i v_j \in E(\Gamma)$, we have $j \leq i(\delta-1)k^2$. Therefore,

$$k^{3k} < i_s = i_1 \cdot \frac{i_2}{i_1} \cdot \ldots \cdot \frac{i_s}{i_{s-1}} \leq 2 \cdot (\delta-1)^{s-1} k^{2(s-1)} < k^{3s},$$

and hence $s \geq k$. Similarly, $\mathrm{dist}_\Gamma(f, U) \geq k$.

It remains to prove (2). As there are no edges between $\{v_1, \ldots, v_{k^{3k}}\}$ and $\{v_{r-k^{3k}+1}, \ldots, v_r\}$, any path from an endpoint of e to an endpoint of f must use a vertex from U. This implies $\mathrm{dist}_\Gamma(e, f) \geq \mathrm{dist}_\Gamma(e, U) \geq k$. □

This puts us in good stead but we still need to find a way to attach the set U for each new gadget to $\delta(H)-1$ vertices in $V(G)$ without interfering with the process on G. To do this, we simply add an independent set I to G and consider the auxiliary H-percolating graph G' with linearly many edges as a graph on vertex set $V(G) \cup I$. Therefore our gadgets will force that $V(G) \cup I$ hosts a complete graph and we can attach the vertex sets U to vertices in I to trigger the percolation on the individual gadgets. One more idea is needed to guarantee that the edges between U and I for different gadgets do not interact with each other to give unwanted copies of H minus an edge. We will use a bipartite graph of large girth to place these edges between gadgets and I, recalling that the *girth* of a graph is the length of the shortest cycle in the graph. Specifically, we need the following.

Lemma 4.8 *Let $k, d \in \mathbf{N}$ be fixed. Provided that $n \in \mathbf{N}$ is sufficiently large, there exists a bipartite graph B with partite sets X, Y of sizes $|X| = n$, $|Y| = 2kn$ such that every vertex in Y has degree d and the girth of B is at least $k+1$.*

Bipartite graphs with many edges and high girth have been studied thoroughly in the literature, see for example [63, 64, 50]. In fact, we will use them again in our construction of *line chains* in Section 7.4. We note that for our purposes we do not require the full strength of those results because the parameters k and d are fixed constants. Indeed, Lemma 4.8 can be obtained easily by considering a random bipartite graph with parts X of size n and Y' of size $(2k+1)n$, with each edge present independently with probability $p = n^{-1+1/(2k)}$. By an application of Chernoff's theorem, a.a.s. all vertices in Y' will have degree at least d and by Markov's inequality, a.a.s. there will be at most n cycles of length at most k. Therefore a.a.s. both these events happen and we can obtain the graph B in Lemma 4.8 by first deleting from Y' any vertex that participates in a cycle of length at most k and then deleting further vertices of Y' and edges to get Y of size exactly kn with each vertex incident to d edges. Finally then, we give the details of the proof of Theorem 4.2

Proof of Theorem 4.2 Let H be a $(2,1)$-inseparable k-vertex graph and G an n-vertex graph with running time $\tau_H(G) = M_H(n)$. We construct a super-graph $\tilde{G} \supseteq G$ such that $\tau_H(\tilde{G}) \geq \tau_H(G)$ and $\tilde{G}$ is H-percolating. Introduce an independent set I of n new vertices, and let G' be an auxiliary graph with vertex set $V(G) \cup I$ such that $|E(G')| \leq 2kn$ and $\langle G' \rangle_H = K_{2n}$. Let Γ be the graph from Lemma 4.7 with $\ell := v(\Gamma)$ vertices, and for each $f \in G^* := G' \setminus \langle G \rangle_H$ introduce a set W^f of $\ell - 4$ new vertices. Fix an edge $e_\tau \in G_\tau \setminus G_{\tau-1}$ and place a copy Γ^f of Γ on $W^f \cup e_\tau \cup f$ such that for some subset $U^f \subseteq W^f$, (1)–(3) of Lemma 4.7 hold for e_τ, f, U^f (with e_τ playing the role of e).

Consider an auxiliary bipartite graph B between I and $\{U^f : f \in G^*\}$ such that

(i) $\deg_B(U^f) = d := |U^f| \cdot (\delta(H) - 1)$ for all $f \in G^*$,

(ii) B has girth at least $k+1$.

Such an auxiliary graph exists by Lemma 4.8 whenever n is sufficiently large, deleting arbitrary vertices from the set Y to get a set of size $e(G^*)$, if necessary. For each $f \in G^*$ choose a partition $N_B(U^f) = \cup_{u \in U^f} N_u$ into $|U^f|$ sets of size $\delta(H)-1$. Define $\tilde{G}$ by

$$V(\tilde{G}) = V(G) \cup I \cup \left(\cup_{f \in G^*} W^f\right),$$
$$E(\tilde{G}) = E(G) \cup \left(\cup_{f \in G^*} \left(E(\Gamma^f - e_\tau) \cup \{vu : u \in U^f, v \in N_u\}\right)\right).$$

With the above choices there is an edge in the auxiliary graph B between $v \in I$ and U^f whenever v has a $\tilde{G}$-neighbour in U^f. Note also that $\tilde{G}$ has

$$v(\tilde{G}) \leq 2n + 2kn(\ell - 4) \leq 4k^{4k}n$$

vertices, as required.

First we show that $\tilde{G}$ percolates. We have $G_\tau \subseteq \tilde{G}_\tau$ since $G \subseteq \tilde{G}$. For each $f \in G^*$, $\Gamma^f \subseteq \tilde{G}_\tau$ and hence $\langle \Gamma^f \rangle_H \subseteq \langle \tilde{G} \rangle_H$. In particular, $f \in E(\langle \tilde{G} \rangle_H)$. Therefore, $G' \subseteq \langle \tilde{G} \rangle_H$. Recall that G' is H-percolating and so $V(G') = I \cup V(G)$ is a clique in $\langle \tilde{G} \rangle_H$. Every vertex in $\cup_{f \in G^*} U^f$ has $\delta(H) - 1$ $\tilde{G}$-neighbours in $V(G')$, so $V(G') \cup \left(\cup_{f \in G^*} U^f\right)$ is a clique in $\langle \tilde{G} \rangle_H$ by Lemma 4.6. Since $\Gamma^f \cup \binom{U^f}{2}$ percolates, $V(\Gamma^f)$ is a clique in $\langle \tilde{G} \rangle_H$. Each of the sets $V(\Gamma^f)$, $f \in G^*$, intersects $V(G') \cup \left(\cup_{f \in G^*} U^f\right)$ in at least $\delta(H) - 1$ vertices. Therefore, repeated applications of Lemma 4.6 guarantee that $\langle \tilde{G} \rangle_H$ is a complete graph.

Finally, we prove that $\tau_H(\tilde{G}) \geq \tau$. We show by induction that for all $0 \leq t \leq \tau$

$$E(\tilde{G}_t) = E(G_t) \cup (E(\tilde{G}) \setminus E(G)).$$

For $t = 0$, this simply holds by definition. For larger t, suppose that we have proved the induction hypothesis up to step $t-1$. This means that the only edges that have been added are the edges in the H-process on G up to time $t-1$. In particular, the neighbourhood of any vertex $v \in I$ in $\tilde{G}_{t-1}$ is the same as its neighbourhood in $\tilde{G}$.

Let F be a copy of H minus an edge in $\tilde{G}_{t-1}$. Suppose first that $V(F)$ intersects I. As $N_{\tilde{G}_{t-1}}(v) = N_{\tilde{G}}(v) \subset \cup_{f \in G^*} U^f$, the connectedness of F and the inequality $\mathrm{dist}_{\Gamma^f}(e_\tau \cup f, U^f) \geq k$ imply that $V(F) \subseteq I \cup \left(\cup_{f \in G^*} W^f\right)$. Every $v \in V(F) \cap I$ has at most one F-neighbour in U^f for each $f \in G^*$, and hence

$$|\{f \in G^* : U^f \cap N_F(v) \neq \varnothing\}| \geq \delta(F) > 2,$$

using that H is $(2,1)$-inseparable in the last inequality here. Moreover, for every $f \in G^*$ such that $U^f \cap V(F) \neq \varnothing$, we must have that there are at least two B-neighbours of U^f in I as otherwise we could disconnect F by removing one vertex, contradicting the $(2,1)$-inseparability of H. Therefore we have that $B[(I \cap V(F)) \cup \{U^f : U^f \cap V(F) \neq \varnothing\}]$ has minimum degree greater than two and therefore contains a cycle. Each vertex in this cycle corresponds to (at least one) vertex of $V(F)$ and these vertices are distinct. Hence the cycle has length at most $v(F) = k$ which contradicts the assumption that the girth of B is at least $k+1$.

We have shown that $I \cap V(F) = \varnothing$, or equivalently, $V(F) \subseteq V(G) \cup \left(\cup_{f \in G^*} W^f\right)$. If $V(F) \cap W^f \neq \varnothing$ for some $f \in G^*$ and $V(F) \not\subseteq V(\Gamma^f)$ then either $f \cap V(F)$

or $e_\tau \cap V(F)$ is a vertex-cut of H' because $\mathrm{dist}_{\Gamma^f}(e_\tau, f) \geq k$. This, however, would contradict the assumption that H is $(2,1)$-inseparable. Consequently, $V(F) \subseteq V(G)$ and $F \subseteq G_{t-1}$ by the induction hypothesis. Therefore any edge that is added during the step t of the F-process on $\tilde{G}$ belongs to $G_t \setminus G_{t-1}$ and we are done. □

We remark that the constant $C = C(H)$ appearing in Theorem 4.2 is probably unnecessary and a similar result should be achievable with a new starting graph $\tilde{G}$ on $(1+o(1))n$ vertices. An approach to do this by packing gadgets onto $o(n)$ vertices using a high girth (approximate) design was suggested by the anonymous referee. We did not pursue this direction, but we believe that it should indeed be possible.

5 (Almost) quadratic running times

In this section, we begin our exploration of $M_H(n)$ for general infection rules H. Our first point of interest will be to what extent the behaviour of large cliques (of size at least five) generalises to other infection rules H. The results of this section are from our paper [47] and are achieved through a program of building a general framework for chain constructions, in particular exploiting the ideas behind the ladder chains and dilation chains that we already encountered. This method turns out to be flexible and we are able to show extremely slow, that is (almost) quadratic, running times for a large array of infection rules H, including many that are far from resembling the large cliques that we set out to generalise.

For one of our almost quadratic examples, namely the wheel graph, we are also able to complement our lower bound with a non-trivial sub-quadratic upper bound. This establishes a new behaviour for maximum running times, exactly the type Conjecture 2.5 predicts for K_5.

5.1 Minimum degree

In our first generalisation of the results of Balogh et al. [14] for cliques, we establish Dirac-type conditions for quadratic and almost quadratic running times.

Theorem 5.1 ([47]) *Suppose $5 \leq k \in \mathbf{N}$ and H is a graph with $v(H) = k$. We have that*

$$M_H(n) \geq \begin{cases} \Omega(n^2) & \text{if } k \geq 6 \text{ and } \delta(H) > 3k/4; \quad (5.1) \\ n^{2-O(1/\sqrt{\log n})} = n^{2-o(1)} & \text{if } \delta(H) \geq k/2+1. \quad (5.2) \end{cases}$$

These two bounds extend the results of [14] for K_k, $k \geq 6$ and for K_5, respectively, to arbitrary infection rules that are dense enough in terms of their minimum degree. Later we will see examples of graphs with minimum degree $k/2$ and running time polynomially separated from quadratic, so the second minimum degree condition is optimal. In fact, as we will see, Theorem 5.1 is best possible not only in guaranteeing almost quadratic running time, but even just super-linear.

A question that remains open is to fully understand the behaviour of quadratic maximum running time in terms of the minimum degree.

Problem 5.2 *Determine the smallest constant $c_{\min} > 0$, such that any infection rule H with $v(H) = k$ sufficiently large and $\delta(H) > c_{\min}k$ has $M_H(n) = \Theta(n^2)$.*

In particular, is there a graph H with $\delta(H) \geq k/2 + 1$ that exhibits a sub-quadratic running time $M_H(n) = o(n^2)$?

5.2 Random graphs

Theorem 5.1 already establishes that cliques of size at least 5 are far from unique in having maximum running time that is (almost) quadratic. Our next result shows that in fact *almost all* infection rules on n vertices have quadratic running time. In fact, unlike Theorem 5.1, where the density is a large constant, for typical infection rules H, a very sparse density turns out to be sufficient to cause quadratic behaviour.

Theorem 5.3 **([47])** *Let $H = G(k,p)$. Then, with probability tending to 1 as $k \to \infty$, we have that*

$$M_H(n) = \begin{cases} O(1) & \text{if } p \leq \left(\frac{1}{2} - o(1)\right) \log k/k; & (5.3)\\ \Omega(n^2) & \text{if } p = \omega(\log k/k). & (5.4)\end{cases}$$

Theorem 5.3 shows a stark phase transition in the behaviour of $H = G(k,p)$ around the connectivity threshold $\log k/k$. We remark that $M_H(n)$ is not necessarily monotone with respect to H and so there is no reason, a priori, to expect such a threshold type result for $H = G(k,p)$.

It is natural to ask whether this threshold is sharper than presented here. The 0-statement, asserting that $M_H(n)$ is constant when p is small, follows from the fact that an isolated edge occurs a.a.s. (see for example [30, Theorem 5.4]). The 1-statement, giving quadratic running time for p much larger than $\log k/k$, is more delicate and in particular, we use that $H = G(k,p)$ is self-stable, i.e. the H-process is stabilises on H as the starting graph. For this we apply a result of Kim, Sudakov and Vu [59]. Whilst it is believable that results on the asymmetry of $G(k,p)$ can be strengthened to give that $H = G(k,p)$ is a.a.s. self-stable as soon as $p > \log k/k$, we did not find such a result in the literature and it is not immediate how one could prove such a statement. Nonetheless, we believe it is likely that our proof can be pushed to give $M_H(n) = \Omega(n^2)$ a.a.s. already when $p > c_1 \log k/k$ for any $c_1 > 1$. The range $\log k/2k < p < \log k/k$ seems to be the most interesting and it is unclear what to expect for the behaviour of $M_H(n)$ with $H = G(k,p)$.

Problem 5.4 *Determine the typical behaviour of $M_H(n)$ for the random infection rule $H = G(k,p)$ in the range $\left(\frac{1}{2} + o(1)\right) \frac{\log k}{k} < p = O(\log k/k)$.*

With the additional condition $p < \frac{\log k}{k}$, the infection rule H in the problem will a.a.s. have one giant component and some isolated vertices, which can be ignored from the perspective of running times. There will also be vertices of degree one which suggests that our chain constructions will be useless for proving lower bounds on $M_H(n)$ as copies of H minus an edge will be abundant and so no analogue of the key property **(*)** from Lemma 2.3 can hold. Conversely, as we will see in Section 7.1, the existence of degree 1 vertices alone is not enough to be able to prove effective upper bounds on running times.

It would also be very interesting to determine the typical behaviour of $M_H(n)$ for H being a random d-regular graph, which we expect to be (almost) quadratic even for $d \geq 3$. Such results would also be helpful to be combined with the methods of Section 7 to generate interesting novel behaviours.

5.3 Constant average degree

The intuitive rule of thumb for the possibility of slow running time is that the infection rule H is dense. The rationale behind this is that dense graphs in chains are easier to hide from interfering with other chains due to their complex internal structure, while sparse graphs have much less such connections. Theorem 5.3 about the random graph $H = G(k, p)$ already indicates the limitations of this intuition, showing that a super-logarithmic average degree is enough for a typical infection rule H to have a slow running time. Our next results go further, identifying two very sparse explicit families with almost quadratic running time.

5.3.1 The square of the Hamilton cycle The first result is about the 4-regular family of the *squares of Hamilton cycles.* To construct the square of the Hamilton cycle one orders the k vertices cyclically and connects any two of them whose distance is at most two.

Theorem 5.5 ([47]) *If H with $k \geq 5$ vertices contains the square of a Hamilton cycle, then*

$$M_H(n) \geq n^{2-O(1/\sqrt{\log n})} = n^{2-o(1)}.$$

Note that Theorem 5.5 also generalises the result on K_5 of Balogh et al. [14] as the square of a cycle of length 5 is K_5. The fact that the result only requires that the infection rule *contains* something is noteworthy. As we will see in Section 7 the function $M_H(n)$ is very far from being monotone in the infection rule H.

5.3.2 The wheel Our second result establishes an almost quadratic lower bound for the maximum running time of a sequence of graphs with average degree strictly less than four. The *wheel graph* W_k is defined to be the graph with $k+1$ vertices obtained by taking a cycle of length k and adding a new vertex that is adjacent to all vertices of the cycle. The striking feature of this example is that we can also couple the lower bound with a non-trivial upper bound. In particular this introduces a completely new behaviour of possible maximum running times, which is sub-quadratic but larger than any polynomial of degree smaller than 2. This can be seen as evidence for the validity of Conjecture 2.5 which predicts a similar behaviour for K_5.

Theorem 5.6 ([47]) *Let $k \geq 7$ be an odd integer. The wheel graph W_k satisfies*

$$M_{W_k}(n) \geq n^{2-o(1)} \quad \text{and} \quad M_{W_k}(n) = o(n^2).$$

In general, upper bounds on $M_H(n)$ are hard to come by and require proving that the process stabilises in a bounded number of rounds *no matter which n-vertex* starting graph G_0 is given. One key idea in our proof is that chains $(H_i, e_i)_{i\in[\tau]}$, which are incredibly useful for lower bounds (see Section 5.4 below), can also be used for upper bounds! Indeed, when proving upper bounds, we have very little information about the graphs G_i in the process whose running time we are trying to bound. However in any H-process $(G_i)_{i\geq 0}$ with running time $\tau := \tau_H(G_0)$ we can embed an H-chain $(H_i, e_i)_{i\in[\tau]}$. Indeed, starting with some edge e_τ added at time τ there is a copy H_τ of H completed by e_τ. The edge $e_{\tau-1}$ is then chosen as an edge of H_τ that is added at time $\tau - 1$, which must exist as otherwise e_τ would be added before time τ. Defining $H_{\tau-1}$ to be a copy of H completed by $e_{\tau-1}$ and continuing this process, we can define the full chain $(H_i, e_i)_{i\in[\tau]}$. Although the starting graph of this chain may not be G_0, or even contained in G_0, the chain still gives us some concrete structure in which to work on for upper bounds.

Another interesting feature of our proof of the upper bound in Theorem 5.6 is that we use the famous *Ruzsa-Szemerédi* $(6,3)$*-problem/theorem* [78] which states that any n-vertex 3-uniform hypergraph with no 6 vertices inducing 3 or more edges, must have $o(n^2)$ edges.

5.4 Proof technique: Chain constructions

All of the lower bounds stated above as well as several further results later in the survey follow from various chain constructions. Part of the challenge for applying chain constructions to infection rules H that are not complete is to find the "correct" way to generalise definitions and properties. In [47], we axiomatised the procedure, giving weak conditions on H-chains $(H_i, e_i)_{i\in[\tau]}$ that allow us to control the running time of the process based on a chain. The resulting notion of *proper* chains is somewhat technical. Suffice it to say here that we derive an analogue to Lemma 2.3 which gives lower bounds on running times for proper chains, with the key property being analogous to the condition **(*)** asking to avoid "unwanted" copies F of H minus an edge, that span several H_i. We also placed the linking chains that we saw for K_5 into this general framework, which allowed us to simply exhibit some collection $\mathcal{H}$ of simple H-chains which satisfies the notion of properness. In particular, we need to avoid "local" unwanted copies F of H minus an edge, that is, those that are induced by one segment chain $(H_i^a, e_i^a)_{i\in[\tau_0]}$ in the collection $\mathcal{H}$ but do not lie in a single $V(H_i)$, and "crossing" copies F of H minus an edge, that span the edges of several different chains in $\mathcal{H}$.

Avoiding local copies F of $H - e$ for some edge $e \in H$ essentially boils down to showing that we cannot place such a copy F on the edges of a simple H-chain $(H_i, e_i)_{i\in[\tau_0]}$ without placing it on $V(H_i)$ for some $i \in [\tau_0]$. For $H = K_5$, we noticed this via Observation 2.4. For general infection rules H, there is no guarantee that we cannot separate a copy F of $H - e$ across a vertex cut of size 2 (given by some edge e_i of the chain). This motivates the definition of $(2,1)$-inseparability from Section 4, which provides a simple condition on H that suffices to avoid the local unwanted copies of $H - e$.

As with K_5, in many cases we can forbid the presence of crossing unwanted copies F of $H - e$ by placing *dilation chains* $\mathcal{H}$ according to an appropriate dilation set A which avoids solutions to certain linear equations and placing one simple H-chain corresponding to each dilation. Through Fact 3.5, we showed that an unwanted copy of K_5^- leads to a triangle spanning more than one chain. In some of our applications in this section, for example for the wheel graph W_k, we cannot guarantee this. We can however work with a slightly weaker concept.

Definition 5.7 A graph F is called *Behrendian* if for any non-monochromatic colouring of the edges in F there is a non-monochromatic cycle $C \subseteq F$ which is the union of two or three monochromatic paths.

The non-monochromatic cycles from Definition 5.7 can still be avoided by taking a dilation set A which is of Behrend type, as in Theorem 3.2, and choosing the constant bounding the size of the coefficients of the equations to be sufficiently large in terms of $k = v(H)$. We therefore arrive at the following theorem.

Theorem 5.8 ([47]) *Let H be a $(2,1)$-inseparable graph such that $H - e$ is Behrendian for every $e \in H$. Then $M_H(n) \geq n^{2-o(1)}$.*

Theorem 5.5 and the second part of Theorem 5.2 are corollaries although checking both the $(2,1)$-inseparability and Behrendian property is not immediate. The wheel does not follow from Theorem 5.8 as it is $(2,1)$-separable. Indeed, it has vertices of degree three that can be disconnected from the rest of the graph by removing an incident edge and the two other adjacent vertices. Nonetheless, we can apply our machinery by explicitly building simple W_k-chains that avoid local unwanted copies of the wheel minus an edge.

For the quadratic lower bounds given in this section, we use *ladder chains* similar to those used for K_6, see Figure 2. We split the vertex set $V(H)$ into $L \cup R$, place a sequence of copies of $H[L]$ on the left with consecutive copies intersecting in a single vertex and similarly on the right. Each simple H-chain $(H_i^a, e_i^a)_{i \in [\tau]}$ in our collection will then be placed with H_i choosing a copy of $H[L]$ from the left and a copy of $H[R]$ according to some *slope* a. For K_6, both $H[L]$ and $H[R]$ were simply copies of K_3 and this greatly simplifies the analysis of this construction. Indeed, it is not hard to show that any unwanted copy of K_6^- will have to use one of the triangles on the left and one on the right. To do this for our applications of dense graphs and random graphs is considerably more intricate. Indeed, for the dense graphs we have very little information about the structure of H and the random graphs $H = G(k,p)$ that we treat can be very sparse.

For the first part of Theorem 5.2, as with Theorem 5.8 above, we actually derive a more general theorem. We say that a graph H is *$(\ell, 1)$-inseparable* if it cannot be disconnected by the removal of an edge and at most ℓ vertices.

Theorem 5.9 ([47]) *If H is $(\lceil v(H)/2 \rceil, 1)$-inseparable, then $M_H(n) = \Omega(n^2)$.*

Further natural examples of k-vertex $(\lceil k/2 \rceil, 1)$-inseparable graphs are complete tripartite graphs with no part larger than $\lfloor k/2 \rfloor - 2$ and t^{th} powers of cycles for $t \geq k/4 + 1$, which are obtained by cyclically ordering the vertices and connecting each vertex to all vertices within distance t in the ordering.

6 Sub-linear running times

After seeing infection rules with long running times, we now jump to the other end of the spectrum and explore infection rules for which the process stabilises fast.

6.1 Trees

We begin with an example.

Example 6.1 For the path P_t with $t \geq 2$ vertices, $M_{P_t}(n) \leq 3$.

Proof If $t = 2$, every edge of K_n is infected immediately, so $M_{P_2}(n) = M_{K_2}(n) = 1$. More generally, unless the starting graph is P_t-stable, the first round creates a copy P of P_t. In the second round, both leaves of P become adjacent to every other vertex (including vertices of P). Moreover, as at this point every vertex in $V(G_0)$ is a leaf of a copy of P_t, in the third round all these vertices also become universal and the process terminates with K_n. □

Infection rules H with a pendant vertex seem to spread the virus very fast: Any vertex in a copy of H with a pendant neighbour in H will become almost universal, that is, connected to every vertex outside the copy of H, in the next round. In particular in every tree T-process, after just the second round every leaf-neighbour within any copy of T created in the first round is already almost universal.

Does this mean that there is an absolute constant that bounds the maximum running time for every tree? This is not quite true; one can witness this with the tree having just a single leaf-neighbour: the star.

Example 6.2 Let $T = K_{1,t-1}$ be the t-vertex star. Then $M_T(n) = t - 1$ for all n sufficiently large.

Proof Let us first see that if a T-process goes on for $t-2$ rounds then it stabilises in the next round. Observe that the centre vertex c_i of a copy of $K_{1,t-1}$ created in the ith round for some $i \leq t-2$, becomes universal at round i. This means that after round $t-2$ every vertex has at least $t-2$ neighbours. So all non-universal vertices are now centres of $K_{1,t-1}$ with one edge missing, and in the next round all vertices become universal and the T-process stabilises by percolating.

For the lower bound let us take G_0 to be the disjoint union of the stars $K_{1,s}$ with centre c_i, $1 \leq s \leq t-2$ and a set W of $n - \binom{t}{2} + 1$ isolated vertices. We aim to show that the T-process $(G_i)_{i\geq 0}$ on G runs for $t-1$ steps before stabilising, that is, at least one edge is infected in round $t-1$ (and not before). We can see by induction that for $1 \leq i \leq t-3$, the graph G_i after round i is the union of the starting graph G_0 and all incident edges to the i vertices $c_{t-i-1}, \ldots, c_{t-2}$. Therefore in G_{t-3} the vertices of W are still independent and have degree $t-3$. Vertex c_1 becomes universal in round $t-2$, raising the degree of the vertices in W to $t-2$. Then in round $t-1$ all edges within W will get infected. That means the process did not stabilise before round $t-1$, so $\tau_H(G_0) \geq t-1$, as desired. □

This example shows that, unlike for the case of paths, the maximum running time does depend on the size of the star. Nevertheless it does not depend on the order n of the starting graph and this is what turns out to extend to any tree T. Our upper bound gives a quadratic dependence on the number of vertices of T.

Theorem 6.3 ([46]) *Let $t \in \mathbf{N}$ and $n \geq 2t$. Every tree T on t vertices satisfies*

$$M_T(n) \leq \frac{1}{8} \cdot (t^2 + 6t + 68).$$

As discussed above, a neighbour of a leaf in a copy of the target tree T in G_1 becomes an almost universal vertex during the second round, and hence the diameter of G_2 already becomes constant, so the statement of Theorem 6.3 perhaps does not come across as unexpected. The actual proof however, that from constant diameter the percolation process finishes in constant time, turned out to be a considerable challenge. Indeed, the analysis needed to handle all the great variety of trees and choices of starting graph is quite delicate. Moreover, extra ideas were needed to achieve an upper bound whose dependence on t is polynomial. An interesting question is to investigate how tight the upper bound of Theorem 6.3 is.

Problem 6.4 ([46]) *For $t \in \mathbf{N}$, determine*

$$M^*(t) := \max\{\limsup_{n\to\infty} M_T(n) : T \text{ is a } t\text{-vertex tree}\}.$$

We avoid simply taking a maximum here as it may be that small values of n interfere with the asymptotics and we take the lim sup instead of a limit as we cannot rule out the possibility that $M_T(n)$ oscillates with n. With this notation, Theorem 6.3 gives that $M^*(t) = O(t^2)$ and stars (Example 6.2) give $M^*(t) \geq t-1$. We expect that stars give the longest maximum running time.

Conjecture 6.5 ([46]) *For all $t \in \mathbf{N}$, we have $M^*(t) = t-1$.*

6.2 Cycles

As discussed in the introduction, the K_3-process always finishes in at most $\lceil \log_2(n-1) \rceil$ rounds, and taking the path P_n to be the starting graph realises this. This is still a quite fast infection spread, but not constant anymore. With the proof of this in mind it should be of no great surprise that the maximum running time of the C_k-process for any cycle length $k \geq 3$ is of order $\log_{k-1} n$. Indeed, the key observation again is that in one step of the process the distance between any two vertices along a shortest path between them is essentially reduced to its $1/(k-1)$-fraction. The next theorem determines the precise value of $M_{C_k}(n)$ for all $k \geq 3$ and all n sufficiently large, giving the first infinite family of graphs H (other than stars and paths) for which the exact running time $M_H(n)$ has been determined. Indeed, the only other non-trivial cases known is when $H = K_4$ [66, 74].

Theorem 6.6 ([45]) *Let $k \geq 3$. For sufficiently large $n \in \mathbf{N}$ we have*

$$M_{C_k}(n) = \begin{cases} \lceil \log_{k-1}(n + k^2 - 4k + 2) \rceil & \text{if } k \text{ is odd;} \\ \lceil \log_{k-1}(2n + k^2 - 5k) \rceil & \text{if } k \text{ is even.} \end{cases} \tag{6.1}$$

What is perhaps unexpected is that the jumps of the monotone increasing function $M_{C_k}(n)$, i.e. those $n \in \mathbf{N}$ with $M_{C_k}(n+1) = M_{C_k}(n) + 1$, behave differently in terms of k depending on the parity of k. For odd k the jumps are close to powers of $k-1$ while for even k the jumps are near one half times powers of $k-1$. Moreover, the precise location of the jumps is determined by a quadratic function of k in both the even and odd cases. This function turns out to be controlled by the *Frobenius number* of the numerical semi-group generated by $k-2$ and k, that is, the largest natural number that cannot be expressed as an integral linear combination of $k-2$ and k with non-negative coefficients.

The difference in behaviour between the odd and even case is also witnessed by the starting graphs that achieve the maximum running time. Whilst the path still provides the maximum running time $M_{C_k}(n)$ when k is odd, for even k the optimal starting graph is slightly different (see Figure 4) and the analysis of the C_k-process on this graph is more delicate.

Figure 4: A starting graph maximising the running time of the C_k-process for $k \in 2\mathbb{N}$.

7 Graph parameters and the running time

In Theorem 6.3 and Theorem 6.6 we saw that the maximum running time of the H-process is constant for trees and logarithmic for cycles, respectively. These are in stark contrast with cliques of order 4, where the running time of the process could be at least linear, and the wide array of graphs encountered in Section 5 where the running time could be (almost) quadratic. Which properties of the infection rule H cause the H-process to always finish quickly? This is what we study in this section, and try to identify necessary and sufficient conditions for sub-linear or linear maximum running time, in terms of relevant graph parameters, involving the degrees, the connectivity and the tree-width of the infection rule. In this section we focus on connected infection rules (see Section 9 for the disconnected case).

7.1 The role of the degrees

In our analysis of trees, the existence of leaves plays a crucial role in showing that the running time is constant. In the analysis of the percolation threshold $p_c(H)$ [13] in $G(n,p)$ the presence of a single pendant vertex in the infection rule H also plays a central role in pushing the threshold towards a sparser range (see Section 4.2). One cannot help but wonder whether the minimum degree being 1 has anything to do with small maximum running time in general. It turns out that, with the exception of cycles, it is indeed *necessary* for a sub-linear running time.

Theorem 7.1 ([46]) *Let H be a connected graph. If $\delta(H) \geq 2$ and $\Delta(H) \geq 3$ then $M_H(n) = \Omega(n)$.*

Corollary 7.2 *If H is connected and $M_H(n) = o(n)$ then $\delta(H) = 1$ or H is a cycle.*

The natural question arises: Is $\delta(H) = 1$ not only (essentially) necessary, but also sufficient for the H-process to stabilise in sub-linear time on an arbitrary starting graph? The following simple example points in this direction. It shows that adding a pendant vertex could reduce the maximum running time from quadratic to constant.

Example 7.3 ([46]) Let $k \geq 3$ and K_k^+ be the $(k+1)$-vertex graph formed by taking a clique of size k and adding a pendant edge. Then $M_{K_k^+}(n) \leq 3$.

Proof Let G_0 be an arbitrary n-vertex starting graph. Unless the K_k^+-process $(G_i)_{i\geq 0}$ immediately stabilises there is some copy of K_k^+ in G_1. Let $U \subset V(G)$ be the set of k vertices that form a clique in this copy of K_k^+. Then in G_2, we have that U and $V(G) \setminus U$ form the two parts of a complete bipartite graph. Thus any vertex in $V(G) \setminus U$ is contained in a k-clique and consequently in G_3 any edge within $V(G) \setminus U$ is added, making G_3 the complete graph. □

Somewhat surprisingly, despite all the above, a pendant vertex is *not* sufficient for a sub-linear or even sub-quadratic maximum running time.

Theorem 7.4 ([46]) *There is a connected H with $\delta(H) = 1$ and $M_H(n) = \Omega(n^2)$.*

This implies that in our quest towards a characterisation of connected graphs H with sub-linear running time we need to identify additional conditions besides the minimum degree being 1. Can limiting the maximum degree help? The graph H in Theorem 7.4 has a vertex of full degree (of degree 15, see Figure 5). The construction in our next result shows that even limiting the maximum degree to 3 is not helpful to guarantee sub-linear maximum running time.

Theorem 7.5 ([46]) *There exists a connected graph H with minimum degree $\delta(H) = 1$ and maximum degree $\Delta(H) = 3$ satisfying $M_H(n) = \Omega(n)$.*

We remark that the maximum degree condition in the theorem cannot be improved to 2 as paths have constant running time by Theorem 6.3. It would be very interesting to determine in general to what extent the limiting of the maximum degree of the infection rule contributes to small maximum running time. The following questions point in this direction.

Question 7.6 *Let H be a connected graph with $\delta(H) = 1$.*

- *What is the largest maximum running time that could occur if $\Delta(H) = 3$? Is it true that $M_H(n) = O(n)$?*
- *What is the smallest possible maximum degree $\Delta(H)$ of H if $M_H(n) = \Omega(n^2)$?*

It is possible to slightly modify the proofs of the above Theorems 7.4 and 7.5 to show that an additional condition in terms of the average degree of H is of no help in guaranteeing faster running time. This modification of Theorem 7.5 for example creates infection rules with minimum degree 1, average degree arbitrarily close to 2, and yet, quadratic maximum running time. On the other hand, connected infection rules with average degree *at most* 2 are unicyclic. These are trees with exactly one extra edge. It is possible, by a small modification of the proof of Theorem 6.3, to show that these, with the exception of cycles, have constant maximum running time.

7.2 Proof technique: Simulation

Let us consider Theorem 7.4 giving a graph H with $\delta(H) = 1$ and $M_H(n)$ quadratic. How to construct an infection rule with these intuitively contradicting properties? The initial high-level idea is to take an infection rule for which we already know how to produce a quadratic running time on some starting graph G, for example the infection rule K_6, and extend it to a larger infection rule H with a pendant vertex, such that running the H-process on an appropriate extension $\tilde{G}$ of

G we can ensure that the original K_6-process is somehow also taking place inside, and hence the H-process on $\tilde{G}$ will be at least as long as the K_6-process on G.

A good plan, a nice plan, easier to dream up though than to actually carry it out. The first idea would be just to attach a pendant edge to K_6, but this was shot down already in Example 7.3 to have maximum running time 3. The main problem with adding a pendant vertex to a "slow infection rule" becomes highly visible: the pendant edge is added immediately to any vertex in any copy of K_6, creating many universal vertices and a speedy finish.

To circumvent this we aim to add a larger structure to K_6 and add a pendant edge to that in such a way that the neighbour of the pendant vertex will always be "locked" into the same vertex of $V(\tilde{G})$. This will partly happen because of the asymmetry we create, partly because of other properties of the graph H. The following construction will work.

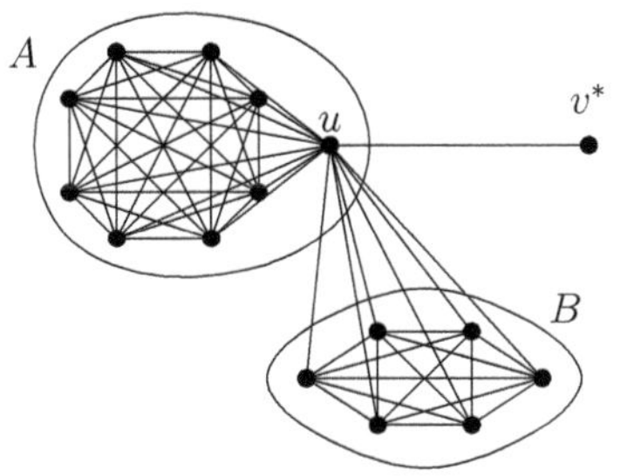

Figure 5: A graph H with minimum degree 1 and quadratic running time.

The graph H consists of three pairwise vertex-disjoint parts: A clique B of size six, a clique A of size nine with a special vertex u which is universal, and a pendant vertex v^* attached to it. As the starting graph $\tilde{G}$ we take the starting graph G of the slow K_6-process defined by Balogh et al. [14], together with a disjoint copy K of K_9, and make one of its vertices w fully connected to $V := V(G)$. Then any copy of $K_6 - e$ in V during the H-process, together with K and an appropriate pendant edge from w back to V forms a copy of $H - e$. So all edges that get infected in the K_6-process on G will also get infected in the H-process on $\tilde{G}$. We still need to make sure though, that the above described are *all* the copies of $H - e$ during the H-process on $\tilde{G}$ and hence the process does not speed up compared to the K_6-process on G, due to an unexpected copy of H minus one of its edges somewhere. The analysis of the K_6-process $(G_i)_{i \geq 0}$ given in [14] implies that no copy of K_7 appears for $\Omega(n^2)$ rounds. Using this and an induction over the rounds, one can show that in any copy of $H - e$ the set A has to be mapped to K and u to w, as required.

The basic idea of this *simulation construction* method is flexible and can be adapted to different settings. The two main ingredients are some *benchmark* infection rule B (this was K_6 above) whose process and running time we intend to simulate and an appropriate *anchor* graph A (this was K_9 above) to which we can connect the benchmark graph B as well as pendent edges to construct H. The starting graph will then be the starting graph for B and a copy of A attached to it. The aim is to show that A can only appear as itself in the copies of H of the H-process so unplanned copies of $H - e$ do not appear. To achieve this we need to construct an anchor graph which is "far enough" from anything that will ever appear in the simulated B-process on G.

For example, to prove Theorem 7.5, these contemplations motivate us towards the construction of the graph depicted in Figure 6. The benchmark infection rule B is C_6 with a chord which has at least linear running time by Theorem 7.1. The anchor graph A is a more complex ad hoc construction, together with carefully chosen attachment vertices u and u' (which are now different from each other to control the maximum degree). The proof that this construction works hinges on an extension [46] of Theorem 7.1 that guarantees that there is some B-process $(G_i)_{i\geq 0}$ with $\tau_B(G_0)$ linear, such that all of the G_i are bipartite. The crucial property of the anchor graph is that it contains many triangles, and so is far from bipartite.

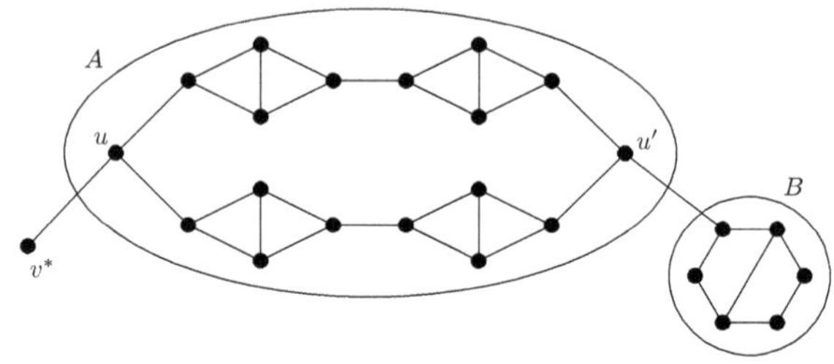

Figure 6: A graph H with $\delta(H) = 1$, $\Delta(H) = 3$ and $M_H(n) = \Omega(n)$.

The small average degree versions of Theorems 7.4 and 7.5 mentioned at the end of the previous section are obtained by adding arbitrarily many pendant edges to the neighbour of the pendant vertex in both constructions. The maximum running time does not change since these vertices are locked into their place by the anchor graph already. We discuss some further applications of this simulation construction method in Section 8.

7.3 The sub-linear regime

The characterisation of infection rules with sub-linear maximum running time remains a tantalising open problem.

Problem 7.7 *Characterise connected graphs H with $M_H(n) = o(n)$.*

To answer this question we need a characterisation of graphs $\hat{H}$ with a vertex $u \in V(\hat{H})$, such that for the graph $\hat{H}^{u+}$ obtained from $\hat{H}$ by attaching a pendant edge at vertex u, has sub-linear running time. Even though we do not do not have a precise conjecture for this question, we are confident that the possible maximum running times are quite limited in this range.

Conjecture 7.8 *If H is connected with $M_H(n) = o(n)$ then $M_H(n) = \Theta(1)$ or $\Theta(\log n)$.*

To gain some more insight, let us run the $\hat{H}^{u+}$ process on some starting graph G. How many rounds can there be when there is a newly infected edge playing the role of the pendant edge in a copy of $\hat{H}^{u+}$? Note that in such a *pendant round* the vertex playing the role of u immediately becomes almost universal, that is, adjacent to all but $v(\hat{H})$ vertices in $V = V(G)$. Now if it so happens that at least $v(\hat{H})^2$ such almost-universal vertices appear at some time i, then G_i contains a clique of size $v(\hat{H})$, and consequently the process will run for at most two more rounds due to the presence of the pendant edge in the infection rule. This shows that if there are more than be $v(\hat{H})^3$ pendant rounds then a very fast finish follows. Otherwise, the neighbour of the pendant vertex is locked into at most constantly many places, like in our simulation construction where it is locked into one. We would like to think of the process between two of these very few pendant rounds as resembling the $\hat{H}$-process. Then either $\delta(\hat{H}) = 1$ and we proceed by induction or $\delta(\hat{H}) \geq 2$ and the maximum running time is either logarithmic or linear time by Theorem 7.1.

The following strengthening of Conjecture 7.8 follows from the above heuristic.

Conjecture 7.9 *For any connected H and $u \in V(H)$ we have*

$$M_{H^{u+}}(n) = O(1) \quad \textit{or} \quad \Theta(M_H(n)).$$

Informally one can say that if $\delta(H) = 1$ then either one can control the almost universal vertices, in which case they should not affect the asymptotics of the maximum running time, or we cannot, which means that at least $v(\hat{H})^2$ appear quickly in the process, causing it to terminate in constant time.

7.4 The role of connectivity

In Theorem 7.1 we saw that all connected graphs H that have sub-linear running time also have minimum degree at most two. When considering graphs with linear running time this condition is no longer necessary, as we know from K_4 for example. Is there an analogue of this theorem for (at most) linear running time? At the end of this subsection we will see that the answer is no, there are graphs with linear maximum running time and arbitrarily large minimum degree.

It turns out however that the maximum running time being linear does instead imply a limit on the connectivity of H. The notion of $(2,1)$-inseparability (Definition 4.1), which played an important role in our chain constructions, also turns out to be the crucial notion in separating linear and super-linear maximum running times. The next result is an analogue of Theorem 7.1 for linear running times and shows that $(2,1)$-inseparable infection rules are super-linear in a strong sense.

Theorem 7.10 **([47])** *For any $(2,1)$-inseparable graph H we have*

$$M_H(n) = \Omega\left(n^{1+\frac{2}{3v(H)-2}}\right).$$

In particular if $M_H(n) = O(n)$ then H is $(2,1)$-separable.

This theorem is proved via yet another type of chain construction, namely *line chains*, where we create a family $\mathcal{H}$ of simple H-chains that is proper by placing one chain on each edge of a linear hypergraph $\mathcal{H}$, that is, a hypergraph in which pairs of edges intersect in at most one vertex (we then link the chains as we did in Section 3). The hypergraph $\mathcal{H}$ we use has large *girth*, meaning there is no small Berge cycle in $\mathcal{H}$, which helps to avoid unwanted copies of H minus an edge. For $\mathcal{H}$ we use bipartite graphs of large girth due to Lazebnik, Ustimenko and Woldar [64].

We remark that $(2,1)$-inseparability is not necessary for super-linear running time. Indeed, we saw in Theorem 5.6 that the wheel graph W_k, which has minimum degree three and hence is $(2,1)$-separable, has almost quadratic running time for any odd $k \geq 7$. Yet, in [46] we give a partial converse to Theorem 7.10 showing that H being $(2,1)$-separable is also sufficient for the H-process to have linear running time, provided H satisfies the extra condition that it *self-percolates*, i.e., the H-process percolates starting from H itself.

Theorem 7.11 ([46]) *If H is $(2,1)$-separable and $\langle H\rangle_H = K_{v(H)}$, then*

$$M_H(n) = O(n).$$

The proof adopts the growing clique idea used for K_4 as well as the existence of chains in all processes as used in Theorem 5.6 for the wheel. Although the condition $\langle H\rangle_H = K_{v(H)}$ is quite restrictive, Theorem 7.11 can still be used to generate a number of interesting examples. For example the graph K_5^-, that is, K_5 minus an edge, percolates on its own vertex set and it is $(2,1)$-separable, hence its maximum running time is at most linear. This is in contrast to K_5 which has (almost) quadratic running time [14]. We note that from the other direction Theorem 7.1 implies $M_{K_5^-}(n) = \Theta(n)$. The following is an infinite sequence of dense examples, showing that the minimum degree condition $k/2+1$ in Theorem 5.2 is best possible to guarantee not only almost quadratic, but also super-linear running times.

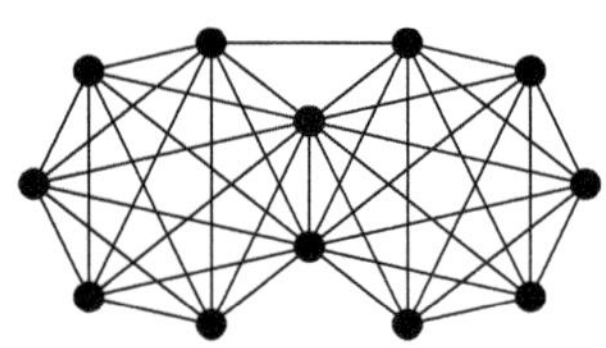

Figure 7: An illustration of the graph H_7'.

Example 7.12 Let $k \geq 3$ and H_k' be the $(2k-2)$-vertex graph composed by 'gluing together' two cliques of size k along a singular edge e and adding one more edge e' between two non-adjacent vertices. (An illustration of H_7' is given in Figure 7). Then $M_{H_k'}(n) = \Theta(n)$.

Proof The H_k'-process on H_k' results in K_{2k-2} after just one step of the process as any missing edge can play the role of e' in a copy of H_k'. Note, moreover that $\kappa(H_k' - e') = 2$ as removing the vertices of the edge e disconnects $H_k' - e'$. Therefore H_k' satisfies the assumptions of Theorem 7.11, and so indeed $M_{H_k'}(n) = \Theta(n)$, where we used Theorem 7.1 for the lower bound on running time here. □

7.5 A conjecture on tree-width

Given Theorem 6.3, another natural parameter to consider is the *tree-width* $\mathrm{tw}(H)$ of the infection rule H. It is well known that $\mathrm{tw}(H) \geq \delta(H) \geq \kappa(H)$. Example 7.3 shows that unlike small minimum degree and small connectivity, small tree-width is *not necessary* for small maximum running time, not even constant. Indeed, for the infection rule K_k^+ with maximum running time 3 we have $\mathrm{tw}(K_k^+) = \mathrm{tw}(K_k) = k-1$. However, it might still be the case that small tree-width is *sufficient* to imply an upper bound on maximum running time. Indeed, trees are the unique graphs H with $\mathrm{tw}(H) = 1$ and thus Theorem 6.3 is simply saying that any graph with tree-width 1 has constant running time. Graphs of tree-width 2 can already have linear running time, by considering the complete bipartite graphs $K_{2,s}$ and noting that $\mathrm{tw}(K_{k,\ell}) = \min\{k, \ell\}$.

Proposition 7.13 ([46]) *For every $s \geq 3$, $M_{K_{2,s}}(n) = \Theta(n)$.*

As the minimum degree of $K_{2,s}$ is two and for $s \geq 3$ the maximum degree is at least three, the lower bound on $M_{K_{2,s}}(n)$ follows from Theorem 7.1. The upper bound in turn establishes that Theorem 7.1 is best possible in the sense that no super-linear maximum running time could be guaranteed in general for graphs with $\delta(H) = 2$ and $\Delta(H) \geq 3$. The upper bound for Proposition 7.13 used a new idea of a partition of vertices which coarsens every 4 rounds in which the process does not stabilise. We believe that linear is the largest running time that an infection rule with tree-width two can have and pose the following conjecture.

Conjecture 7.14 *Any graph H with* $\mathrm{tw}(H) = 2$ *has* $M_H(n) = O(n)$.

Finally we remark that using bounds on tree-width to give effective upper bounds on running time only has the hope to work for very small values of tree-width, namely tree-width 1, as in Theorem 6.3 and tree-width 2 as in Conjecture 7.14. Indeed, the wheel graph W_7 has tree-width 3 and almost quadratic running time by Theorem 5.6.

8 Bipartite infection rules

In the frustrating world of upper bounds our next one is refreshingly simple, yet meaningfully general. Recall that the *extremal number* (or *Turán number*) of a graph, denoted $\mathrm{ex}(n, H)$, is the maximum number of edges of an n-vertex graph that does not contain H as a subgraph.

Proposition 8.1 ([47]) *Let H be a graph with at least two edges. Then*

$$M_H(n) \leq 2\mathrm{ex}(n, H).$$

Proof Let $(G_i)_{i \geq 0}$ be an H-process with $\tau = \tau_H(G_0) = M_H(n) \geq 2$. For $1 \leq i \leq \tau$, let $e_i \in G_i \setminus G_{i-1}$ be an edge added at time i. Now consider the graph G' with

$$E(G') = \{e_i : i \in [\tau], i = 1 \bmod 2\}.$$

We claim that G' has no copy of H. Indeed, if there is some copy F of H in G' and $i_* \in [\tau]$ is the maximal index such that $e_{i_*} \in E(F)$, then $F - e_{i_*} \subseteq G_{i_*-2}$, and so e_{i_*} would be added by time $i_* - 1$, contradicting the definition of e_{i_*}. Thus, $\tau/2 \le e(G') \le \mathrm{ex}(n, H)$, as required. □

As a consequence of Proposition 8.1 and the classical Kővári-Sós-Turán theorem [58], we obtain that for bipartite graphs H, the maximum running time $M_H(n)$ is polynomially separated from quadratic.

Corollary 8.2 *Let H be a bipartite graph such that the two partite sets of H have size r and s, respectively, where $1 \le r \le s$. Then*

$$M_H(n) = O(n^{2-1/r}).$$

At first sight, the upper bound of Proposition 8.1 seems awfully generous. For it to be tight, we should be able to construct a pair of H-free graphs that are both optimally dense up to a constant factor, such that their edges appear one by one in the even and odd rounds of an H-process, respectively. Dense H-free graphs are notoriously hard to come by for bipartite H [50], so good luck locating a magical starting graph producing two of them through the H-process!

Indeed, for many graphs, like trees and unicyclic graphs, we already know from Theorems 6.3 and 6.6 that this upper bound is not tight, by far in fact. For complete bipartite graphs with a part of size two we can also do much better, as their maximum running time is linear by Proposition 7.13, while the Turán number is of order $n^{3/2}$. Before completely disrespecting Proposition 8.1 though, we should reflect on what happens for non-bipartite graphs. The generous upper bound *is* tight [14] up to a constant factor for cliques of order at least 6, and in fact also for "most" graphs by Theorem 5.3. So maybe we are also in for a surprise for large enough bipartite H?

8.1 Small graphs

First we settle that the maximum running time of $K_{3,3}$ is significantly higher than that of $K_{2,s}$. This will follow from an analogue of Theorem 5.2, that provides a general lower bound for bipartite infection rules with large proportional minimum degree on both sides.

Theorem 8.3 ([47]) *Let $3 \le r \le s$ and suppose H is a bipartite graph with parts X, Y with $|X| = r$, $|Y| = s$ and such that $d(x) \ge \frac{s}{2}+1$ for all $x \in X$ and $d(y) \ge \frac{r}{2}+1$ for all $y \in Y$. Then*

$$M_H(n) \ge n^{3/2-o(1)}.$$

The bound on the minimum degrees is best possible. Indeed, for an arbitrary k divisible by 4 consider two copies of $K_{k/4,k/4}$ connected with a single edge. This graph is $(2,1)$-separable and self-percolating and hence has at most linear running time by Theorem 7.10. The minimum degree on both sides of the bipartition is $k/4$, exactly half of the sides.

The proof of Theorem 8.3 follows a bipartite adaptation of our dilated chain construction and employs generalisations of Sidon sets as the set of dilations. In fact, arithmetic sets with the exact properties and density that we needed were not known to exist and so in [47] we also adapted ideas of Ruzsa [77] to construct the necessary sets. Observe that Theorem 8.3 is applicable for both $K_{3,3}$ and the cube graph Q_3. Unfortunately the known upper bounds of the respective Turán numbers do not provide matching exponents. For the cube Q_3 we can remove the $o(1)$ from the exponent of the lower bound in Theorem 8.3 using a bipartite version of the ladder chain construction with Sidon sets.

Theorem 8.4 ([47]) *For the running time of the cube Q_3 we have*

$$\Omega(n^{3/2}) \leq M_{Q_3}(n) \leq O(n^{8/5}).$$

Here the upper bound follows from Proposition 8.1 and the best known upper bounds on the Turán number of the cube, due to Erdős and Simonovits [43]. Actually, for $\mathrm{ex}(n, Q_3)$ the best known bounds are the same as for $M_{Q_3}(n)$.

The Turán number of $K_{3,3}$ is known asymptotically and the exponent is 5/3 [50], matching the Kővari-Sós-Turán upper bound. While the exponent 3/2 in the lower bound still falls short of the upper bound of Corollary 8.2, both for $H = K_{3,3}$ and Q_3 we believe it to be essentially tight.

Conjecture 8.5 *For $K_{3,3}$ and Q_3 we have maximum running times*

$$M_{K_{3,3}}(n) = n^{3/2+o(1)} \qquad \text{and} \qquad M_{Q_3}(n) = n^{3/2+o(1)}.$$

Both for $K_{3,3}$ and Q_3, it would even be extremely interesting to improve on any of the known bounds, even if only just by a log-factor.

Remark One could use the slow $K_{3,3}$ process in the simulation construction method (Section 7.2) to create graphs with minimum degree 1, maximum degree 4, and maximum running time $n^{3/2}$. In the proof of Theorem 7.5, one simply replaces the benchmark graph (C_6 plus a chord) with $K_{3,3}$.

The obtained infection rule has a single vertex of degree 4. In order to get down to maximum degree 3, one needs to delete an edge from $K_{3,3}$; unfortunately this reduces the maximum running time to linear. One can also aim to create a larger almost 3-regular bipartite infection rule randomly and hope that it has a longer running time. We are not convinced though that such a graph would have super-linear running time either.

8.2 Large complete bipartite graphs

For larger complete bipartite graphs, using bipartite dilation chains together with a probabilistic alteration tweak, we can improve the lower bound significantly.

Theorem 8.6 ([47]) *For $3 \leq r \leq s$, the maximum running time $M_{K_{r,s}}(n)$ is bounded from below by*

$$M_{K_{r,s}}(n) \geq n^{2-\frac{1}{r}-\frac{1}{s-1}-o(1)}. \tag{8.1}$$

The bound in Theorem 8.6 becomes stronger than that of Theorem 8.3 when $s > \frac{3r-2}{r-2}$, in particular for $K_{r,r}$ with $r \geq 5$. With r fixed and s growing, the bound of Theorem 8.6 approaches the upper bound given by Corollary 8.2.

Recall that before this section we have only encountered infection rules with maximum running times constant, logarithmic, linear and (almost) quadratic. What other running times are possible? Using Theorem 8.6 together with Proposition 8.1, we can establish an infinite sequence of infection rules H, each with a distinct maximum running time strictly between $n^{3/2}$ and quadratic.

Corollary 8.7 ([47]) *For every $3 \leq r \in \mathbf{N}$ there exists $s_r \in \mathbf{N}$ such that*

$$\omega\left(n^{2-1/(r-1)}\right) \leq M_{K_{r,s_r}}(n) \leq O(n^{2-1/r}).$$

We expect that there are in fact infinitely many exponents $\alpha \in \mathbb{Q}$ possible such that $M_H(n) = \Theta(n^\alpha)$. However due to our lack of understanding of $M_H(n)$, we do not know that the graphs in Corollary 8.7 offer such examples. In fact we do not even know that maximum running times are approximately polynomial.

Conjecture 8.8 *For every infection rule H, we have that the limit*

$$\lim_{n\to\infty} \frac{\log M_H(n)}{\log n}$$

exists.

In general, it is unclear which of our bounds are closer to the truth for bipartite graphs and the picture is complicated by the fact that the extremal number is not known for many bipartite graphs H of interest, see for example [50]. Despite the ample warning signs involving the general case for cliques, we gamble that Proposition 8.1 is never tight for bipartite graphs.

Conjecture 8.9 *For all bipartite graphs H, we have that*

$$M_H(n) = o(\mathrm{ex}(n, H)).$$

It could well be that a stronger conjecture is true and there is always a polynomial separation between $M_H(n)$ and $\mathrm{ex}(n, H)$ for all bipartite graphs H. In a different direction, for general H we believe the following is true.

Conjecture 8.10 *For all infection rules H with at least 3 edges, we have*

$$M_H(n) \leq O\left(\min_e \mathrm{ex}(n, H - e)\right).$$

This conjecture formulates the limitations intrinsic in all our chain constructions, where the edges e_i form a graph free of any copy of H minus an edge. It can also be considered a challenge to devise novel constructions beyond the current ideas. If one believes the conjecture of Erdős [39] about the Turán number of r-degenerate graphs then Conjecture 8.10 implies for the symmetric complete bipartite graph that $M_{K_{r,r}}(n) \leq n^{2-\frac{1}{r-1}-o(1)}$.

9 Disconnected infection rules

Up to this point we have restricted our attention to connected infection rules, mainly because we do not expect to understand the behaviour of a disconnected infection rule H if we have not worked out the individual behaviour of its components, and the interpretation of graph percolation as a virus spreading lends itself to connected H. Nevertheless the general questions we have studied for connected infection rules are mathematically just as valid when H is disconnected, and we close this survey by discussing this setting.

9.1 Forests and 2-regular graphs

A consequence of the proof of Theorem 6.3 is that the T-process for any tree T will percolate for every starting graph with at least $2v(T)$ vertices that is not T-stable, even if it is just a union of T minus an edge and isolated vertices. Together with the constant maximum running time for trees this implies that $M_F(n) = O(1)$ for any forest F. When studying maximum running times of graph bootstrap processes we are usually interested in starting graphs whose order is much larger than the order of the infection rule. For forests, investigating the F-process on graphs with $v(F)$ vertices is of interest due to connections with Problem 6.4, asking for the largest possible running time for t-vertex trees. In particular, an upper bound of the form $M_F(f) \leq c \cdot f$ for Question 9.1 below would result in a linear upper bound for trees. We refer to [46] for details.

Question 9.1 *For $f \in \mathbf{N}$, what is the maximum value of $M_F(f)$ over all f-vertex forests F?*

As for unions of cycles, i.e. 2-regular graphs, recall that when the infection rule is C_k any two vertices of distance $k-1$ become adjacent in each round of the process, which led to the maximum running time on n-vertex graphs essentially being $\log_{k-1}(n)$. In the case that H is the disjoint union of cycles the maximum running time is controlled by the cycle that provides the fastest reduction of the diameter, which is just the largest cycle in H.

Theorem 9.2 ([46]) *If $s \geq 2$ and $H := C_{k_1} \sqcup \ldots \sqcup C_{k_s}$ is the disjoint union of cycles of lengths $k_1 \geq \ldots \geq k_s$, then for sufficiently large n we have that*

$$\log_{k_1-1}(n) - 1 \leq M_H(n) < \log_{k_1-1}(n) + k_1^3 s^4.$$

The idea underlying the proof is to take a copy of $C_{k_1} \sqcup \ldots \sqcup C_{k_s}$ that appears in the first step of the process, fix all cycles of that copy but the largest one, and consider the C_{k_1}-process on the part of the host graph that avoids the fixed cycles. This approach is an instance of a more general strategy that we will explore next.

9.2 Bounds obtained from simulation

For the rest of our discussion, we concentrate on disconnected infection rules with just two connected components. Given a graph $H = H_1 \sqcup H_2$, which is the disjoint union of H_1 and H_2 to what extent is the running time $M_H(n)$ determined by $M_{H_1}(n)$ and $M_{H_2}(n)$? In particular, can we hope for some non-trivial bounds in terms of $M_{H_1}(n)$ and $M_{H_2}(n)$?

A general strategy to obtain lower bounds on $M_{H_1 \sqcup H_2}(n)$ is to take a starting graph G for H_1 together with an isolated copy of H_2 such that $\langle G \rangle_{H_1}$ does not contain any copies of H_2 minus an edge. In the H-process on $G \sqcup H_2$ each completed copy of H will then consist of the fixed copy of H_2 and a new copy of H_1 with vertices in $V(G)$. This is essentially the simulation construction introduced in Section 7. For example, consider $H_1 = K_6$ and H_2 a sufficiently large wheel, say $H_2 = W_{101}$, with the starting graph G constructed in [14] to give $M_{K_6}(n) = \Theta(n^2)$. It is not immediately obvious, but possible to verify, that the final graph $\langle G \rangle_{K_6}$ is free of copies of W_{101} minus an edge. Therefore $\tau_{K_6 \sqcup W_{101}}(G \sqcup W_{101}) = \tau_{K_6}(G) = \Theta(n^2) = \omega(M_{W_{101}}(n))$. In this example $M_{H_1 \sqcup H_2}(n)$ has the same asymptotic growth as the larger of $M_{H_1}(n)$ and $M_{H_2}(n)$.

Our next example (Figure 8) demonstrates that $M_H(n)$ may lie strictly between $M_{H_1}(n)$ and $M_{H_2}(n)$.

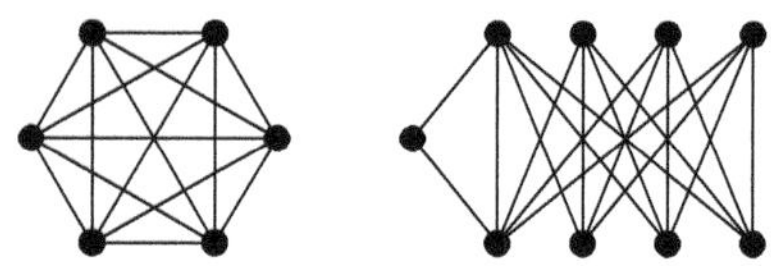

Figure 8: A graph $H = H_1 \sqcup H_2$ whose running time lies strictly between those of its components.

As in our previous example we choose $H_1 = K_6$. The component H_2 is obtained by starting with $K_{4,4}$ and adding a new vertex with precisely one neighbour in each part. Theorem 7.1 implies $M_{H_2}(n) = \Omega(n)$, and it is not difficult to check that due to the vertex of degree two and the symmetries of $K_{4,4}$ it is $(2,1)$-separable and self-percolating, so $M_{H_2}(n) = O(n)$ by Theorem 7.11. On the other hand, $M_{H_1}(n) = \Theta(n^2)$. In [47, Section 8] it is shown that the running time of H is polynomially separated from those of its components: $n^{1+\alpha} \leq M_H(n) \leq n^{2-\alpha}$ for some $\alpha > 0$. The upper bound here uses the fact that once we have more than $\mathrm{ex}(n, K_{4,4})$ edges, we can find many copies of $K_{4,4}$. Once these appear we can show a growing clique behaviour that forces the process to stabilise in at most linearly more steps. The lower bound is obtained via simulation as described above. To construct a starting graph G for K_6 that is free of H_2 minus an edge we use line chains generated by a hypergraph of high girth, similar to the proof of Theorem 7.10.

9.3 Interacting infection rules

In both examples of the previous section we reduced the disconnected infection rule to a connected one by confining one component to a single isolated copy. Although in the upper bound for the graph in Figure 8 it is crucial that new copies of both components appear throughout the process, we are essentially assuming a two-stage process. First we consider only new copies of K_6, until we can be sure that several copies of $K_{4,4}$ appear, that is, when the number of new edges reaches

$\text{ex}(n, K_{4,4})$. From that point onwards the first component takes over and causes the process to terminate within a linear number of steps. None of the bounds required any interplay between the two connected infection rules.

An example where the individual infection rules of the components do work together is given in Figure 9: The triangle component causes vertices of distance two in the host graph to become adjacent at every step (apart from the six vertices that are reserved for the right component), while the right component introduces edges between the endpoints of any two edge-disjoint triangles. This leads to a maximum running time of $O(1)$. Individually however, the component have maximum running time logarithmic, and linear respectively (using Theorems 7.1 and 7.11).

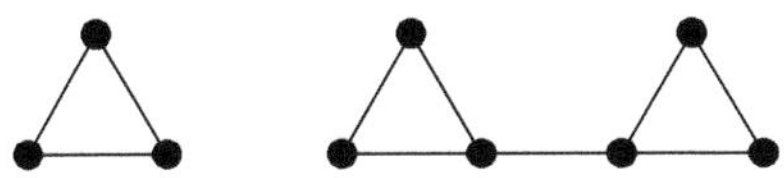

Figure 9: A disconnected graph with maximum running time $O(1)$.

This example further tells us that Corollary 7.2 breaks down for disconnected infection rules. We remark though that Theorem 7.1 can be extended to disconnected H by the stronger requirement that every component of H has minimum degree at least 2 and maximum degree at least 3.

We have seen examples where $M_H(n)$ is much smaller than either of $M_{H_1}(n)$ and $M_{H_2}(n)$, asymptotically equal to one of $M_{H_1}(n)$ and $M_{H_2}(n)$, or strictly between the two. However we are not aware of any choice of H that satisfies $M_H(n) = \omega(M_{H_1}(n))$ and $M_H(n) = \omega(M_{H_2}(n))$.

Question 9.3 *Is there a disconnected infection rule* $H = H_1 \sqcup H_2$ *such that* $M_H(n) = \omega(M_{H_1}(n))$ *and* $M_H(n) = \omega(M_{H_2}(n))$*?*

Acknowledgements

The third author is deeply grateful to József Balogh, Gal Kronenberg, and Alexey Pokrovskiy for their collaboration on [14], ideas from that paper were instrumental in shaping this survey. We also thank the anonymous referee for their careful reading and helpful suggestions.

References

[1] J. Adler and U. Lev, *Bootstrap percolation: visualizations and applications*, Brazilian Journal of Physics **33** (2003), 641–644.

[2] J. Adler, D. Stauffer, and A. Aharony, *Comparison of bootstrap percolation models*, Journal of Physics A: Mathematical and General **22** (1989), no. 7, L297–L301.

[3] M. Aizenman and J. L. Lebowitz, *Metastability effects in bootstrap percolation*, Journal of Physics A: Mathematical and General **21** (1988), no. 19, 3801.

[4] M. Akhmejanova, I. Vorobyev, and M. Zhukovskii, *Weak saturation numbers of large complete bipartite graphs*, arXiv preprint arXiv:2508.19435 (2025).

[5] N. Alon, *An extremal problem for sets with applications to graph theory*, Journal of Combinatorial Theory, Series A **40** (1985), no. 1, 82–89.

[6] O. Angel and B. Kolesnik, *Sharp thresholds for contagious sets in random graphs*, The Annals of Applied Probability **28** (2018), no. 2, 1052–1098.

[7] R. Ascoli and X. He, *Rational values of the weak saturation limit*, arXiv preprint arXiv:2501.15686 (2025).

[8] P. Balister, B. Bollobás, and P. Smith, *The time of bootstrap percolation in two dimensions*, Probability Theory and Related Fields **166** (2016), no. 1-2, 321–364.

[9] J. Balogh and B. Bollobás, *Bootstrap percolation on the hypercube*, Probability Theory and Related Fields **134** (2006), no. 4, 624–648.

[10] J. Balogh, B. Bollobás, and R. Morris, *Bootstrap percolation in three dimensions*, The Annals of Probability (2009), 1329–1380.

[11] ———, *Majority bootstrap percolation on the hypercube*, Combinatorics, Probability and Computing **18** (2009), no. 1-2, 17–51.

[12] ———, *Bootstrap percolation in high dimensions*, Combinatorics, Probability and Computing **19** (2010), no. 5-6, 643–692.

[13] ———, *Graph bootstrap percolation*, Random Structures & Algorithms **41** (2012), no. 4, 413–440.

[14] J. Balogh, G. Kronenberg, A. Pokrovskiy, and T. Szabó, *The maximum length of K_r-Bootstrap Percolation*, arXiv preprint arXiv:1907.04559 (2019), To appear in Proceedings of the American Mathematical Society.

[15] Z. Bartha and B. Kolesnik, *Weakly saturated random graphs*, Random Structures & Algorithms **65** (2024), no. 1, 131–148.

[16] Z. Bartha, B. Kolesnik, and G. Kronenberg, *H-percolation with a random H*, Electronic Communications in Probability **29** (2024), 1–5.

[17] Z. Bartha, B. Kolesnik, G. Kronenberg, and Y. Peled, *Sharp Fuss-Catalan thresholds in graph bootstrap percolation*, arXiv preprint arXiv:2510.26724 (2025).

[18] E. Bayraktar and S. Chakraborty, *$K_{r,s}$ graph bootstrap percolation*, Preprint (2019), arXiv:1904.12764.

[19] F. A. Behrend, *On sets of integers which contain no three terms in arithmetical progression*, Proceedings of the National Academy of Sciences of the United States of America **32** (1946), no. 12, 331.

[20] F. Benevides and M. Przykucki, *On slowly percolating sets of minimal size in bootstrap percolation*, Electronic Journal of Combinatorics **20** (2013), no. 2, P46.

[21] ———, *Maximum percolation time in two-dimensional bootstrap percolation*, SIAM Journal on Discrete Mathematics **29** (2015), no. 1, 224–251.

[22] M. Bidgoli, A. Mohammadian, and B. Tayfeh-Rezaie, *On $K_{2,t}$-Bootstrap Percolation*, Graphs and Combinatorics **37** (2021), no. 3, 731–741.

[23] M. Biskup and R. Schonmann, *Metastable behavior for bootstrap percolation on regular trees*, Journal of Statistical Physics **136** (2009), no. 4, 667–676.

[24] B. Bollobás, *Weakly k-saturated graphs*, Beiträge zur Graphentheorie (Kolloquium, Manebach, 1967), 1968, pp. 25–31.

[25] ———, *Extremal graph theory*, Academic Press New York, 1978.

[26] B. Bollobás, C. Holmgren, P. Smith, and A. Uzzell, *The time of bootstrap percolation with dense initial sets*, Annals of Probability **42** (2014), no. 4, 1337–1373.

[27] B. Bollobás, M. Przykucki, O. Riordan, and J. Sahasrabudhe, *On the maximum running time in graph bootstrap percolation*, Electronic Journal of Combinatorics **24** (2017), no. 2, P2.16.

[28] B. Bollobás, P. Smith, and A. J. Uzzell, *The time of bootstrap percolation with dense initial sets for all thresholds*, Random Structures & Algorithms **47** (2015), no. 1, 1–29.

[29] B. Bollobás and A. G. Thomason, *Threshold functions*, Combinatorica **7** (1987), no. 1, 35–38.

[30] B. Bollobás, *Random Graphs*, 2nd ed., Cambridge University Press, 2001.

[31] M. Borowiecki and E. Sidorowicz, *Weakly p-saturated graphs*, Discussiones Mathematicae Graph Theory **22** (2002), no. 1, 17–29.

[32] N. Cancrini, F. Martinelli, C. Roberto, and C. Toninelli, *Kinetically constrained spin models*, Probability theory and related fields **140** (2008), no. 3, 459–504.

[33] R. Cerf and F. Manzo, *The threshold regime of finite volume bootstrap percolation*, Stochastic Processes and their Applications **101** (2002), no. 1, 69–82.

[34] J. Chalupa, P. L. Leath, and G. R. Reich, *Bootstrap percolation on a Bethe lattice*, Journal of Physics C: Solid State Physics **12** (1979), no. 1, L31.

[35] W. Chen, X. Liu, and X. Yang, *A note on weak saturation number of trees*, arXiv preprint arXiv:2502.15626 (2025).

[36] Y. Cui and L. Pu, *Weak saturation numbers of $K_{2,t}$ and $K_p \cup K_q$*, AKCE International Journal of Graphs and Combinatorics **16** (2019), no. 3, 237–240.

[37] B. L. Currie, J. R. Faudree, R. J. Faudree, and J. R. Schmitt, *A survey of minimum saturated graphs*, Electronic Journal of Combinatorics (2012), DS19–Oct 11, 2021.

[38] P. Dukes, J. Noel, and A. Romer, *Extremal bounds for 3-neighbor bootstrap percolation in dimensions two and three*, SIAM Journal on Discrete Mathematics **37** (2023), no. 3, 2088–2125.

[39] P. Erdős, *Some recent results on extremal problems in graph theory*, Theory of Graphs (Rome, 1966), Gordon and Breach, New York, 1967, pp. 117–123.

[40] P. Erdős, Z. Füredi, and Z. Tuza, *Saturated r-uniform hypergraphs*, Discrete mathematics **98** (1991), no. 2, 95–104.

[41] P. Erdős, A. Hajnal, and J. Moon, *A problem in graph theory*, The American Mathematical Monthly **71** (1964), no. 10, 1107–1110.

[42] P. Erdős and A. Rényi, *On random graphs I*, Publ. Math. Debrecen **6** (1959), 290–297.

[43] P. Erdős and M. Simonovits, *Some extremal problems in graph theory*, Combinatorial Theory and its Applications I, North-Holland Amsterdam, 1970, pp. 378–392.

[44] A. Espuny Díaz, B. Janzer, G. Kronenberg, and J. Lada, *Long running time for hypergraph bootstrap percolation*, European Journal of Combinatorics **115** (2024), 103783.

[45] D. Fabian, P. Morris, and T. Szabó, *Slow graph bootstrap percolation I: Cycles*, arXiv preprint arXiv:2308.00498 (2023).

[46] ———, *Slow graph bootstrap percolation II: Accelerating properties*, Journal of Combinatorial Theory, Series B **172** (2025), 44–79.

[47] ———, *Slow graph bootstrap percolation III: Chain constructions*, arXiv preprint arXiv:2508.03835 (2025).

[48] R. J. Faudree, R. J. Gould, and M. S. Jacobson, *Weak saturation numbers for sparse graphs*, Discussiones Mathematicae Graph Theory **33** (2013), no. 4, 677–693.

[49] P. Frankl, *An extremal problem for two families of sets*, European Journal of Combinatorics **3** (1982), no. 2, 125–127.

[50] Z. Füredi and M. Simonovits, *The history of degenerate (bipartite) extremal graph problems*, Erdős Centennial, Springer, 2013, pp. 169–264.

[51] K. Gunderson, S. Koch, and M. Przykucki, *The time of graph bootstrap percolation*, Random Structures & Algorithms **51** (2017), no. 1, 143–168.

[52] I. Hartarsky and L. Lichev, *The maximal running time of hypergraph bootstrap percolation*, SIAM Journal on Discrete Mathematics **38** (2024), no. 2, 1462–1471.

[53] A. E. Holroyd, *Sharp metastability threshold for two-dimensional bootstrap percolation*, Probability Theory and Related Fields **125** (2003), no. 2, 195–224.

[54] S. Janson, T. Łuczak, T. Turova, and T. Vallier, *Bootstrap percolation on the random graph $G_{n,p}$*, The Annals of Applied Probability **22** (2012), no. 5, 1989–2047.

[55] G. Kalai, *Intersection patterns of convex sets*, Israel J. Math. **48** (1984), 161–174.

[56] ______, *Weakly saturated graphs are rigid*, Convexity and graph theory, North-Holland Mathematics Studies, vol. 87, 1984, pp. 189–190.

[57] ______, *Hyperconnectivity of graphs*, Graphs and Combinatorics **1** (1985), no. 1, 65–79.

[58] P. Kővári, V. T. Sós, and P. Turán, *On a problem of Zarankiewicz*, Colloquium Mathematicum, vol. 3, Polska Akademia Nauk, 1954, pp. 50–57.

[59] J. H. Kim, B. Sudakov, and V. H. Vu, *On the asymmetry of random regular graphs and random graphs*, Random Structures & Algorithms **21** (2002), no. 3-4, 216–224.

[60] B. Kolesnik, *The sharp K_4-percolation threshold on the Erdős–Rényi random graph*, Electronic Journal of Probability **27** (2022), 1–23.

[61] D. Korándi, Y. Peled, and B. Sudakov, *A random triadic process*, SIAM Journal on Discrete Mathematics **30** (2016), no. 1, 1–19.

[62] G. Kronenberg, T. Martins, and N. Morrison, *Weak saturation numbers of complete bipartite graphs in the clique*, Journal of Combinatorial Theory, Series A **178** (2021).

[63] F. Lazebnik, V. A. Ustimenko, and A. J. Woldar, *A new series of dense graphs of high girth*, Bulletin of the American Mathematical Society **32** (1995), no. 1, 73–79.

[64] F. Lazebnik and R. Viglione, *On the connectivity of certain graphs of high girth*, Discrete Mathematics **277** (2004), no. 1-3, 309–319.

[65] L. Lovász, *Flats in matroids and geometric graphs*, Combinatorial Surveys (Proc. 6th British Combinatorial Conference), 1977, pp. 45–86.

[66] K. Matzke, *The saturation time of graph bootstrap percolation*, arXiv preprint arXiv:1510.06156 (2015).

[67] R. Morris, *Bootstrap percolation, and other automata*, European Journal of Combinatorics **66** (2017), 250–263.

[68] ______, *Monotone cellular automata*, Surveys in Combinatorics 2017, London Mathematical Society Lecture Note Series, Cambridge University Press, 2017, pp. 312–371.

[69] N. Morrison and J. A. Noel, *Extremal bounds for bootstrap percolation in the hypercube*, Journal of Combinatorial Theory, Series A **156** (2018), 61–84.

[70] ———, *A sharp threshold for bootstrap percolation in a random hypergraph*, Electronic Journal of Probability **26** (2021), 1–85.

[71] J. A. Noel and A. Ranganathan, *On the Running Time of Hypergraph Bootstrap Percolation*, Electronic Journal of Combinatorics **30** (2023), no. 2, P2.46.

[72] O. Pikhurko, *Uniform families and count matroids*, Graphs and Combinatorics **17** (2001), no. 4, 729–740.

[73] ———, *Weakly saturated hypergraphs and exterior algebra*, Combinatorics, Probability and Computing **10** (2001), no. 5, 435–451.

[74] M. Przykucki, *Maximal percolation time in hypercubes under 2-bootstrap percolation*, Electronic Journal of Combinatorics **19** (2012), no. 2, P41.

[75] M. Przykucki and T. Shelton, *Smallest percolating sets in bootstrap percolation on grids*, arXiv preprint arXiv:1907.01940 (2019).

[76] K. F. Roth, *On certain sets of integers*, J. London Math. Soc **28** (1953), no. 1, 104–109.

[77] I. Ruzsa, *Solving a linear equation in a set of integers I*, Acta Arithmetica **65** (1993), no. 3, 259–282 (en).

[78] I. Ruzsa and E. Szemerédi, *Triple systems with no six points carrying three triangles*, Combinatorics (Keszthely, 1976), Coll. Math. Soc. J. Bolyai **18** (1978), no. 939-945, 2.

[79] A. Shapira, *Behrend-type constructions for sets of linear equations*, Acta Arithmetica **122** (2006), no. 1, 17–33.

[80] N. Terekhov and M. Zhukovskii, *Weak saturation in graphs: a combinatorial approach*, Journal of Combinatorial Theory, Series B **172** (2025), 146–167.

[81] Z. Tuza, *Extremal problems on saturated graphs and hypergraphs*, Ars Combinatorica, Eleventh British Combinatorial Conference (London 1987) **25b** (1988), 105–113.

[82] S. Ulam, *Random processes and transformations*, Proceedings of the International Congress on Mathematics 1950, 1952, pp. 264–275.

[83] J. von Neumann, *Theory of self-reproducing automata*, University of Illinois Press, Urbana, 1966.

Ulm
Germany
dfabian@mailbox.org

Departament de Matemàtiques
Universitat Politècnica de Catalunya (UPC)
Carrer de Pau Gargallo 14, 08028 Barcelona
Spain
pmorrismaths@gmail.com

Department of Mathematics and Computer Science
Freie Universität Berlin
Arnimallee 3, 14195 Berlin
Germany
szabo@math.fu-berlin.de

Rigidity of Graphs and Frameworks: A Matroid Theoretic Approach

James Cruickshank, Bill Jackson, Tibor Jordán, Shin-ichi Tanigawa

Abstract

A d-dimensional (bar-and-joint) framework (G, p) consists of a graph $G = (V, E)$ and a realisation $p : V \to \mathbf{R}^d$. It is rigid if every continuous motion of the vertices which preserves the lengths of the edges is induced by an isometry of $\mathbf{R}^d$. The study of rigid frameworks has increased rapidly since the 1970's, stimulated by numerous applications in areas such as civil and mechanical engineering, CAD, molecular conformation, sensor network localisation, and low rank matrix completion. We will describe some of the main results in combinatorial rigidity theory and their applications to other areas of combinatorics, putting an emphasis on links to matroid theory.

1 Introduction

The study of the rigidity of frameworks can be traced back to a claim made by Euler in 1776 that "A closed spatial figure allows not changes, as long as it is not ripped apart." Giving a formal proof for Euler's claim turned out to be a challenging problem. A celebrated result of Cauchy in 1813 implies that every triangulated convex spherical surface in $\mathbf{R}^3$ is rigid. Gluck [41] showed that every generic triangulated spherical surface in $\mathbf{R}^3$ is rigid. Rather surprisingly, Connelly [19,20] disproved Euler's original claim by constructing a flexible triangulated spherical surface.

We can use the language of rigidity theory to give a more precise description of these results. A d-dimensional *(bar-joint) framework* is defined as a pair (G, p) consisting of a graph G and a *realisation* $p : V(G) \to \mathbf{R}^d$. We will also say that (G, p) is a *realisation* of G in $\mathbf{R}^d$. Two realisations (G, p) and (G, q) of G in $\mathbf{R}^d$ are *congruent* if (G, p) can be obtained from (G, q) by an isometry of $\mathbf{R}^d$, i.e., a combination of translations, rotations and reflections. The framework (G, p) is *globally rigid* if every framework which has the same edge lengths as (G, p) is congruent to (G, p). It is *rigid* if every continuous motion of the vertices of (G, p) in $\mathbf{R}^d$ which preserves the lengths of its edges results in a framework which is congruent to (G, p), and otherwise it is *flexible*. These concepts are illustrated in Figure 1.

It is straightforward to show that a realisation of a graph on the real line is rigid if and only if the graph is connected: necessity is obvious, and sufficiency follows by an easy induction on the number of vertices. In contrast, Abbot [1] showed that it is NP-hard to determine whether a given d-dimensional framework is rigid when $d \geq 2$, and Saxe [94] showed the same holds for global rigidity when $d \geq 1$. These problems become more tractable, however, if we restrict attention to *generic frameworks*, i.e. frameworks (G, p) in which the multiset of coordinates of all points p_v, $v \in V$, is algebraically independent over $\mathbb{Q}$. Gluck [41] showed that the rigidity of a generic framework depends only on its underlying graph. This allows us to define a graph G as being *rigid in* $\mathbf{R}^d$ if some (or equivalently every) generic realisation of G in $\mathbf{R}^d$ is rigid. Analogous, but much deeper, results of Connelly [23] and Gortler, Healy, and Thurston [43] imply that the global rigidity of a generic framework depends

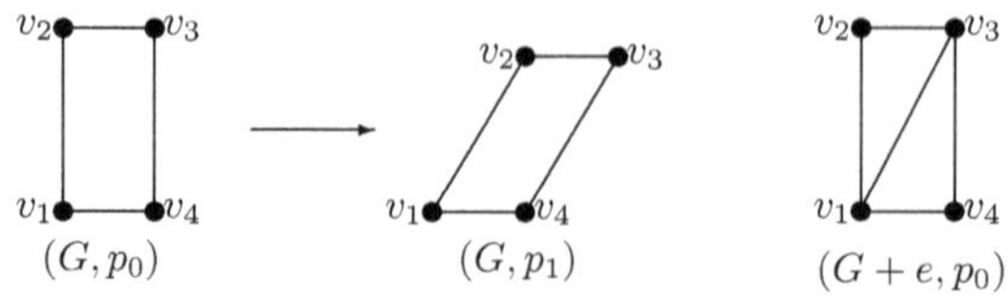

Figure 1: The 2-dimensional frameworks (G, p_0) and (G, p_1) are not rigid since (G, p_1) can be obtained from (G, p_0) by a continuous motion of its vertices in $\mathbf{R}^2$ which preserves all edge lengths, but changes the distance between v_1 and v_3. The 2-dimensional framework $(G + e, p_0)$ is rigid. It is not globally rigid, however, since we can obtain a non-congruent framework with the same edge lengths by reflecting v_4 in the line through $v_1 v_3$.

only on its underlying graph; so, we can define a graph G as being *globally rigid* in $\mathbf{R}^d$ if some (or equivalently every) generic realisation of G in $\mathbf{R}^d$ is globally rigid. The problem of finding a combinatorial characterisation of rigid or globally rigid graphs in $\mathbf{R}^d$ has been solved when $d = 1, 2$, but is a major open problem in discrete geometry for $d \geq 3$.

Combinatorial rigidity theory has expanded rapidly since the 1970's. It has branched out to include several other types of frameworks and has found applications in many diverse areas. The aim of this survey is to introduce the reader to some foundational results, proof techniques, and open problems in rigidity, and to describe some striking applications to other areas of combinatorics. We refer the reader wishing to discover more to the notes [59,115], text books [24,46], and survey articles on rigidity in [42]. We will assume a knowledge of basic concepts in matroid theory and refer the reader to [91] as a source text for these.

2 Terminology and preliminary results

2.1 Graphs and matroids

We will use the term *multigraph* for graphs which may contain loops or multiple edges and will reserve the term *graph* for graphs without loops and multiple edges. Given a finite set V, we use K_V to denote the complete graph with vertex set V and $\binom{V}{2}$ to denote its edge set. We say that a multigraph $G = (V, E)$ is (s,t)-*sparse* for two integers s, t if $|F| \leq s|V(F)| - t$ for all $\emptyset \neq F \subseteq E$ and (s,t)-*tight* if it also satisfies $|E| = s|V| - t$. Given $X \subseteq V$, we use $i_G(X)$ to denote the number of edges in $G[X]$, the subgraph of G induced by X. Thus, G is (s,t)-sparse if and only if $i(X) \leq s|X| - t$ for all $X \subseteq V$ which induce at least one edge in G. We will use the notation G/uv to denote the graph obtained from G by contracting an edge $e = uv$.

The following result of Edmonds [35] implies that the edge sets of the (s,t)-sparse subgraphs of K_V are the independent sets of a matroid on $\binom{V}{2}$, which is non-trivial when $t \leq 2s - 1$.

Theorem 2.1 *Let E be a finite set and $f : 2^E \to \mathbb{Z}$ be a non-decreasing, submodular* *function. Put*

$$\mathcal{I}_f := \{F \subseteq E : |I| \le f(I) \text{ for any } I \subseteq F \text{ with } I \neq \emptyset\}.$$

Then $\mathcal{M}_f := (E, \mathcal{I}_f)$ *is a matroid with rank function* $\hat{f} : 2^E \to \mathbb{Z}$ *given by*

$$\hat{f}(F) := \min\left\{ |F_0| + \sum_{i=1}^{k} f(F_i) : F_0 \subseteq F \text{ and } \{F_1, \ldots, F_k\} \text{ is a partition of } F \backslash F_0 \right\}.$$

We refer to the matroid $\mathcal{M}_f$ given by Edmonds' theorem as the *matroid induced by* f. For any finite set V and integers s, t, the (s,t)*-sparsity matroid of* K_V is the matroid on $\binom{V}{2}$ induced by the function $f : F \mapsto s|V(F)| - t$ for all $F \subseteq \binom{V}{2}$.

Nash-Williams [87] showed that the (k,k)-sparsity matroid of K_V is the matroid union of k copies of the cycle matroid of K_V for any integer $k \ge 1$. This implies, in particular, that it can be realised as the row matroid of a $|\binom{V}{2}| \times k|V|$ matrix A defined as follows. We first choose a generic map $q : \binom{V}{2} \to \mathbf{R}^k$ and a reference ordering of the vertices in V. We then define the row of A indexed by the edge $e = uv$ with $u < v$ to be

$$\begin{array}{c} \\ e{=}uv \end{array} \begin{array}{c} \\ \left[\begin{array}{ccccc} & u & & v & \\ 0\ldots0 & q_e & 0\ldots0 & -q_e & 0\ldots0 \end{array}\right]. \end{array}$$

2.2 Infinitesimal rigidity and rigidity matroids

A basic step in deciding whether a given d-dimensional realisation of a graph $G = (V, E)$ is rigid is to first linearise the problem. We consider the *measurement map* $f_G : \mathbf{R}^{d|V|} \to \mathbf{R}^{|E|}$, which maps each $p \in \mathbf{R}^{d|V|}$ onto the vector of squared edge lengths of the framework (G,p). The *rigidity matrix* $R(G,p)$ of the framework is the Jacobian matrix of f_G evaluated at p. It is the $|E| \times d|V(G)|$ matrix, in which each row is indexed by an edge, sets of d consecutive columns are indexed by the vertices, and the row indexed by the edge $e = uv$ has the form:

$$\begin{array}{c} \\ e{=}uv \end{array} \begin{array}{c} \\ \left[\begin{array}{ccccc} & u & & v & \\ 0\ldots0 & p_u - p_v & 0\ldots0 & p_v - p_u & 0\ldots0 \end{array}\right]. \end{array}$$

The space of *infinitesimal motions* of (G,p) is given by the right kernel of $R(G,p)$. It consists of the vectors $\dot{p} : V \to \mathbf{R}^d$ with the property that $\dot{p}_u - \dot{p}_v$ is orthogonal to $p_u - p_v$ for all $uv \in E$. Maxwell pointed out in his fundamental work on statics [81] that the space of infinitesimal motions will have a subspace of *trivial motions* of dimension $\binom{d+1}{2}$, generated by the infinitesimal translations and rotations of $\mathbf{R}^d$, whenever $p(V)$ affinely spans $\mathbf{R}^d$. He deduced that

$$\operatorname{rank} R(G,p) \le d|V| - \binom{d+1}{2} \tag{2.1}$$

when $|V| \ge d+1$, and that (G,p) will be rigid whenever equality holds (since this will imply that every infinitesimal motion of (G,p) is an infinitesimal isometry of $\mathbf{R}^d$). We say that (G,p) is *infinitesimally rigid* if $|V| \ge d+1$ and equality holds in (2.1), or $|V| \le d$ and $\operatorname{rank} R(G,p) = \binom{|V|}{2}$. The 2-dimensional framework on the right of Figure 2 is an example of a rigid framework which is not infinitesimally rigid (since it has a non-trivial infinitesimal motion which fixes each vertex of degree three and gives each vertex of degree two an arbitrary 'infinitesimal velocity' in the direction perpendicular to the line through its neighbours).

Asimow and Roth [5] used elementary differential geometry to show that infinitesimal rigidity is equivalent to rigidity whenever p is a regular point of the measurement map. Since the entries in $R(G,p)$ are polynomial functions of the coordinates of p with integer coefficients, this gives

Theorem 2.2 *A generic framework is rigid in* $\mathbf{R}^d$ *if and only if it is infinitesimally rigid in* $\mathbf{R}^d$.

Equivalently, a graph is (generically) rigid in $\mathbf{R}^d$ if and only if it has an infinitesimally rigid realisation in $\mathbf{R}^d$. These observations have important algorithmic consequences: they tell us that the problem of deciding whether a graph is rigid in $\mathbf{R}^d$ belongs to NP and, in combination with the Schwarz-Zippel Lemma [95, 116], give rise to a randomised polynomial algorithm for solving this problem. Unfortunately, this tells us nothing about the structure of graphs which are rigid in $\mathbf{R}^d$. We will use techniques from graph theory, matroid theory and geometry to obtain structural information on such graphs.

The *rigidity matroid* $\mathcal{R}(G,p)$ of a framework (G,p) is the row matroid of the rigidity matrix $R(G,p)$. Thus, $\mathcal{R}(G,p)$ is a matroid on E in which a set of edges $F \subseteq E$ is independent if the rows of $R(G,p)$ indexed by F are linearly independent. Since the rows of $R(G,p)$ are nonzero scalar multiples of the rows of an oriented edge/vertex incidence matrix of G when p is an injection from V to $\mathbf{R}$, the rigidity matroid of any such 1-dimensional realisation of G is just the cycle matroid of G.

The *d-dimensional rigidity matroid* $\mathcal{R}_d(G)$ of a graph G is defined to be the rigidity matroid $\mathcal{R}(G,p)$ of any generic realisation of G in $\mathbf{R}^d$. Thus, G is rigid in $\mathbf{R}^d$ if and only if the rank of $\mathcal{R}_d(G)$ is $d|V| - \binom{d+1}{2}$ when $|V| \geq d+1$. Since every independent set in a matroid can be extended to a base, an algorithm to decide whether any given set of edges of G is independent in $\mathcal{R}_d(G)$ would enable us to determine whether G is rigid in $\mathbf{R}^d$.

We will simplify terminology by describing subgraphs H of G using properties of their edge set in $\mathcal{R}_d(G)$. For example, we will say H is $\mathcal{R}_d$-independent to mean $E(H)$ is independent in $\mathcal{R}_d(G)$ and will use $r_d(H)$ to denote the rank of $E(H)$ in $\mathcal{R}_d(G)$. We can apply (2.1) to the induced subgraphs of G to obtain the following graph theoretic necessary condition for $\mathcal{R}_d$-independence.

Lemma 2.3 (Maxwell's independence criterion) *Suppose* G *is an* $\mathcal{R}_d$*-independent graph. Then* $i_G(X) \leq d|X| - \binom{d+1}{2}$ *for all* $X \subseteq V$ *with* $|X| \geq d+1$.

Maxwell's criterion is also sufficient to imply $\mathcal{R}_d$-independence when $d = 1$ since $\mathcal{R}_1(G)$ is the cycle matroid of G, and Maxwell's criterion is equivalent to saying that G is a forest. We will see in the following sections that it remains sufficient when $d = 2$ (and hence the 2-dimensional rigidity matroid is equal to the $(2,3)$-sparsity matroid), but is not sufficient to imply $\mathcal{R}_d$-independence when $d \geq 3$.

Figure 2: Two 2-dimensional stressed frameworks with an associated stress matrix of maximum possible rank. The framework on the right is obtained from the one on the left by subdivisions. This operation preserves the maximum rank property of a stressed framework. The framework on the right is not infinitesimally rigid, and its underlying graph does not have enough edges to be generically (globally) rigid in $\mathbf{R}^2$.

2.3 Global rigidity and stress matrices

The space of *equilibrium stresses* of a d-dimensional framework (G,p) is given by the left kernel of $R(G,p)$. It consists of the vectors $\omega : E \to \mathbf{R}$ with the property that, for all $v \in V$, $\sum_{uv \in E} \omega_{uv}(p_u - p_v) = \mathbf{0}$ (see Figure 2). Given an equilibrium stress ω of (G,p), we can define the corresponding *stress matrix* Ω as the $|V| \times |V|$ symmetric matrix in which the entry in row u and column v is $-\omega_{uv}$ if $u \neq v$ and $uv \in E$, 0 if $u \neq v$ and $uv \notin E$, and the diagonal entries are chosen so that the row and column sums of Ω are each equal to zero. Connelly showed in his fundamental work on global rigidity [21,23] that the rank of Ω is at most $|V|-d-1$, and (G,p) is globally rigid in the case when equality holds and p is generic. Gortler, Healy, and Thurston [43] subsequently verified a conjecture of Connelly by showing that the converse of the second part of the previous statement also holds (when $|V| \geq d+2$). This gives us

Theorem 2.4 *A generic d-dimensional framework (G,p) is globally rigid if and only if G is a complete graph on at most $d+1$ vertices, or (G,p) has a stress matrix of rank $|V|-d-1$.*

Theorem 2.4 plays an analogous role for generic global rigidity as Theorem 2.2 does for generic rigidity. In particular, it implies that the global rigidity of a generic d-dimensional framework (G,p) depends only on the graph G, and that G is (generically) globally rigid in $\mathbf{R}^d$ if and only if it has a realisation in $\mathbf{R}^d$ which is infinitesimally rigid and has a stress matrix of rank $|V|-d-1$. This, in turn, gives rise to a randomised polynomial algorithm for deciding whether G is (generically) globally rigid in $\mathbf{R}^d$, see [26,43]. An example of a graph $G=(V,E)$, which is not generically globally rigid in $\mathbf{R}^2$, but still has a realisation with a stress matrix of rank $|V|-3$, is given on the right of Figure 2.

Hendrickson [48] gave the following graph theoretic necessary conditions for a graph to be generically globally rigid. We say that a graph G is *redundantly rigid in* $\mathbf{R}^d$ if $G - e$ is rigid in $\mathbf{R}^d$ for all edges e of G.

Theorem 2.5 *Suppose G is a graph which is globally rigid in $\mathbf{R}^d$. Then either G is a complete graph on at most $d+1$ vertices, or G is $(d+1)$-connected and redundantly rigid in $\mathbf{R}^d$.*

Necessity of the connectivity condition follows since, if G has a vertex-cut of size less than $d+1$, then, for any generic realisation, we can reflect one side of the cut in a hyperplane containing the vertices of the cut. This operation will preserve the length of all the edges, but will change the distance between any two vertices on opposite sides of the cut. Hendrickson verifies necessity of redundant rigidity by showing that, if G has at least $d+2$ vertices and has an edge $e = uv$ for which $G-e$ is not rigid, then, for any generic realisation (G,p), we can continuously deform $(G-e,p)$ in such a way that we preserve all its edge lengths and eventually arrive at a non-congruent framework $(G-e,p')$ which satisfies $\|p(u)-p(v)\| = \|p'(u)-p'(v)\|$.

We will see that Hendrickson's necessary conditions for global rigidity are also sufficient when $d = 1, 2$, but not for $d \geq 3$. Another necessary condition for global rigidity in $\mathbf{R}^d$ has recently been obtained by Garamvölgyi, Gortler and Jordán [38].

Theorem 2.6 *Let G be a globally rigid graph in $\mathbf{R}^d$ on at least $d+2$ vertices. Then $\mathcal{R}_d(G)$ is a connected matroid.*

This necessary condition is implied by Hendrickson's conditions when $d = 1, 2$, but not for $d \geq 3$. The combination of all three conditions is still not sufficient to imply global rigidity in $\mathbf{R}^d$ when $d \geq 3$.

The following sufficient condition for global rigidity is given in [99]. It can be viewed as a partial converse to Theorem 2.5.

Theorem 2.7 *If $G = (V, E)$ is a graph and $G - v$ is rigid in $\mathbf{R}^d$ for all $v \in V$, then G is globally rigid in $\mathbf{R}^d$.*

2.4 Inductive constructions

Proofs that a given family of graphs are rigid, or globally rigid, in $\mathbf{R}^d$ are often inductive in nature and proceed by using a suitably chosen graph operation to reduce to a smaller graph in the family, applying induction to deduce that the smaller graph is rigid, and then using the property that the inverse of our chosen operation preserves rigidity. We will give examples of such 'rigidity preserving operations' in this subsection. The first two operations were introduced by Henneberg in [49].

2.4.1 0-extensions Given a graph H, the d-dimensional 0-*extension* operation creates a new graph G by adding a vertex v of degree at most d to H. The d-dimensional 0-extension operation preserves $\mathcal{R}_d$-independence: this follows from the following more general statement for the rigidity matrix of an arbitrary framework, which can be verified by using the fact that the only non-zero entries in the columns of this matrix labelled by v occur in the $d_G(v)$ rows labelled by the edges incident to v.

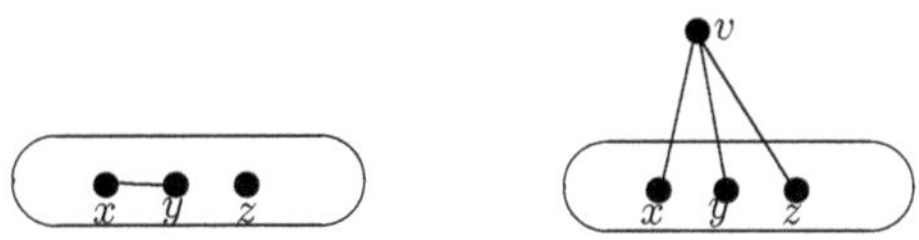

Figure 3: The 2-dimensional 1-extension operation.

Lemma 2.8 *Suppose (G,p) is a framework in $\mathbf{R}^d$, and v is a vertex in G such that $\{p_v - p_u : u \in N_G(v)\}$ is linearly independent in $\mathbf{R}^d$. Then* $\operatorname{rank} R(G,p) = \operatorname{rank} R(G-v, p|_{V-v}) + d_G(v)$.

2.4.2 1-extensions Given a graph H, the d-dimensional 1-*extension* operation creates a new graph G from H by deleting an edge xy and then adding a vertex v and $d+1$ new edges from v to x, y and $d-1$ other vertices of H, see Figure 3.

Lemma 2.9 *The d-dimensional 1-extension operation preserves both rigidity and global rigidity in* $\mathbf{R}^d$.

Sketch proof Rigidity can be verified by using a 'special position' argument. Suppose H is rigid in $\mathbf{R}^d$, G is a 1-extension of H and $H = G - v + xy$. Choose a generic framework (H,p) in $\mathbf{R}^d$ and extend it to a framework (G,p') by placing v on the line through p_x, p_y. Then (H,p) is infinitesimally rigid in $\mathbf{R}^d$ and we can use Lemma 2.8 to deduce that $(G+xy,p')$ is also infinitesimally rigid in $\mathbf{R}^d$. On the other hand, the fact that the points p'_v, p'_x, p'_y are collinear implies that the rows of $R(G+xy,p')$ indexed by xy, vx, vy are minimally linearly dependent, and hence $R(G+xy,p')$ and $R(G,p')$ have the same rank. This gives us an infinitesimally rigid realisation of G and implies that G is rigid in $\mathbf{R}^d$.

Connelly [23] applied a similar special position argument to show that the 1-extension operation preserves the property of having a 'maximum rank stress matrix' and hence preserves global rigidity in $\mathbf{R}^d$ by Theorem 2.4. We refer the reader to [23] for more details. □

2.4.3 Vertex splitting Suppose z is a vertex of a graph H. The *(d-dimensional) vertex splitting operation* constructs a new graph G from H by deleting z and then adding two new vertices u and v with $N_G(u) \cup N_G(v) = N_H(z) \cup \{u,v\}$ and $|N_G(u) \cap N_G(v)| \geq d-1$, see Figure 4. (We can view vertex splitting as an inverse operation to contracting the edge uv.) Whiteley [113] showed that vertex splitting preserves generic rigidity in $\mathbf{R}^d$.

Lemma 2.10 *Suppose G is a graph, $uv \in E$ and $|N_G(u) \cap N_G(v)| \geq d-1$. If G/uv is rigid in $\mathbf{R}^d$, then G is rigid in $\mathbf{R}^d$.*

Sketch proof Whiteley's proof uses yet another special position argument. He chooses a generic framework $(G/uv, p)$ and extends it to a framework (G,p') by putting $p'_u = p'_v = p_z$, where z is the vertex created by contracting uv. He then applies a limiting argument to the rigidity matrix to deduce that it has full rank for

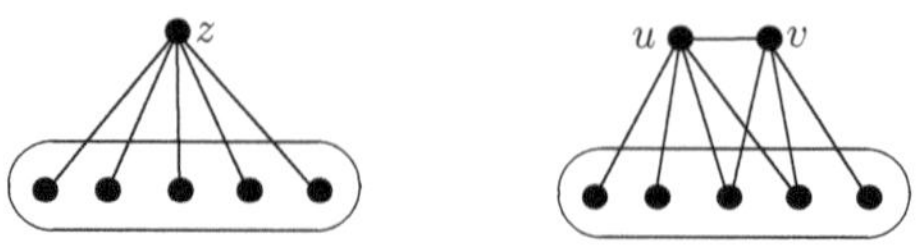

Figure 4: The 3-dimensional vertex splitting operation.

all realisations sufficiently close to this special position. We refer the reader to [113] for more details. □

Whiteley conjectures that an analogous result holds for global rigidity in $\mathbf{R}^d$ as long as the vertex splitting is *non-trivial*, i.e., the degrees of u and v in G are at least $d+1$. Connelly shows in his online notes [18] that vertex splitting preserves global rigidity whenever G has an infinitesimally rigid realisation in $\mathbf{R}^d$ in which u and v are coincident. A more precise statement and proof of his result is given in [28]. Jordán and Tanigawa verify another special case of Whiteley's conjecture in [63] by showing that a graph is globally rigid in $\mathbf{R}^d$ whenever it can be constructed from a globally rigid graph of maximum degree at most $d+2$ by a sequence of non-trivial vertex splits.

2.4.4 Gluing and substitution Our next two results describe situations in which we can construct a graph which is (globally) rigid in $\mathbf{R}^d$ by 'gluing together' two smaller graphs or replacing one subgraph by another subgraph.

Lemma 2.11 *Let G be a graph, and G_1, G_2 be subgraphs of G with $V(G_1) \cap V(G_2) = X$, $G = G_1 \cup G_2$ and $V(G_1) \backslash V(G_2) \neq \emptyset \neq V(G_2) \backslash V(G_1)$.*
(a) Suppose G_1, G_2 are rigid in $\mathbf{R}^d$. Then G is rigid in $\mathbf{R}^d$ if and only if $|X| \geq d$.
(b) Suppose G_1, G_2 are globally rigid in $\mathbf{R}^d$. Then G is globally rigid in $\mathbf{R}^d$ if and only if $|X| \geq d+1$.

Sketch proof Necessity is obvious since, for any generic realisation (G, p) in $\mathbf{R}^d$, we can rotate a component of $G - X$ about a $(d-2)$-dimensional affine subspace containing $p(X)$ if $|X| < d$, and reflect it in a $(d-1)$-dimensional affine subspace containing $p(X)$ if $|X| = d$. We refer the reader to [115] for proofs of sufficiency. □

Lemma 2.12 *Let $G_i = (V_i, E_i)$ be a graph for $i = 1, 2, 3$, with $V_1 \cap V_2 = X = V_1 \cap V_3$.*
(a) If $G_1 \cup G_2$ and G_3 are both rigid in $\mathbf{R}^d$ and $|X| \geq d$, then $G_1 \cup G_3$ is rigid in $\mathbf{R}^d$.
(b) If $G_1 \cup G_2$ and G_3 are both globally rigid in $\mathbf{R}^d$ and $|X| \geq d+1$, then $G_1 \cup G_3$ is globally rigid in $\mathbf{R}^d$.

Sketch proof (a) Choose generic realisations p of $G_1 \cup G_2$ and q of $G_1 \cup G_3$ in $\mathbf{R}^d$ such that $p|_{G_1} = q|_{G_1}$. Let $\dot{q}$ be an infinitesimal motion of $(G_1 \cup G_3, q)$. Since G_3 is rigid in $\mathbf{R}^d$, we may assume (by composing $\dot{q}$ with a suitable trivial infinitesimal

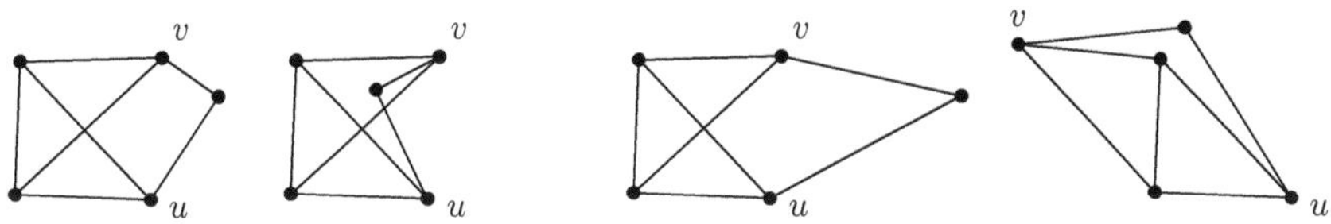

Figure 5: Two pairs of generic realisations of a graph G in $\mathbf{R}^2$. The vertex pair $\{u, v\}$ is globally linked in the two frameworks on the left since these are the only two non-congruent realisations of G in $\mathbf{R}^2$ with these edge lengths and the distance between u and v is the same in both of them. On the other hand, $\{u, v\}$ is not globally linked in the two frameworks on the right since they have the same edge lengths and the distance between u and v is different. Thus $\{u, v\}$ is weakly globally linked, but not globally linked in G in $\mathbf{R}^2$.

motion) that $\dot{q}|_{G_3} = 0$. Then $\dot{q}$ corresponds to an infinitesimal motion $\dot{p}$ of $(G_1 \cup G_2, p)$ with $\dot{p}|_{G_1} = \dot{q}|_{G_1}$ and $\dot{p}|_{G_2} = 0$. Since $G_1 \cup G_2$ is rigid and $|X| \geq d$, this gives $\dot{p} = 0$. Hence $\dot{q} = 0$ and $G_1 \cup G_3$ is rigid.

Part (b) can be proved similarly. □

2.5 Linked pairs of vertices

Let (G, p) be a d-dimensional framework. We say a pair of vertices $\{u, v\}$ of G are *linked* in (G, p) if every continuous motion of its vertices in $\mathbf{R}^d$, which preserves the lengths of its edges, preserves the distance between u and v. Thus, (G, p) is rigid if and only if all its pairs of vertices are linked. In the case when p is generic, $\{u, v\}$ is linked in (G, p) if and only if $r_d(G + uv) = r_d(G)$; this follows from the fact that every infinitesimal motion of a generic framework can be extended to a finite motion. Hence '$\{u, v\}$-linkedness' is a generic property, and we can define a pair of vertices $\{u, v\}$ of G to be *linked (in G) in* $\mathbf{R}^d$ if they are linked in every (or equivalently some) generic realisation of G in $\mathbf{R}^d$.

Similarly, we say that $\{u, v\}$ is *globally linked* in (G, p) if, for every framework (G, q) in $\mathbf{R}^d$ which has the same edge lengths as (G, p), we have $\|q(u) - q(v)\| = \|p(u) - p(v)\|$. It is not so straightforward to define the generic version of this concept since $\{u, v\}$ can be globally linked in one generic realisation of G in $\mathbf{R}^d$ and not in another, see Figure 5. Hence, we say $\{u, v\}$ is *globally linked (in G) in* $\mathbf{R}^d$ if $\{u, v\}$ is globally inked in (G, p) for *all* generic $p : V \to \mathbf{R}^d$, and $\{u, v\}$ is *weakly globally linked (in G) in* $\mathbf{R}^d$ if $\{u, v\}$ is globally linked in (G, p) for *some* generic $p : V \to \mathbf{R}^d$. It follows immediately that G is globally rigid in $\mathbf{R}^d$ if and only if every pair of vertices of G is globally linked in $\mathbf{R}^d$. Somewhat surprisingly, Jordán and Villányi [64] have recently shown that global rigidity can also be characterised by weak global linkedness.

Theorem 2.13 *A graph G is globally rigid in* $\mathbf{R}^d$ *if and only if every pair of vertices of G is weakly globally linked in* $\mathbf{R}^d$.

3 Rigidity and global rigidity in dimensions at most two

3.1 Rigidity

We have already seen that Maxwell's independence criterion characterises rigidity on the line. Pollaczek-Geiringer [92] showed that it also characterises generic rigidity in the plane in 1927, but her result went largely unnoticed, until it was rediscovered by Laman [74] in 1970. Pollaczek-Geiringer's original paper was brought to the attention of the current day rigidity community by Brigitte Servatius in 2010.

Theorem 3.1 *A graph with at least two vertices is rigid in* $\mathbf{R}^2$ *if and only if it has a spanning subgraph* $G = (V, E)$*, which satisfies* $|E| = 2|V| - 3$ *and* $i_G(X) \leq 2|X| - 3$ *for all* $X \subseteq V$ *with* $|X| \geq 2$.

Necessity follows from Lemma 2.3. To prove sufficiency it will suffice to show that every $(2,3)$-tight graph is rigid in $\mathbf{R}^2$. We will sketch two proofs of this statement to illustrate proof techniques which are commonly used to verify rigidity.

First proof of Theorem 3.1 Let $G = (V, E)$ be a $(2,3)$-tight graph. We show that G is rigid in $\mathbf{R}^2$ using the inductive strategy outlined in Section 2.4. Since G is $(2,3)$-tight, G has a vertex v of degree two or three. If $d_G(v) = 2$ then $G - v$ is $(2,3)$-tight, so is rigid in $\mathbf{R}^2$ by induction, and we can now apply Lemma 2.8 to deduce that G is rigid in $\mathbf{R}^2$. Hence, we may assume that $N_G(v) = \{x, y, z\}$. It will suffice to show that $G - v + uw$ is $(2,3)$-tight for some $u, w \in \{x, y, z\}$, since we can then deduce that G is rigid in $\mathbf{R}^2$ by applying induction and Lemma 2.9 to $G - v + uw$.

Suppose for a contradiction that $G - v + uw$ is not $(2,3)$-tight for all $u, w \in \{x, y, z\}$. Let M be the $(2,3)$-sparsity matroid of K_V. We have $r_M(G) = |E|$ and $r_M(G - v) = |E| - 3$ since G is M-independent. The assumption that $G - v + uw$ is M-dependent for all $u, w \in \{x, y, z\}$ now gives $r_M(G - v + xy + yz + xz) = r_M(G - v)$. Since $E(K_4)$ is a circuit in M, this implies that $r_M(G + xy + yz + xz) \leq r_M(G - v) + 2 = r_M(G) - 1$, and contradicts the fact that r_M is non-decreasing. □

Our second proof of Theorem 3.1 is a version of a proof by Lovász and Yemini [79], which was given by Whiteley in [114].

Second proof of Theorem 3.1 Let $G = (V, E)$ be a $(2,3)$-tight graph. We will show that G is rigid in $\mathbf{R}^2$ by constructing an infinitesimally rigid realisation of G. Choose an edge $xy \in E$, and let $G^{xy} = (V, E^{xy})$ be the multigraph obtained by adding a new copy of xy to G. The hypothesis that G is $(2,3)$-tight implies that G^{xy} is $(2,2)$-tight. Choose a generic map $q : E^{xy} \to \mathbf{R}^2$ and a reference ordering for the vertices in V. Let A^{xy} be the $|E^{xy}| \times 2|V|$-matrix, in which the entry in the row indexed by the edge $e = uw$ with $u < w$ and columns indexed by the vertex v is $q(e)$ if $v = u$, $-q(e)$ if $v = w$, and $\mathbf{0}$ if $v \neq u, w$. Then $\operatorname{rank} A^{xy} = |E^{xy}|$ by the discussion in the last paragraph of Section 2.1. Thus, if A is the matrix obtained by deleting row xy from A^{xy}, then we have $\operatorname{rank} A^{xy} = \operatorname{rank} A + 1$. This implies that there exists a vector $\alpha^{xy} \in \ker A$ with $\alpha_x^{xy} \neq \alpha_y^{xy}$. By taking a suitable linear combination of the vectors α^{xy}, we can construct a vector $\alpha \in \ker A$ with the property that $\alpha_x \neq \alpha_y$ for all $xy \in E$. We can now define our desired framework

(G,p) by putting $p_v = \alpha_v^\perp$ for each $v \in V$, where $(a,b)^\perp := (-b,a)$. Then each row of $R(G,p)$ will be a non-zero scalar multiple of the corresponding row of A. Hence, $\operatorname{rank} R(G,p) = \operatorname{rank} A = 2|V|-3$, and (G,p) is infinitesimally rigid. □

Once we know that the 2-dimensional rigidity matroid is equal to the $(2,3)$-sparsity matroid, Theorem 2.1 immediately gives us an expression for the rank function of this matroid. Lovász and Yemini [79] showed that this expression can be simplified to

$$r_2(G) = \min_{\mathcal{X}} \left\{ \sum_{X\in\mathcal{X}} (2|X|-3) \right\}, \tag{3.1}$$

where the minimum is taken over all families $\mathcal{X}$ of subsets of V, each of cardinality at least two, which cover E and are 1-*thin*, i.e., they satisfy $|X_i \cap X_j| \leq 1$ for all distinct $X_i, X_j \in \mathcal{X}$. In addition, they show that the right hand side of (3.1) achieves its minimum value when we take $\mathcal{X}$ to be the vertex sets of the 2-*dimensional rigid components of* G, i.e., the maximal subgraphs of G which are rigid in $\mathbf{R}^2$.

3.2 Global rigidity

It is not difficult to see that a graph $G = (V,E)$ is globally rigid on the line if and only if G is a complete graph on at most two vertices or is 2-connected. Necessity follows from Lemma 2.11. (Note that the redundant rigidity condition of Theorem 2.5 follows from the connectivity condition in this 1-dimensional case.) We can prove sufficiency by induction on $|E|$. We may assume that $|V| \geq 4$ and that G is minimally 2-connected, i.e., $G-e$ is not 2-connected for all $e \in E$. Choose an ear decomposition of G, i.e., a sequence of cycles $(C_1, C_2, \dots, C_t)$ which cover E and are such that $E(C_i)\setminus \bigcup_{j=1}^{i-1} E(C_j)$ induces a path in G which is internally vertex-disjoint from $\bigcup_{j=1}^{i-1} C_j$ for all $2 \leq i \leq t$. Since G is minimally 2-connected, the last 'ear' of this ear decomposition is an induced path of length at least two. Let v be an internal vertex of this last ear, and x,y be the neighbours of v in G. Then $G-v+xy$ is 2-connected and hence is globally rigid on the line by induction. We can now apply Lemma 2.9 to deduce that G is globally rigid on the line.

A similar inductive proof technique was used in [51] to show that Hendrickson's necessary conditions for global rigidity are also sufficient when $d = 2$, based on the generalisation of ear decompositions from graphs to matroids given by Coullard and Hellerstein [27].

Theorem 3.2 *A graph G is globally rigid in $\mathbf{R}^2$ if and only if G is a complete graph on at most 3 vertices or G is 3-connected and redundantly rigid in $\mathbf{R}^2$.*

The key idea in the proof of sufficiency in [51] is to show that the 2-dimensional rigidity matroid of a 3-connected, redundantly rigid graph $G = (V,E)$ is connected and then take an ear decomposition of $\mathcal{R}_2(G)$. We can assume inductively that $G-e$ is not both 3-connected and redundantly rigid, and also, by a previous result of Berg and Jordán [9], that G is not an $\mathcal{R}_2$-circuit. We next show that the subgraph of G induced by the last ear in the ear decomposition of $\mathcal{R}_2(G)$ contains a vertex v of degree 3 such that $\mathcal{R}_2(G-v+xy)$ is connected (and hence $G-v+xy$ is redundantly rigid in $\mathbf{R}^2$) for some $x,y \in N_G(v)$. The argument is rather technical, and an additional technical argument is needed to show that we can find such a vertex v so

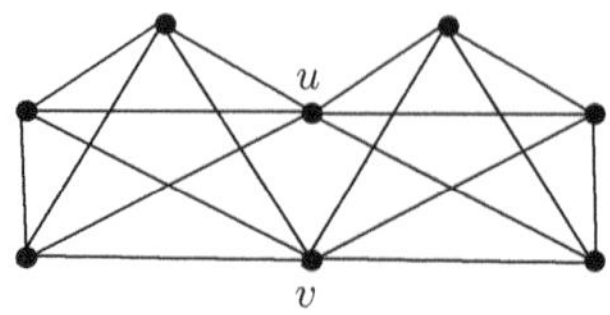

Figure 6: The double banana graph.

that $G-v+xy$ is 3-connected. Once we have such a vertex, we can apply induction to $G-v+xy$ and then use Lemma 2.9 to deduce that G is globally rigid.

A simplified version of this proof appeared in [99]. A different, and even simpler, proof based on Theorem 2.13 has recently been given by Jordán and Villányi in [64]. In fact, they go further and obtain a complete characterisation of when two vertices in a graph are weakly globally linked in $\mathbf{R}^2$. The related problem of characterising when two vertices in a graph are globally linked in $\mathbf{R}^2$ is one of the few remaining open problems for generic rigidity in $\mathbf{R}^2$. A conjectured characterisation, together with a proof of sufficiency and other partial results, can be found in [54, 55].

4 Rigidity and global rigidity in higher dimensions

The problem of characterising graphs which are generically rigid or globally rigid in $\mathbf{R}^d$ is open for all $d \geq 3$. We will investigate how the rank formula for the 2-dimensional rigidity matroid given in (3.1) may be extended to higher dimensions. There are many results which show that special families of graphs are rigid in higher dimensions. We refer the reader to the comprehensive survey articles on rigidity in [42] for information on these results.

4.1 Rigidity in 3-space

We first note that Maxwell's necessary condition, Lemma 2.3, is not sufficient to imply $\mathcal{R}_3$-independence. The so called *double banana graph* of Figure 6 satisfies Maxwell's condition for $\mathcal{R}_3$-independence. Since it has 18 edges, it would be rigid in $\mathbf{R}^3$ if it were $\mathcal{R}_3$-independent, and this would contradict Lemma 2.11(a).

To motivate a conjectured characterisation of rigidity in $\mathbf{R}^3$, we return to the expression for the rank function of the 2-dimensional rigidity matroid of a graph $G=(V,E)$ given in (3.1) and the result of [79] that the right hand side of (3.1) achieves its minimum value when we take the 1-thin cover $\mathcal{X}$ to be the vertex sets of the 2-dimensional rigid components of G. It is easy to see that $r_2(G)$ is at most $\sum_{X\in\mathcal{X}}(2|X|-3)$ for any cover $\mathcal{X}$ of E with vertex sets of size at least two: this follows from the fact that any base of $\mathcal{R}_2(G)$ can contain at most $2|X|-3$ edges in $G[X]$ for each $X \in \mathcal{X}$.

A similar reasoning tells us that

$$r_3(G) \leq \min\left\{|F| + \sum_{X\in\mathcal{X}}(3|X|-6)\right\}, \tag{4.1}$$

where the minimum is taken over all $F \subseteq E$ and all covers $\mathcal{X}$ of $E \backslash F$ with sets of size at least three.

Equality does not hold in (4.1) when we take $G = (V, E)$ to be the double banana graph in Figure 6 since, for each $X \subseteq V$ with $|X| \geq 3$, we have $i_G(X) \leq 3|X| - 6$, and hence the right hand side of (4.1) is at least $|E| = 18$. We can, however, obtain a tight upper bound on $r_3(G)$ if we take account of 'hinges' in the cover. Consider the graph $G + uv$, and let G_1 and G_2 be the two copies of K_5 in $G + uv$. Let B be a base of $\mathcal{R}_2(G + uv)$ which contains uv, and let $B_i = B \cap E(G_i)$ for $i = 1, 2$. Then $|B_i| \leq 3|V(G_i)| - 6 = 9$ for each $i = 1, 2$. Since $uv \in B_1 \cap B_2$, this gives $r_3(G) = |B| \leq |B_1| + |B_2| - 1 = 17$. (We can also show that $r_2(G) \geq 17$ by choosing an arbitrary edge of G and showing that $G - e$ can be constructed recursively from a path of length two by applying the 3-dimensional 0- and 1-extension operations. Or equivalently, that $G - e$ can be reduced to a path of length two by their inverse operations.)

We can generalise the argument in the preceding paragraph to obtain a better upper bound on $r_3(G)$ for an arbitrary graph $G = (V, E)$. Given a cover $\mathcal{X}$ of E, we say that $\mathcal{X}$ is *t-thin* if any two distinct sets in $\mathcal{X}$ intersect in at most t vertices, and define the *set of hinges of* $\mathcal{X}$ to be

$$H(\mathcal{X}) = \left\{ \{x, y\} \in \binom{V}{2} : \{x, y\} \subseteq X_1 \cap X_2 \text{ for two distinct sets } X_1, X_2 \in \mathcal{X} \right\}.$$

For each hinge $h \in H(\mathcal{X}\}$, the *degree of* h *in* $\mathcal{X}$, $d_{\mathcal{X}}(h)$, is the number of sets in $\mathcal{X}$ which contain h. The *$\mathcal{R}_3$-closure of* G, $\mathrm{cl}_3(G)$, is the graph obtained from G by adding an edge between all non-adjacent pairs of vertices which are linked in $\mathbf{R}^3$ (equivalently, adding all edges uv which satisfy $r_3(G + uv) = r_3(G)$ to G). We will refer to the maximal cliques in $\mathrm{cl}_3(G)$ as the *rigid clusters* of G. See Figure 7. Dress et al. [32] conjecture that the cover of E given by the rigid clusters of G determines $r_3(G)$.

Conjecture 4.1 (Dress Conjecture) *For any graph G*

$$r_3(G) = |F| + \sum_{X \in \mathcal{X}} (3|X| - 6) - \sum_{h \in H(\mathcal{X})} (d_{\mathcal{X}}(h) - 1), \tag{4.2}$$

where F is the set of edges of G which belong to rigid clusters of cardinality two, and $\mathcal{X}$ is the set of rigid clusters of cardinality at least three.

This conjecture can be illustrated using the graph G on the left of Figure 7, which is taken from Tay [100]. This graph has 7 rigid clusters, each of cardinality 5. The corresponding cover $\mathcal{X}$ has 7 hinges, all of degree two, and we have

$$\sum_{X \in \mathcal{X}} (3|X| - 6) - \sum_{h \in H(\mathcal{X})} (d_{\mathcal{X}}(h) - 1) = 7 \times 9 - 7 = 56,$$

which is indeed equal to $r_3(G)$.

The Dress Conjecture is of strong theoretical interest, but would not immediately allow us to calculate $r_3(G)$ since it does not tell us how to determine $\mathrm{cl}_3(G)$. Dress et al. offered a second conjecture in [32], which would imply that the problem of

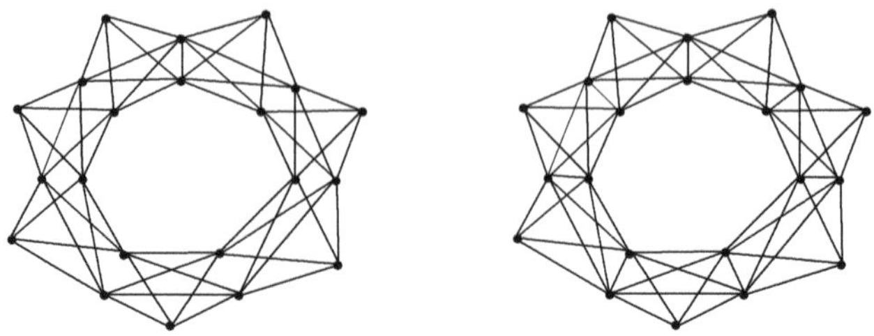

Figure 7: A graph and its $\mathcal{R}_3$-closure. The closure has 7 maximal cliques, each of cardinality five, giving rise to 7 rigid clusters in the original graph.

determining whether G is rigid in $\mathbf{R}^3$ belongs to NP $\cap$ coNP. They conjectured that $r_3(G)$ is equal to $\min\{|F| + \text{val}(\mathcal{X})\}$, where the minimum is taken over all $F \subseteq E$ and all 2-thin covers $\mathcal{X}$ of $E \backslash F$ with vertex sets of cardinality at least three, and

$$\text{val}(\mathcal{X}) := \sum_{X \in \mathcal{X}} (3|X| - 6) - \sum_{h \in H(\mathcal{X})} (d_{\mathcal{X}}(h) - 1)\}.$$

Unfortunately, this conjecture turned out to be spectacularly false: it was shown in [52] that $\min\{|F| + \text{val}(\mathcal{X})\}$ is negative for all graphs with 56 vertices.

We can try to repair the second Dress conjecture by restricting the pairs $(F, \mathcal{X})$, over which the minimum is taken, to more closely resemble the cover of E given by the rigid clusters of G. Lemma 2.11 implies that the set $\mathcal{X}$ of rigid clusters of G is 2-thin. In addition, Theorem 4.3 below gives some evidence that $\mathcal{X}$ is *$\mathcal{R}_3$-degenerate*, i.e., $\mathcal{X}$ can be ordered as $(X_1, X_2, \ldots, X_t)$ in such a way that, for all $2 \le i \le t$, the set of hinges of the subfamily $\{X_1, X_2, \ldots, X_i\}$, which are contained in X_i, is $\mathcal{R}_3$-independent (when viewed as a set of edges of K_V).

It had already been shown in [53, Lemma 5 and Conjecture 3.2] that, for any graph $G = (V, E)$,

$$r_3(G) \le \min\{|F| + \text{val}(\mathcal{X})\}, \tag{4.3}$$

where the minimum is taken over all $F \subseteq E$ and all $\mathcal{R}_3$-degenerate covers $\mathcal{X}$ of $E \backslash F$ with sets of cardinality at least three, and conjectured that equality holds.[1] Theorem 4.3 below suggests that this conjecture may remain true if we restrict our attention to a very simple type of $\mathcal{R}_3$-degenerate cover. We say that a vertex cover $\mathcal{X}$ of E is *k-shellable* if $\mathcal{X}$ can be ordered as $(X_1, X_2, \ldots, X_t)$ in such a way that, for all $2 \le i \le t$, $|X_i \cap \bigcup_{j=1}^{i-1} X_j| \le k$.

Conjecture 4.2 *For any graph G*

$$r_3(G) = \min\{|F| + \text{val}(\mathcal{X})\}, \tag{4.4}$$

where the minimum is taken over all $F \subseteq E$ and all 2-thin, 4-shellable covers $\mathcal{X}$ of $E \backslash F$ with sets of cardinality at least five.

[1]The statement of [53, Lemma 5] is in terms of 'iterated covers' and contains the additional hypothesis that the cover is 2-thin. The formulation above is given as in [17, Lemma 6.2].

Note that every 4-shellable cover is $\mathcal{R}_3$-degenerate since every graph on at most four vertices is $\mathcal{R}_3$-independent. Note also that the rigid cluster cover $\mathcal{X}$ of the graph $G = (V, E)$ on the left of Figure 7 is 4-shellable since every rigid cluster of G contains exactly two hinges of $\mathcal{X}$. Thus, (4.3) immediately implies that $r_3(G) \leq \text{val}(\mathcal{X}) = 56$. This tells us in particular that G is not rigid in $\mathbf{R}^3$ since $r_3(G) < 3|V| - 6 = 57$.

Clinch, Jackson and Tanigawa [17, Theorems 6.3 and 6.1] show that Conjectures 4.1 and 4.2 both hold if we replace the 3-dimensional rigidity matroid by another matroid on $\binom{V}{2}$, the $\mathcal{C}_2^1$-cofactor matroid, which features in the study of bivariate splines. More generally, for any $d \geq 1$ and any graph $G = (V, E)$, the *$\mathcal{C}_{d-1}^{d-2}$-cofactor matroid of G*, $\mathcal{C}_{d-1}^{d-2}(G)$, is the row matroid of the $|E| \times d|V|$ matrix obtained by choosing a reference ordering for V and a generic map $p : V \to \mathbf{R}^2$, and taking the row of the matrix indexed by $e = uv$ with $u < v$ to be:

$$\begin{array}{c} \\ e{=}uv \end{array} \begin{array}{c} \begin{matrix} \quad\quad\quad\quad\quad u & \quad\quad\quad\quad\quad\quad\quad\quad\quad v \end{matrix} \\ \left[\ 0 \ldots 0 \quad g_d(p_u - p_v) \quad 0 \ldots 0 \quad -g_d(p_u - p_v) \quad 0 \ldots 0\ \right], \end{array}$$

where $g_d : (x, y) \mapsto (x^{d-1}, x^{d-2}y, \ldots, y^{d-1})$. Let $c_{d-1}^{d-2}(G)$ denote the rank of $\mathcal{C}_{d-1}^{d-2}(G)$. A *$\mathcal{C}_{d-1}^{d-2}$-rigid-cluster* of G is a maximum clique in the '$\mathcal{C}_{d-1}^{d-2}$-closure' of G.

Theorem 4.3 *For any graph G,*

$$c_2^1(G) = \min\{|F| + \text{val}(\mathcal{X})\}, \tag{4.5}$$

where the minimum is taken over all $F \subseteq E$ and all 2-thin, 4-shellable covers $\mathcal{X}$ of $E \backslash F$ with sets of cardinality at least five. In addition, equality holds in (4.5) when we take $\mathcal{X}$ to be the family of C_2^1-rigid-clusters of G of size at least five and F to be the set of edges of G which are not covered by $\mathcal{X}$.

The $\mathcal{C}_2^1$-cofactor matroid has many properties in common with the 3-dimensional rigidity matroid, and the two matroids are conjectured to be identical by Whiteley in [115]. Whiteley's conjecture, and hence also Conjectures 4.1 and 4.4, would follow from the following conjecture from [115] by a simple inductive argument.

Conjecture 4.4 *Suppose $G = (V, E)$ is a graph, and $v \in V$ with $N_G(v) = \{v_1, v_2, v_3, v_4, v_5\}$.*
(a) If $v_1v_2, v_3v_4 \notin E$ and $G - v + v_1v_2 + v_3v_4$ is $\mathcal{R}_3$-independent, then G is $\mathcal{R}_3$-independent.
(b) If $v_1v_2, v_2v_3, v_iv_j, v_jv_k \notin E$ for some $v_i, v_j, v_k \in N_G(v)$ with $v_j \neq v_2$, and $G - v + v_1v_2 + v_2v_3$ and $G - v + v_iv_j + v_jv_k$ are both $\mathcal{R}_3$-independent, then G is $\mathcal{R}_3$-independent.

We refer to the operation which constructs G from $G - v + v_1v_2 + v_3v_4$ in (a) as the *X-replacement operation*, and the operation which constructs G from $G - v + v_1v_2 + v_2v_3$ and $G - v + v_iv_j + v_jv_k$ in (b) as the *double-V-replacement operation*. Whiteley [115] uses a simple special position argument to show that the X-replacement operation preserves C_2^1-independence. This fact is used in [16] to deduce that the double-V-replacement operation also preserves C_2^1-independence.

Theorem 4.3 follows from a different expression for the rank function of the C_2^1-cofactor matroid [17, Theorem 5.7]. Given a graph $G = (V, E)$, a *K_5-cover of G* is a family of edge sets of copies of K_5 in K_V, which cover E. A K_5-cover is *proper* if it has an ordering $(C_1, C_2, \ldots, C_t)$ such that $C_i \not\subseteq \bigcup_{j=1}^{i-1} C_j$ for all $2 \leq i \leq t$. Note that the rigid cluster cover of the graph on the left of Figure 7 is a proper K_5-cover.

Theorem 4.5 *For any graph G,*

$$c_2^1(G) = \min\left\{|F| + \left|\bigcup_{C\in\mathcal{Y}} C\right| - |\mathcal{Y}|\right\}, \tag{4.6}$$

where the minimum is taken over all $F \subseteq E$ and all proper K_5-covers $\mathcal{Y}$ of $G - F$.

This result can also be viewed as a characterisation of $c_2^1(G)$ in terms of '$\mathcal{C}_2^1$-degenerate covers' of E with vertex sets of cardinality 5 by putting $\mathcal{X} = \{V(C) : C \in \mathcal{Y}\}$ and noting that $\operatorname{val} \mathcal{X} = \left|\bigcup_{C\in\mathcal{Y}} C\right| - |\mathcal{Y}|$.

Theorems 4.3 and 4.5 each imply that the problem of determining whether a graph $G = (V, E)$ satisfies $c_2^1(G) = 3|V| - 6$, belongs to NP $\cap$ coNP. On the other hand, there is no known polynomial algorithm for determining the value of the functions on the right hand side of (4.5) or (4.6).

4.2 Rigidity in dimensions at least four

The problem of characterising when a graph is rigid in $\mathbf{R}^d$ seems to be significantly more difficult when $d \geq 4$. This is signified by the results of Bolker and Roth [11] which imply that $K_{d+2,d+2}$ is a closed, non-rigid circuit in the d-dimensional rigidity matroid whenever $d \geq 4$. This in turn implies that the d-dimensional X-replacement operation does not preserve $\mathcal{R}_d$-independence when $d \geq 4$ (since $K_{d+2,d+2}$ can be constructed from K_2 by a sequence of d-dimensional 0- and 1-extension operations and an X-replacement). In addition, the d-dimensional version of Conjecture 4.1 fails, since every $\mathcal{R}_d$-rigid cluster of $K_{d+2,d+2}$ is a copy of K_2.

On the other hand, it is at least conceivable that a modified version of Theorem 4.5 may hold for $\mathcal{R}_d$. A *$\{K_{d+2}, K_{d+2,d+2}\}$-cover of a graph $G = (V, E)$* is a family of edge sets of subgraphs of K_V, each of which is a copy of K_{d+2} or $K_{d+2,d+2}$. The following conjecture appeared in the special case when $d = 4$ as [17, Conjecture 8.2] and was then extended to arbitrary d in [56, Conjecture 6].

Conjecture 4.6 *For any graph G,*

$$r_d(G) = \min\left\{|F| + \left|\bigcup_{C\in\mathcal{Y}} C\right| - |\mathcal{Y}|\right\}, \tag{4.7}$$

where the minimum is taken over all $F \subseteq E$ and all proper $\{K_{d+2}, K_{d+2,d+2}\}$-covers $\mathcal{Y}$ of $G - F$.

4.3 Global rigidity

Connelly [22] showed that the family of complete bipartite graphs provide examples which show that the necessary conditions for generic global rigidity in $\mathbf{R}^d$ given in Theorem 2.5 are not sufficient when $d \geq 3$. The following characterization of globally rigid complete bipartite graphs is stated without proof in [65, Theorem 63.2.2]. A proof can be found in [60].

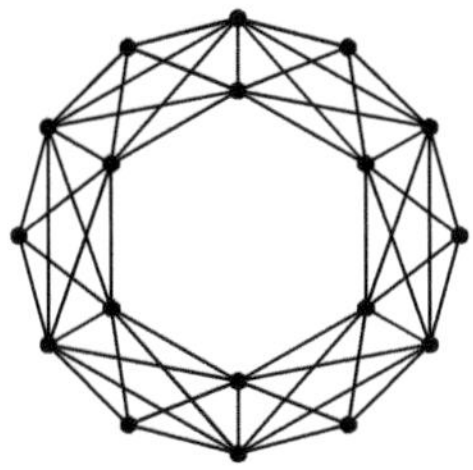

Figure 8: A 4-connected graph which is redundantly rigid but not globally rigid in $\mathbf{R}^3$.

Theorem 4.7 *The complete bipartite graph $K_{s,t}$ is globally rigid in $\mathbf{R}^d$ if and only if $s,t \geq d+1$ and $s+t \geq \binom{d+2}{2}+1$.*

This implies that $K_{5,5}$ is not globally rigid in $\mathbf{R}^3$ and, more generally, gives a finite family of graphs which satisfy the necessary conditions for global rigidity given in Theorem 2.5 and are not globally rigid in $\mathbf{R}^d$ for any fixed d. Infinite families of such graphs were constructed by Frank and Jiang [37] for each $d \geq 5$ and subsequently by Jordán, Király and Tanigawa [62] for each $d \geq 3$. The smallest graph in their family for $d = 3$ is shown in Figure 8. To see that this graph G is not globally rigid in $\mathbf{R}^3$, consider the graph G' obtained from G by replacing one of the K_5-subgraphs H by a K_4-subgraph H' with the same vertices of attachment as H. Let F be the set of four edges of H' which do not belong to any of the K_5-subgraphs of G', and let $e \in F$. Then we can use (4.3) and the 2-thin, 4-shellable cover $\mathcal{X}$ of $G' - F$ consisting of the five K_5-subgraphs to show that $G' - e$ is not rigid in $\mathbf{R}^3$: we have $r_3(G'-e) \leq |F-e| + \mathrm{val}\,(\mathcal{X}) = 3 + 5 \times 9 - 4 = 44 = 3|V(G')| - 7$. This implies that G' is not redundantly rigid in $\mathbf{R}^3$ and hence is not globally rigid in $\mathbf{R}^3$ by Theorem 2.5. We can now use Lemma 2.12(b) to deduce that G is not globally rigid in $\mathbf{R}^3$.

Lemma 2.12(b) also tells us that we can construct an infinite family of 4-connected graphs which are redundantly rigid but not globally rigid in $\mathbf{R}^3$ by replacing a K_5-subgraph of the graph G in Figure 8 by any graph on at least five vertices which is globally rigid in $\mathbf{R}^3$.

The above discussion suggests the following new necessary condition for global rigidity in $\mathbf{R}^3$.

Lemma 4.8 *Let $G = (V,E)$ be a graph on at least five vertices which is globally rigid in $\mathbf{R}^3$. Suppose there exists a set $F \subseteq E$ and a 3-thin, 4-shellable cover $\mathcal{X}$ of $E \backslash F$ with sets of cardinality at least five such that $|F| + \mathrm{val}\,(\mathcal{X}) = 3|V| - 6$. Then $|F| = \emptyset$ and $|\mathcal{X}| = 1$.*

Proof Theorem 2.5 and the hypothesis that G is a globally rigid graph with at least five vertices imply that G is 4-connected and redundantly rigid in $\mathbf{R}^3$.

We first show that $F = \emptyset$. Suppose, for a contradiction, that $F \neq \emptyset$, and choose $e \in F$. Then $|F-e| + \mathrm{val}\,(\mathcal{X}) = 3|V| - 7$ and hence $G - e$ is not rigid in $\mathbf{R}^3$ by (4.3), contradicting the redundant rigidity of G. Hence $F = \emptyset$.

To prove the second part of the lemma, we assume, for a contradiction, that $|\mathcal{X}| \geq 2$. Let $(X_1, X_2, \ldots, X_t)$ be a '4-shellable ordering' of $\mathcal{X}$, and put $U = X_t \cap \bigcup_{i=1}^{t-1} X_i$. Then $|U| = 4$ since G is 4-connected and X_t is the last set in the 4-shellable ordering of $\mathcal{X}$.

Let F' be the set of all edges of K_U which are not hinges of $\mathcal{X}$, and let G' be obtained from G by replacing the subgraph induced by X_t by K_U. Then $\mathcal{X}' := \mathcal{X} - X_t$ is a 3-thin, 4-shellable cover of $G' - F'$ and $|F'| + \mathrm{val}\,(\mathcal{X}') = 3|V(G')| - 6$. If $F' \neq \emptyset$, then we could use the first part of the lemma to deduce that G' is not globally rigid in $\mathbf{R}^3$. This would contradict Lemma 2.12(b), since G is globally rigid, and G' is obtained from G by replacing $G[X_t]$ by a copy of K_4. Hence $F' = \emptyset$ and every edge of K_U is a hinge of $\mathcal{X}$. Since $\mathcal{X}$ is 3-thin, this implies that $|\mathcal{X}| \geq 3$, and hence $|\mathcal{X}'| \geq 2$. We can now deduce by induction on $|V|$ that G' is not globally rigid in $\mathbf{R}^3$. This again contradicts Lemma 2.12(b) and completes the proof of the lemma.
□

Note that the lemma becomes false if we remove the hypothesis that $\mathcal{X}$ is 3-thin. Consider the graph G obtained by taking the union of two copies of K_5 with four vertices in common. Then G is globally rigid in $\mathbf{R}^3$, and the cover $\mathcal{X}$ of G consisting of the vertex sets of each of the K_5-subgraphs is 4-shellable and satisfies $\mathrm{val}\,(\mathcal{X}) = 3|V| - 6$.

It is conceivable that the necessary condition for global rigidity given by Lemma 4.8 is also sufficient for all graphs other than $K_{5,5}$.

Conjecture 4.9 *Let $G = (V, E)$ be a graph on at least five vertices, which is not a copy of $K_{5,5}$. Then G is globally rigid in $\mathbf{R}^3$ if and only if, for all $F \subseteq E$ and all 3-thin, 4-shellable covers of $E \backslash F$,*

$$|F| + \mathrm{val}\,(\mathcal{X}) \geq 3|V| - 6$$

with equality only when $F = \emptyset$ and $\mathcal{X} = \{V\}$.

If true, this conjecture would imply that the complete bipartite graph $K_{5,5}$ is the only 5-connected graph which is redundantly rigid but not globally rigid in $\mathbf{R}^3$, and hence give an affirmative answer to [38, Problem 4.2].

To our knowledge, there are no conjectured combinatorial characterisations of graphs which are globally rigid in $\mathbf{R}^d$ when $d \geq 4$. This makes the following recent result of Villányi [109] even more impressive.

Theorem 4.10 *Every $d(d+1)$-connected graph is globally rigid in $\mathbf{R}^d$.*

This result answers a conjecture of Connelly, Jordán and Whiteley [25] and also resolves a longstanding conjecture of Lovász and Yemini [79] for rigidity in $\mathbf{R}^d$. The connectivity hypothesis in the theorem is best possible since there exist $(d(d+1)-1)$-connected graphs which are not even rigid in $\mathbf{R}^d$. Villányi's proof of Theorem 4.10 is surprisingly elementary: he first uses an ingenious application of the probabilistic method based on Lemmas 2.8 and 2.9 to verify rigidity, and then uses Theorem 2.13 to extend this to global rigidity.

Another exciting new development in generic (global) rigidity is the result by Lew et al. [77], who show that the threshold probabilities for rigidity and global rigidity of Erdős–Rényi random graphs coincide with that for having minimum degree at least d and $d+1$, respectively. Their proof cleverly exploits properties of the rigidity matroid of the Erdös-Rényi random graph.

5 Applications to other areas of combinatorics

5.1 Orientations and packings of graphs

Tools from rigidity theory, along with matroid and graph theoretic methods, have recently been used to solve two long standing open problems in graph theory concerning highly connected orientations and removable spanning trees, respectively. It follows from a classical theorem of Nash-Williams [85] that every $2k$-edge-connected graph has a k-arc-connected orientation. In 1985, Thomassen [106] asked whether a similar statement is true for vertex-connectivity. He conjectured that, for every positive integer k, there exists a (smallest) integer $f(k)$ such that every $f(k)$-connected graph has a k-connected orientation. It is easy to see that if $f(k)$ exists, then $f(k) \geq 2k$. It follows from the above mentioned result of Nash-Williams that $f(1) = 2$. Jordán [58] showed that $f(2) \leq 18$. His proof uses the following result on packing rigid subgraphs, together with results on 2-connected orientations of Eulerian graphs. Thoughout this section, we will say that a graph G is *d-rigid* if it is (generically) rigid in $\mathbf{R}^d$, and that G is *minimally d-rigid* if it is d-rigid, but $G - e$ is not d-rigid for all edges e of G.

Theorem 5.1 *Every $6t$-connected graph contains t edge-disjoint 2-rigid (and hence 2-connected) spanning subgraphs.*

The proof of Theorem 5.1 used (3.1) together with Edmonds' formula for the rank function of unions of matroids [34]. It also showed that a packing of 2-rigid spanning subgraphs can be found in polynomial time.

Subsequently, Thomassen [107] proved that $f(2) = 4$. Durand de Gevigney [33] showed that, for $k \geq 3$, deciding whether a graph has a k-connected orientation is NP-hard.

The proof of Theorem 4.10 showed that sufficient connectivity conditions for d rigidity can be obtained without knowing the rank function of $\mathcal{R}_d(G)$. A similar approach was used by Garamvölgyi, Jordán, Király and Villányi [40] to obtain the following packing result, which can be considered as a result on the matroid union of t copies of the d-dimensional generic rigidity matroid.

Theorem 5.2 *Every $(t \cdot 10d(d+1))$-connected graph contains t edge-disjoint d-rigid (and hence d-connected) spanning subgraphs.*

The existence of a constant $h(d)$ such that every $(t \cdot h(d))$-connected graph contains t edge-disjoint d-connected spanning subgraphs was conjectured in [39]. The bound on $h(d)$ in Theorem 5.2 is certainly not tight. It is conceivable, that $h(d) \leq d(d+1)$. For $t = 1$, this is Theorem 4.10.

Theorem 5.2 was combined with our next two lemmas in [40] to solve Thomassen's orientation problem. Given a graph $G = (V, E)$ and a function $g : V \to \mathbb{Z}_+$, we shall use the notation $g(X) = \sum_{v \in X} g(v)$ for subsets $X \subseteq V$. The in-degree of a subset X of vertices in a digraph D is denoted by $\rho_D(X)$. Hakimi [47] proved that G has an orientation $\vec{G}$ with $\rho_{\vec{G}}(v) = g(v)$ for all $v \in V$ if and only if (i) $i_G(X) \leq g(X)$ for all nonempty $X \subseteq V$, and (ii) $|E| = g(V)$ hold.

We may use Hakimi's result and the sparsity count of Lemma 2.3 to show the existence of certain in-degree specified orientations of minimally d-rigid graphs. Given a minimally d-rigid graph $G = (V, E)$ with $|V| \geq \binom{d+1}{2}$ and a subset $R \subseteq V$ with $|R| = \binom{d+1}{2}$, we define the in-degree specification function $g_{d,R}$ by putting $g_{d,R}(v) = d$ for all $v \in V - R$ and $g_{d,R}(r) = d - 1$ for all $r \in R$. We say that an orientation $\vec{G}$ of G is a (d, R)-*orientation* if its in-degrees respect the specification $g_{d,R}$.

Lemma 5.3 *Let $G = (V, E)$ be a minimally d-rigid graph with $|V| \geq \binom{d+1}{2}$, and let $R \subseteq V$ be a set of vertices with $|R| = \binom{d+1}{2}$. Then G has a (d, R)-orientation.*

The second lemma provides a lower bound on the number of in-neighbors of certain subsets in (d, R)-orientations, establishing a link between degree-specified orientations and high vertex-connectivity.

Lemma 5.4 *Let d and k be integers with $k \geq 2$ and $d \geq 4k-4$. Let $G = (V, E)$ be a minimally d-rigid graph with $|V| \geq \binom{d+1}{2}$, $R \subseteq V$ a set of vertices with $|R| = \binom{d+1}{2}$, and let $\vec{G}$ be a (d, R)-orientation of G. Finally, let $X \subseteq V$ be a set of vertices. If $|X \cap R| \leq \frac{\binom{d+1}{2}}{2}$, then X has at least k in-neighbors in $\vec{G}$.*

These lemmas were combined with Theorem 5.2 in [40] to give:

Theorem 5.5 *Every $(320 \cdot k^2)$-connected graph has a k-connected orientation.*

Sketch proof Suppose that $k \geq 2$, and let $d = 4k-4$. By Theorem 5.2, $G = (V, E)$ has two edge-disjoint minimally d-rigid spanning subgraphs G_1 and G_2.

Let us fix a set $R \subseteq V$ of vertices with $|R| = \binom{d+1}{2}$. The desired orientation $\vec{G}$ of G is defined by first defining two orientations of G_1 and G_2, and then orienting the remaining edges arbitrarily. The orientation of G_1 is chosen to be a (d, R)-orientation. The orientation of G_2 is chosen to be a *reversed* (d, R)-orientation. By Lemma 5.3, these orientations exist. It is not hard to show, using Lemma 5.4, that the union $\vec{G}$ of these oriented spanning subgraphs is k-connected. □

The bound on $f(k)$ given by Theorem 5.5 is probably far from being tight. In particular, it is still open whether $f(k) = 2k$ holds.

One may also consider the rooted version of the above orientation problem. A digraph D is *rooted k-connected* (from a given root vertex r), if it contains k internally disjoint directed paths from r to each vertex $v \in V(D) - r$. Rooted k-connected graphs are defined analogously. It was recently conjectured in [61] that sufficiently highly rooted connected graphs have rooted k-connected orientations. It was shown in [61] that every rooted 6-connected graph has a rooted 2-connected orientation, and such an orientation can be found in polynomial time. The proof of this result also employs methods from rigidity theory. In particular, a key step of the proof is to show that if G is rooted 6-connected from r and $G - r$ is 2-edge connected, then G is redundantly rigid in $\mathbf{R}^2$, which strengthens a result of [79].

The packing Theorem 5.2 immediately implies the solution of another conjecture on graph connectivity concerning removable spanning trees. Kriesell [84, Problem

444] conjectured in 2003 that for every positive integer k there exists a (smallest) integer $g(k)$ such that every $g(k)$-connected graph G contains a spanning tree T for which $G - E(T)$ is k-connected.

As with Thomassen's conjecture, the edge-connected version of Kriesell's conjecture is a well-known result: it follows from a classical theorem of Nash-Williams [86] and Tutte [108] that every $(2k+2)$-edge-connected graph G contains a spanning tree T such that $G - E(T)$ is k-edge-connected. In particular, we have $g(1) = 4$. Theorem 5.1 implies that $g(2) \leq 12$. The $k = 3$ case was settled by Garamvölgyi, Jordán, and Király [39] using the matroid union of t copies of the $\mathcal{C}_2^1$-cofactor matroid and Theorem 4.3.

Theorem 5.6 *Every $12t$-connected graph contains t edge-disjoint $\mathcal{C}_2^1$-rigid (and hence 3-connected) spanning subgraphs. In particular, $g(3) \leq 24$.*

By applying Theorem 5.2 with $t = 2$ and choosing a spanning tree of one of the two d-rigid spanning subgraphs, we get $g(d) \leq 20d(d+1)$. To improve upon this result, one can consider the union of the d-dimensional rigidity matroid and the graphic matroid. It was shown in [40] that this leads to the bound $g(d) \leq d^2+3d+5$.

Theorem 5.7 *Every (d^2+3d+5)-connected graph has a spanning tree T for which $G - E(T)$ is d-connected.*

The bound given by Theorem 5.7 is still not optimal. Perhaps every $(d(d+1)+2)$-connected graph contains edge-disjoint copies of a d-rigid spanning subgraph and a spanning tree. Again, this bound would be tight.

An immediate corollary of Theorem 5.7 is that, if G is sufficiently highly connected, then, for all $s, t \in V(G)$, there exists a path P from s to t in G such that $G - E(P)$ is k-connected. The existence of such paths was verified earlier in [71], assuming that G is $(1600k^4 + k + 2)$-connected. This connectivity requirement can be substantially weakened using Theorem 5.7.

5.2 Rigidity and simplicial complexes

An *abstract simplicial complex* Δ is a finite family of sets with the property that $\tau \subseteq \sigma \in \Delta$ implies that $\tau \in \Delta$. An element $\sigma \in \Delta$ is called a *face of dimension* $|\sigma| - 1$, or a $(|\sigma| - 1)$-*face* for short, and we define $\dim(\Delta) = \max\{\dim \sigma : \sigma \in \Delta\}$. A *facet* of Δ is a maximal face, and we say that Δ is a *simplicial k-complex* if all its facets have dimension k. We write $\mathcal{F}_j(\Delta)$ for the set of j-faces of Δ and put $f_j(\Delta) = |\mathcal{F}_j(\Delta)|$. The *$f$-vector of* Δ is $f(\Delta) = (f_0, f_1, \ldots)$. The *graph of* Δ is $G(\Delta) = (\mathcal{F}_0(\Delta), \mathcal{F}_1(\Delta))$, and so elements of $\mathcal{F}_0(\Delta)$, respectively $\mathcal{F}_1(\Delta)$, are called *vertices*, respectively *edges*, of Δ. Abusing notation, we will write v instead of $\{v\}$ for the vertex whose single element is v and, similarly, uv instead of $\{u, v\}$. The *link* of a face $\sigma \in \Delta$ is defined by $\mathrm{lk}_\Delta(\sigma) = \{\tau \setminus \sigma : \tau \in \Delta, \tau \supseteq \sigma\}$.

5.2.1 Simplicial circuits and Fogelsanger's Theorem Given a finite set X and positive integer k, the *k-simplicial matroid* $\mathcal{S}_k(X)$ is the row matroid of the $\binom{|X|}{k+1} \times \binom{|X|}{k}$ matrix I, defined over $\mathbf{Z}_2$, in which the rows are indexed by $\binom{X}{k+1}$, the columns are indexed by $\binom{X}{k}$, and the (σ, τ)-entry is 1 if $\tau \subset \sigma$, and 0 otherwise. Clearly,

$\mathcal{S}_1([n])$ is the graphic matroid on K_n. See [110, Chapter 10] for more information on simplicial matroids.[2]

A *simplicial k-circuit* is a simplicial k-complex Δ such that $\mathcal{F}_k(\Delta)$ is a circuit in $\mathcal{S}_k(\mathcal{F}_0(\Delta))$. For example, a cycle graph is a simplicial 1-circuit. More explicitly, we have the following characterisation.

Lemma 5.8 *A simplicial k-complex Δ is a simplicial k-circuit if and only if every $(k-1)$-face of Δ is contained in an even number of k-faces, and no proper subcomplex of Δ has this property.*

In particular, for $d \geq 2$, the boundary complex of simplicial d-polytope[3] is a simplicial $(d-1)$-circuit. A more general example is the following. A *k-pseudomanifold* is a simplicial k-complex Π such that

1. every $(k-1)$-face of Π belongs to exactly two k-faces of Π, and
2. Π is *strongly connected*, i.e., if $\sigma, \sigma' \in \mathcal{F}_k(\Pi)$, then there is a sequence of k-faces $\sigma = \tau_1, \cdots, \tau_s = \sigma'$ such that $|\tau_i \cap \tau_{i+1}| = k$ for $i = 1, \cdots, s-1$.

In particular, every triangulation of a k-dimensional manifold is a k-pseudomanifold.[4] It follows easily from Lemma 5.8 that a k-pseudomanifold is a simplicial k-circuit for all $k \geq 1$, although the converse is not necessarily true when $k \geq 2$.

Theorem 5.9 (Fogelsanger's Rigidity Theorem) *Let Δ be a simplicial k-circuit for some $k \geq 2$. Then $G(\Delta)$ is rigid in $\mathbf{R}^{k+1}$.*

The proof is an inductive argument based on edge contraction/vertex splitting. Given two vertices u, v of a simplicial complex Δ, we define Δ/uv to be the complex obtained from Δ by deleting every face that contains $\{u, v\}$ and then replacing every face σ that contains v by $\sigma \setminus \{v\} \cup \{u\}$. An immediate difficulty for the induction is that edge contraction does not necessarily preserve the property of being a simplicial circuit. Fogelsanger discovered a key decomposition argument to circumvent this problem.

Lemma 5.10 *Let Δ be a simplicial k-circuit for some $k \geq 2$, $uv \in \mathcal{F}_1(\Delta)$, and suppose that Δ/uv is not a simplicial k-circuit. Then there exist simplicial k-circuits $\Delta_1, \cdots, \Delta_m$ such that*

(F1) $G(\Delta) = \bigcup_{i=1}^m G(\Delta_i)$,

(F2) $uv \in \Delta_i$ *for* $1 \leq i \leq m$, *and, either* Δ_i *is the boundary complex of a* $(k+1)$-*simplex, or* Δ_i/uv *is a simplicial k-circuit,*

(F3) for $2 \leq j \leq m$, $\mathcal{F}_k(\Delta_j) \cap \left(\bigcup_{i=1}^{j-1} \mathcal{F}_k(\Delta_i)\right) \neq \emptyset$.

[2] In Welsh's notation, the $(k+1)$-simplicial matroid is what we call $\mathcal{S}_k(X)$. Our notation is more in line with the topological convention that the dimension of a face σ is $|\sigma| - 1$.

[3] We will use the term *d-polytope* to mean the convex hull of a finite set of points which affinely span $\mathbf{R}^d$. It is *simplicial* if every $(d-1)$-face is a $(d-1)$-simplex.

[4] Recall that a triangulation of a manifold is a simplicial complex which has a homeomorphism from its geometric realisation to the given manifold.

In fact, Fogelsanger proved more general versions of Theorem 5.9 and Lemma 5.10 for *minimal homology cycles* over an arbitrary field. These coincide with our simplicial circuits when the field is $\mathbf{Z}_2$. See [36] for details.

Next, we outline the proof of Theorem 5.9,

Sketch proof Pick $uv \in \mathcal{F}_1(\Delta)$. If Δ/uv is again a simplicial k-circuit then, by induction on the number of vertices, $G(\Delta)/uv = G(\Delta/uv)$ is rigid in $\mathbf{R}^{k+1}$. Now, since Δ is a simplicial k-circuit, $\{u, v\}$ is contained in at least two k-faces. This implies that u and v have at least k common neighbours in $G(\Delta)$, and so, by Lemma 2.10, G is rigid in $\mathbf{R}^{k+1}$.

Thus, we can assume that Δ/uv is not a simplicial k-circuit. Let $\Delta_1, \cdots, \Delta_m$ be the simplicial k-circuits, whose existence is asserted by Lemma 5.10. By (F2) and using the induction argument of the previous paragraph, we see that $G(\Delta_i)$ is rigid in $\mathbf{R}^{k+1}$ for each i. Now, (F3) implies that, for $2 \leq j \leq m$, $\bigcup_{i=1}^{j-1} G(\Delta_i)$ and $G(\Delta_j)$ have at least $k+1$ common vertices, so an induction argument using Lemma 2.11 and (F1) proves that $G(\Delta)$ is rigid in $\mathbf{R}^{k+1}$. □

The following result on the global rigidity of simplicial k-circuits was recently obtained in [29].

Theorem 5.11 *Let Δ be a simplicial k-circuit for $k \geq 2$.*

1. *If $k \geq 3$, then $G(\Delta)$ is globally rigid in $\mathbf{R}^{k+1}$ if and only if $G(\Delta)$ is $(k+2)$-connected or is a copy of K_{k+2}.*
2. *If $k = 2$, then $G(\Delta)$ is globally rigid in $\mathbf{R}^3$ if and only if $G(\Delta)$ is 4-connected and not planar, or $G(\Delta)$ is a copy of K_4.*

The proof of Theorem 5.11 is again an induction based on vertex splitting and on Fogelsanger's decomposition. However, there are significant extra complications that arise from controlling the $(k+2)$-connectedness in the inductive arguments, and also from the fact that, as mentioned in Section 2.4, the natural analogue of Lemma 2.10 for global rigidity is still a conjecture. For more details see [29].

Theorems 5.9 and 5.11 have some important consequences for the theory of f-vectors of simplicial k-polytopes and, more generally, of k-pseudomanifolds. It is natural to wonder what is the smallest possible number of j-faces in a simplicial polytope with a given number of vertices. For $0 \leq j \leq k \leq n-2$, define

$$\phi_j(k,n) = \begin{cases} \binom{k+1}{j} n - j\binom{k+2}{j+1} & \text{if } j < k, \\ kn - (k-1)(k+2) & \text{if } j = k. \end{cases}$$

The number $\phi_j(k,n)$ is the numbers of j-faces in a *stacked k-sphere*. This family of simplicial k-complexes can be defined recursively as follows. The smallest stacked k-sphere is the boundary complex of the $(k+1)$-dimensional simplex. For $n \geq k+3$, a stacked k-sphere with n vertices is a simplicial complex obtained by deleting one of the facets of a stacked k-sphere with $n-1$ vertices and adding a new vertex, whose link is the boundary complex of the deleted facet. An easy induction argument proves that if Σ is a stacked k-sphere with n vertices, then $f_j(\Sigma) = \phi_j(k,n)$. The following theorem is due to Barnette [6, 7].

Theorem 5.12 (The Lower Bound Theorem) *Let $k \geq 2$ be an integer and Δ be the boundary complex of a simplicial $(k+1)$-polytope with n vertices. Then, for $0 \leq j \leq k$,*

$$f_j(\Delta) \geq \phi_j(k,n). \tag{5.1}$$

Moreover, if $k \geq 3$ and equality holds in (5.1) for some $1 \leq j \leq k$, then Δ is a stacked k-sphere. If $k = 2$, then equality holds if and only if Δ is a triangulation of the 2-sphere.

Sketch proof We outline a proof of the inequality using rigidity that is due to Kalai [67]. The use of global rigidity to characterise the case of equality is taken from [29].

It is well known that the link of a vertex of Δ is the boundary complex of a simplicial k-polytope. Based on this fact, McMullen, Perles and Walkup [82] gave an inductive argument to show that we can reduce the proof of the theorem to the case $j = 1$. In this case, (5.1) becomes

$$f_1(\Delta) \geq (k+1)n - \binom{k+2}{2}. \tag{5.2}$$

Let $p : \mathcal{F}_0(\Delta) \to \mathbf{R}^{k+1}$ be the embedding of the vertices that arises naturally from the given polytope. Note that sufficiently small perturbations of a simplicial polytope do not change its boundary complex, so we may assume that p is generic. By Theorem 5.9, the graph $G(\Delta)$ is rigid. Now, (5.2) follows immediately from the fact that equality holds in (2.1) for infinitesimally rigid frameworks.

To deal with the case of equality when $k \geq 3$, we observe that, if equality holds in (5.2), then $G(\Delta)$ is minimally rigid. If $f_0(\Delta) \geq k+2$, then Theorem 2.5 implies that $G(\Delta)$ is not globally rigid. Theorem 5.11 now tells us that $G(\Delta)$ is not $(k+2)$-connected. It can be be shown that, in this case, Δ is obtained by gluing two smaller complexes together along a common facet, whose set of vertices is a $(k+1)$-vertex separator for $G(\Delta)$, and removing the common facet. It also follows that each of these two smaller complexes is the boundary complex of a polytope, and that equality in (5.2) holds in both cases. So, by induction, each of these complexes is a stacked sphere, and therefore, Δ is also a stacked sphere. The case of equality for $k = 2$ follows by a similar argument. □

Some historical comments. Rigidity of frameworks associated to polytopes is a very classical topic, dating at least to Cauchy's Rigidity Theorem for 3-dimensional convex polyhedra [15]. Dehn proved an infinitesimal rigidity version of Cauchy's Theorem [31], and later Alexandrov extended Cauchy's result to higher dimensions [3]. Gluck extended Dehn's result to generic rigidity in $\mathbf{R}^3$ of plane triangulations [41]. However, in a surprising development, Connelly showed that there is a non-convex polyhedron, i.e., an embedded simplicial 2-sphere in $\mathbf{R}^3$, that is continuously flexible [20]. Whiteley extended Alexandrov's theorem by showing that, for all $d \geq 3$, if we triangulate the 2-faces of a non-simplicial polytope without adding new vertices, then the resulting framework is infinitesimally rigid [111]. Kalai [67] proved Theorem 5.9 for triangulations of manifolds of dimension at least 3 and used it to extend the Lower Bound Theorem to triangulated manifolds. Tay subsequently extended the Lower Bound Theorem to the class of normal pseudomanifolds [101]. Following Kalai's paper [67], rigidity properties and lower bound theorems have been investigated for several other classes of complexes, including balanced complexes [73, 90], cubical complexes [70], and symmetric complexes [30, 72]. As far as we are aware, the following is still open.

Conjecture 5.13 *Let $k \geq 2$ be an integer, and Δ be a simplicial k-circuit with n vertices. Then $f_j(\Delta) \geq \phi_j(k,n)$ for all $j \geq 0$. Moreover, if equality holds for some $1 \leq j \leq k$, then Δ is a simplicial k-sphere.*

5.2.2 Connections with commutative algebra In this section, we will outline some connections between rigidity theory and combinatorial commutative algebra. We will assume that the reader is familiar with the basic concepts and terminology of commutative algebra. Since we are focused on the connection with rigidity, all of the rings that we discuss in this section will be $\mathbf{R}$-algebras. However, many of the algebraic concepts and results are also valid over other fields. Let $\mathbf{R}[X] = \mathbf{R}[x_1, \cdots, x_n]$ be the polynomial ring in n commuting variables. For $\tau \subset [n]$, we will write x_τ for the *squarefree monomial* $\Pi_{i \in \tau} x_i$. The mapping $\tau \mapsto x_\tau$ is a bijection between nonempty subsets of $[n]$ and the set of squarefree monomials in $\mathbf{R}[X]$.

To a simplicial complex Δ with $\mathcal{F}_0(\Delta) = [n]$, we associate an ideal of $\mathbf{R}[X]$ defined by

$$I_\Delta = (x_\tau : \tau \subset [n], \tau \notin \Delta).$$

In words, I_Δ is the ideal of $\mathbf{R}[X]$ generated by squarefree monomials corresponding to non-faces of Δ. The Stanley-Reisner ring of Δ is defined by

$$\mathbf{R}[\Delta] = \frac{\mathbf{R}[X]}{I_\Delta}.$$

It is a commutative graded algebra, whose j-homogeneous component $\mathbf{R}[\Delta]_j$ is the vector space with a basis consisting of all monomials (including non-squarefree monomials) of total degree j, whose support is in Δ. The Stanley-Reisner ring is one of the central objects in combinatorial commutative algebra, see [50, 83, 98] for more comprehensive treatments.

How does this connect with rigidity theory? Suppose that $p : [n] \to \mathbf{R}^d$. For $1 \leq j \leq d$, let $\theta_j = \sum_{i=1}^n p(i)_j x_i \in \mathbf{R}[\Delta]_1$, and write θ for the sequence $\theta_1, \ldots, \theta_d$. Let $\ell = x_1 + \cdots + x_n$, $B = B(\Delta, \theta) = \frac{\mathbf{R}[\Delta]}{(\theta)}$, and $C = C(\Delta, \theta) = \frac{\mathbf{R}[\Delta]}{(\theta, \ell)}$. Then B and C are graded $\mathbf{R}$-algebras. Let B_j, respectively C_j, be the j-homogeneous component of B, respectively C. The connection with classical rigidity theory is given by the following result of Lee [76]. Recall the definition of a stress matrix from Section 2.3.

Theorem 5.14 *There is a natural isomorphism between C_2 and the vector space of all stress matrices of $R(G(\Delta), p)$.*

Sketch proof The basis $\{x_i x_j + (I_\Delta)_2 : ij \in \mathcal{F}_2(\Delta)\}$ induces an inner product on $\mathbf{R}[\Delta]_2$. It can be shown that

$$(\theta,\ell)_2^{\perp} = \left\{ \left(\sum_{1\le i,j\le n} \Omega_{ij} x_i x_j \right) + (I_\Delta)_2 : \Omega \text{ is a stress matrix of } (G(\Delta),p) \right\}.$$

□

In fact, there are similar interpretations of C_j for $2 \le j \le \dim \Delta + 1$ in terms of the skeletal rigidity matrices associated to the pair (Δ, p), see [76] for further details. In the case $j = 1$, we have B_1, respectively C_1, is naturally isomorphic to the space of linear, respectively affine, dependencies among the points $p(1), \ldots, p(n)$ and so, it has dimension $n - d$, respectively $n - d - 1$, when p is in general position.

Up to now our discussion has only required elementary concepts of commutative algebra. The remainder of this section will make reference to some more advanced notions from commutative algebra, see [13, 83] for more detail on these algebraic aspects. For our purposes, it suffices to mention the following facts.

Suppose that $p : \mathcal{F}_0(\Delta) \to \mathbf{R}^d$ is generic with $\dim \Delta = d - 1$. It is not too difficult to show that $B_j = 0$ for $j \ge d + 1$. In particular, this means that in this case B is an *artinian reduction* of $\mathbf{R}[\Delta]$.

The Cohen-Macaulay property is extensively studied in commutative algebra. Macaulay has shown $\mathbf{R}[\Delta]$ is a Cohen-Macaulay ring if and only if

$$\dim B_i = h_i(\Delta) = \sum_{j=0}^{i} (-1)^{i-j} \binom{d-j}{i-j} f_{j-1}(\Delta). \tag{5.3}$$

See [97] for an account of this. Also, we note that it follows from a theorem of Reisner that the Stanley-Reisner ring of a simplicial sphere is Cohen-Macaulay, see [13, Corollary 5.3.9].

Using (5.3) and Theorem 5.14, we obtain the following.

Lemma 5.15 *Suppose that* $\mathbf{R}[\Delta]$ *is Cohen-Macaulay and* $\dim \Delta = d-1 \ge 1$. *Then* $G(\Delta)$ *is rigid in* $\mathbf{R}^d$ *if and only if multiplication by* ℓ *is an injective map* $B_1 \to B_2$.

An element m of a graded Artinian algebra $\mathcal{A}$ is said to have the *weak Lefschetz property* if $m \in \mathcal{A}_1$ and the multiplication map $x \mapsto mx$ from $\mathcal{A}_i$ to $\mathcal{A}_{i+1}$ has maximal rank for all i, i.e., this map is either injective or surjective. The weak Lefschetz property is of great interest in commutative algebra and algebraic geometry. Lemma 5.15 indicates that there is a connection between rigidity theory and the weak Lefschetz property for an artinian reduction of the Stanley-Reisner ring. This connection has been developed much further, see [102–105]. In particular, we mention Adiprasito's recent proof of the g-theorem, which characterises the f-vectors of rational homology spheres. This breakthrough result proves, among other things, that artinian reductions of the corresponding Stanley-Reisner rings have the weak Lefschetz property. The proof makes crucial use of the coning property of skeletal rigidity [2].

Finally, in this section, we briefly discuss, without giving details, the theory of algebraic shifting. For a simplicial complex Δ with vertex set $[n]$, Kalai [69] constructs a complex Δ^s with the following properties.

(1) $f_i(\Delta^s) = f_i(\Delta)$ for all $i \ge 0$.

(2) If $j \in \sigma \in \Delta^s$ and $i < j$ such that $i \notin \Delta^s$, then $\sigma \setminus \{j\} \cup \{i\} \in \Delta^s$.

(3) For $d \geq 1$, $\operatorname{rank} \mathcal{R}_d(G(\Delta^s)) = \operatorname{rank} \mathcal{R}_d(G(\Delta))$.

Roughly speaking, the algebraic shifting operator $\Delta \to \Delta^s$ corresponds at the level of Stanley-Reisner ideals to the generic initial ideal construction - there are some technicalities to deal with the fact that generic initial ideals of squarefree monomial ideals need not be squarefree. Algebraic shifting has many other interesting algebraic and combinatorial properties. For example, it is not difficult to deduce from (2) and (3) above that a graph G, considered as a 1-dimensional simplicial complex, with vertex set $[n]$ for some $n \geq d+1$, is rigid in $\mathbf{R}^d$ if and only if $\{d, n\} \in G^s$. We note that this does not yield a more efficient algorithm for determining rigidity since computing the generic initial ideal of a squarefree monomial ideal is computationally as difficult as computing the rank of the generic rigidity matrix.

The basic construction of algebraic shifting was introduced in [66] in the context of the hyperconnectivity matroid of a graph, which is a relative of the rigidity matroid. Indeed, Kalai shows that analogous shifting operators can be constructed for any suitably symmetric matroid, and uses this to study growth functions of symmetric matroids, see [68]. Nevo developed the theory further and gave some applications to f-vectors of graded posets in his PhD thesis [88]. We also note some recent work on volume rigidity and algebraic shifting [14].

5.3 Abstract rigidity and matroid maximality problems

We will describe Jack Graver's ingenious concept of abstract rigidity matroids, which now plays a central role in combinatorial rigidity and can be used to connect rigidity theory to other topics in combinatorics, such as weakly saturated sequences of graphs, tropical ideals and products of matroids.

5.3.1 Abstract rigidity A key observation of Graver [45] for the introduction of abstract rigidity is that the gluing property, one of the characteristic properties of graph rigidity given in Lemma 2.11(a), can be described purely in terms of the closure operator of the rigidity matroid.

Let M be a matroid on the edge set of the complete graph K_n with $n \geq d+1$ and closure operator cl_M. Then M is an *abstract d-rigidity matroid* if M satisfies the following two axioms (which correspond to the 'only-if' and 'if' directions of Lemma 2.11(a), respectively):

(G1) If $E_1, E_2 \subseteq E(K_n)$ with $V(E_1) = V_1$, $V(E_2) = V_2$ and $|V_1 \cap V_2| \leq d-1$, then $\mathrm{cl}_M(E_1 \cup E_2) \subseteq \binom{V_1}{2} \cup \binom{V_2}{2}$;

(G2) If $E_1, E_2 \subseteq E(K_n)$ with $\mathrm{cl}_M(E_1) = \binom{V_1}{2}$, $\mathrm{cl}_M(E_2) = \binom{V_2}{2}$ and $|V_1 \cap V_2| \geq d$, then $\mathrm{cl}_M(E_1 \cup E_2) = \binom{V_1 \cup V_2}{2}$.

It is known that Lemma 2.11(a) holds for any framework (G, p) such that p is in general position in $\mathbf{R}^d$. Hence the rigidity matroid of any d-dimensional framework (K_n, p) in general position is an example of an abstract d-rigidity matroid, and this matroid may be distinct from the generic d-dimensional rigidity matroid of K_n.

Although the axioms of abstract d-rigidity appear to be a rather specialised concept in rigidity theory, they actually define a very natural class of matroids on $E(K_n)$ if we consider some equivalent characterizations of abstract rigidity. To describe these, we first list some basic properties of a matroid M on $E(K_n)$, which all hold when $M = \mathcal{R}_d(K_n)$ and $n \geq d+1$.

(R1) The rank of M is $dn - \binom{d+1}{2}$.

(R2) Every copy of K_{d+2} is a circuit in M.

(R3) Every copy of $K_{1,n-d}$ is a cocircuit in M.

(R4) If $G \subseteq K_n$, v is a vertex of G of degree d, and $G - v$ is M-independent, then G is M-independent.

Theorem 5.16 (Graver-Servatius-Servatius [46] and Nguyen [89]) *Let n, d be positive integers with $n \geq d+1$, and M be a matroid on $E(K_n)$. Then the following statements are equivalent.*

- *M is an abstract d-rigidity matroid.*
- *M satisfies (R1) and (R2).*
- *M satisfies (R1) and (R3).*
- *M satisfies (R1) and (R4).*
- *M satisfies (R2) and (R3).*
- *M satisfies (R2) and (R4).*

Let us consider the case when $d = 1$ as an example. The graphic matroid of K_n is the generic 1-dimensional rigidity matroid, and hence it is an abstract 1-rigidity matroid. Indeed, the graphic matroid has rank $n-1$, implying (R1), and every copy of K_3 is a circuit, implying (R2). Conversely, suppose M is an abstract 1-rigidity matroid of K_n. Then, by (R4), any forest is M-independent, and by (R1), any cycle is M-dependent. Hence, M is the graphic matroid of K_n. Thus, the graphic matroid of K_n is the unique abstract 1-rigidity matroid of K_n. We will see that the class of abstract d-rigidity matroids is much richer when $d \geq 2$.

Whiteley [115] observed that the generic C_{d-1}^{d-2}-cofactor matroid of K_n introduced in Section 4.1 is another important example of an abstract d-rigidity matroid.

Proposition 5.17 *The generic C_{d-1}^{d-2}-cofactor matroid of K_n is an abstract d-rigidity matroid.*

Several fundamental properties of generic rigidity matroids are valid for any abstract rigidity matroid. For example, the following lemma shows that Maxwell's condition (Lemma 2.3) holds at the level of abstract rigidity, see [16, Lemma 4.1].

Lemma 5.18 *Let M be a matroid on the edge set of K_n. Suppose that M satisfies (R2). Then, for any M-independent subgraph $G = (V, E)$ of K_n, $i_G(X) \leq d|X| - \binom{d+1}{2}$ for all $X \subseteq V$ with $|X| \geq d+1$.*

5.3.2 The Graver-Whiteley Maximality Conjecture Graver's main motivation for defining abstract d-rigidity was to gain a better understanding of the d-dimensional rigidity problem by characterising the generic d-dimensional rigidity matroid as an extremal instance in the family of all abstract d-rigidity matroids.

For a family $\mathcal{M}$ of matroids on the same ground set, the *weak order* $\preceq$ is defined as the partial order on $\mathcal{M}$ such that, for any $M_1, M_2 \in \mathcal{M}$, $M_1 \preceq M_2$ if every independent set in M_1 is independent in M_2.

Graver [45] conjectured that the generic d-dimensional rigidity matroid of K_n is the unique maximal matroid in the family of abstract d-rigidity matroids of K_n under the weak order, and verified his conjecture when $d = 2$.

Theorem 5.19 *The generic* 2*-dimensional rigidity matroid of* K_n *is the unique maximal matroid in the family of all abstract* 2*-rigidity matroids of* K_n *for all* $n \geq 3$.

Indeed, Theorem 5.19 is a direct consequence of Geiringer's theorem since every independent set in any abstract 2-rigidity matroid satisfies Maxwell's condition by Lemma 5.18, and such a set is independent in the generic 2-dimensional-rigidity matroid by Theorem 3.1.

Graver's conjecture turned out to be false for $d \geq 4$. The first counterexample was given by N. J. Thurston (cf. [46]) for $d = 4$. Whiteley subsequently used the complete bipartite graph $K_{d+2,d+2}$ to show that the generic C^{d-2}_{d-1}-cofactor matroid provides a counterexample for all $d \geq 4$.

Lemma 5.20 *For all* $d \geq 4$, $E(K_{d+2,d+2})$ *is a circuit in the generic* d*-dimensional rigidity matroid and is independent in the generic* C^{d-2}_{d-1}*-cofactor matroid.*

The rigidity part of Lemma 5.20 follows from the Bolker-Roth [11] characterization of the rigidity of complete bipartite graphs. The cofactor part follows from the fact that the X-replacement operation preserves independence in the generic C^{d-2}_{d-1}-cofactor matroid [115], cf. the discussion in the first paragraph of Section 4.2.

Lemma 5.20 motivates the following modified conjecture.

Conjecture 5.21 (Graver-Whiteley Maximality Conjecture) *The generic* C^{d-2}_{d-1}*-cofactor matroid of* K_n *is the unique maximal matroid in the family of abstract* d*-rigidity matroids of* K_n *for all* $n \geq d+1 \geq 2$.

If true, this conjecture would give useful structural information about the generic C^{d-2}_{d-1}-cofactor matroid. Indeed, it is shown in [17] that the partial result that there exists a unique maximal abstract d-rigidity matroid M of K_n would imply that the following covering property holds for M: for each cyclic flat F of M, every $e \in F$ is contained in a copy of K_{d+2} in F.

Clinch, Jackson, and Tanigawa [16] verified Conjecture 5.21 for $d = 3$.

Theorem 5.22 *The generic* C^1_2*-cofactor matroid of* K_n *is the unique maximal matroid in the family of abstract* 3*-rigidity matroids of* K_n *for all* $n \geq 4$.

They then used the above mentioned covering property from [17] to prove Theorem 4.5. Despite these positive results, Graver's original conjecture still remains open when $d = 3$.

Conjecture 5.23 (Graver's maximality conjecture for $d = 3$) *The generic 3-rigidity matroid of K_n is the unique maximal matroid in the family of abstract 3-rigidity matroids of K_n for all $n \geq 4$.*

The truth of Conjecture 5.23 would give a combinatorial characterisation of generic rigidity in $\mathbf{R}^3$ by Theorems 4.3 and 5.22.

As pointed out by Crespo and Santos [93], when $d \geq 4$ we do not even know if $\mathcal{R}_d(K_n) \prec C_{d-1}^{d-2}(K_n)$ or the two matroids are incomparable in the weak order.

5.3.3 Duality, symmetric powers, and Mason's problem We will describe a connection between abstract rigidity and the symmetric powers of a matroid introduced by Lovász [78] and Mason [80].

Let K_V° be the complete graph with vertex set V and with a loop at each vertex, and $K_{1,t}^\circ$ be the star with $t+1$ vertices and a loop at its central vertex. We denote the loop at a vertex v by the multiset $\{v, v\}$.

Suppose M is a loopless matroid on a finite set V of rank t, and N is a matroid on $E(K_V^\circ)$. Following Anderson [4], we say that N is a *second symmetric quasi-power* of M if

(S1) the rank in N of each copy of $K_{1,t}$ and $K_{1,t}^\circ$ in K_V° is equal to the rank of the neighbours of its 'central vertex' in M. More precisely, $r_M(X) = r_N(vX)$ for all $X \subseteq V$ and all $v \in V$, where $vX = \{\{v, x\} : x \in X\}$.

And N is a a *second symmetric power* of M if it also satisfies

(S2) the rank of N is $\binom{t+1}{2}$.

A given matroid M may have several or no second symmetric powers, see [4, 80] for more details. In the special case when M is linear, however, we can always construct a canonical example of a symmetric second power by using the symmetric power of the ambient space of its linear representation. Specifically, if M has a linear representation over a field $\mathbb{F}$ by associating a vector $p_v \in \mathbb{F}^t$ to each $v \in V$, then we can construct a linear matroid on $E(K_n^\circ)$ by associating the symmetric power $p_u p_v$ with each $uv \in E(K_n^\circ)$.[5] The resulting matroid N will satisfy (S2) by the fact that the second symmetric power of $\mathbb{F}^t$ has dimension $\binom{t+1}{2}$, and N will also satisfy (S1) by bilinearity.

For example, the uniform matroid U_n^t of rank t with n elements can be linearly represented by assigning a generic vector $p_v \in \mathbf{R}^t$ to each $v \in V$, when $|V| = n$. Thus, we can obtain a second symmetric power of U_n^t by assigning $p_u p_v$ to each $uv \in E(K_n^\circ)$. We will refer to this matroid as the *generic second power of* $U_{n,t}$ and refer to its restriction to $E(K_n)$ (by removing the loops $\{v, v\}$ of K_n°) as the *generic symmetric t-tensor matroid of* K_n. A recent result of Brakensiek et al. [12] tells us that the generic symmetric t-tensor matroid of K_n is the dual matroid of the d-dimensional generic rigidity matroid of K_n when we take $d = n - 1 - t$.

Theorem 5.24 *Let n, d, t be positive integers with $d + t = n - 1$. Then the generic symmetric t-tensor matroid of K_n is the dual of the generic d-dimensional rigidity matroid.*

[5] We will be concerned with the case when $\mathbb{F} = \mathbf{R}$. In this case, or any other case when $\mathbb{F}$ has characteristic zero, we can take the symmetric power to be the symmetric matrix $p_u p_v^T + p_v p_u^T$.

Motived by Theorem 5.24, we can also consider the dual of the other matroids in the family of abstract d-rigidity matroids. We will see that this gives rise to other second symmetric powers of U_n^t restricted to $E(K_n)$. Specifically, consider axiom (S1) in the case when $M = U_n^t$ and we take the restriction of N to $E(K_n)$. Then (S1) can be simplified to:

(S1') Every copy of $K_{1,t+1}$ is a circuit in N.

We can now define the family of *abstract symmetric t-tensor matroids* to be the family of matroids on $E(K_n)$ satisfying (S1') and (S2).

For $d + t = n - 1$, a matroid N on $E(K_n)$ satisfies (S2) if and only if its dual N^* satisfies (R1), whereas N satisfies (S1') if and only if N^* satisfies (R3). We can combine this observation with Theorem 5.16 to obtain the following duality relation from [57].

Lemma 5.25 *Let n, d, t be positive integers with $d + t = n - 1$. Then a matroid N on $E(K_n)$ is an abstract symmetric t-tensor matroid if and only if its dual N^* is an abstract d-rigidity matroid.*

Lemma 5.25 enables us to make dual formulations of the conjectures in Section 5.3.2 in terms of symmetric t-tensor matroids. The dual formulation of Graver's maximality conjecture can be verified using techniques from rigidity theory when t is small.

Theorem 5.26 *Let n, t be positive integers with $2 \leq t \leq n - 1$ and $t \leq 5$. Then the generic symmetric t-tensor matroid of K_n is the unique maximal element in the family of abstract symmetric t-tensor matroids of K_n. Thus, Graver's maximality conjecture for generic rigidity is true whenever $d \geq n - 6$.*

Theorem 5.26 is proved in [57] by providing a sparsity condition, which characterises independence in the generic symmetric t-tensor matroid when $t \leq 5$. It is also possible to derive Theorem 5.26 from the characterisation of all $\mathcal{R}_d$-circuits on n vertices when $n \leq d + 6$, given in [44].

Mason [80] asked if there is always a unique maximal element in the family of second symmetric powers of a matroid. This property does not hold for the related family of *second powers* of a matroid by Las Vergnas [75] or the larger family of *symmetric quasi-powers* of a matroid by [57], but, to the best of our knowledge, Mason's question for second symmetric powers is still open. We formulate the special case of this question for uniform matroids as the following conjecture.

Conjecture 5.27 *Let n, t be positive integers. Then the family of second symmetric powers of U_n^t has a unique maximal element.*

In view of Theorem 5.26, it is tempting to conjecture that the generic second symmetric power of U_n^t will be the unique maximal element in Conjecture 5.27, but this is false when $t \geq 7$. We will use rigidity theory to show this, but we first need to fill a small gap between second symmetric powers of $U_{n,t}$, which are defined on $E(K_n^\circ)$, and abstract d-rigidity matroids, which are defined on $E(K_n)$.

Suppose N is a second symmetric power of U_n^t. Then its restriction to $E(K_n)$ always gives an abstract symmetric t-tensor matroid by definition. We can use matroid duality to give an inverse construction. Suppose N is an abstract symmetric t-tensor matroid on $E(K_n)$, and let N^* be its dual. Then N^* is an abstract d-rigidity matroid with $d = n - t - 1$ by Lemma 5.25. We can extend N^* to a matroid $\hat{N}^*$ on $E(K_n^\circ)$ by appending each loop edge of K_n° as a loop element in the matroidal sense. Let B_n be the bicircular matroid on $E(K_n^\circ)$, and let $\hat{N}^* \vee B_n$ be the matroid union of $\hat{N}^*$ and B_n. Then $\hat{N}^* \vee B_n$ has rank $(d+1)n - \binom{d+1}{2}$, and every copy of $K_{1,n-d+1}$ and $K_{1,n-d}^\circ$ is a cocircuit. Therefore, its dual $(\hat{N}^* \vee B_n)^*$ satisfies (S1) and (S2) for U_n^t, and hence is a second symmetric power of U_n^t.

We can now apply the above construction, taking N to be the dual of the generic C_{d-1}^{d-2}-cofactor matroid of K_n. Then $M = (\hat{N}^* \vee B_n)^*$ is a second symmetric power of U_n^t. In addition, we have $M \not\preceq M'$ when M' is the generic second symmetric power of U_n^t, $t \geq 7$ and $t + 5 \leq n \leq 2t - 2$. This follows from Theorem 5.24 and Lemma 5.20 by taking the dual of the restrictions of M and M' to $E(K_n)$. We can now deduce:

Theorem 5.28 *The generic second symmetric power of U_n^t is not the unique maximal element in the family of all second symmetric powers of U_n^t whenever $t \geq 7$ and $t + 5 \leq n \leq 2t - 2$.*

5.3.4 Birigidity As a bipartite analogue of rigidity matroids, Kalai, Nevo, and Novik [70] introduced birigidity matroids. Let $G = (V_1 \cup V_2, E)$ be a bipartite graph, and consider point-configurations $p_1 : V_1 \to \mathbf{R}^{d_1}$ and $p_2 : V_2 \to \mathbf{R}^{d_2}$. The *birigidity matrix* of (G, p_1, p_2) is defined as the $|E| \times (d_2|V_1| + d_1|V_2|)$ matrix, in which the row associated with an edge $e = uv$ is given by

$$\begin{array}{c} \\ e{=}uv \end{array} \begin{array}{c} \begin{matrix} & u & & v & \end{matrix} \\ \begin{bmatrix} 0\ldots0 & p_v & 0\ldots0 & p_u & 0\ldots0 \end{bmatrix}. \end{array}$$

The row matroid of the birigidity matrix defines a matroid on E, which is called the *birigidity matroid* of (G, p_1, p_2). The *generic* (d_1, d_2)*-birigidity matroid* of G, $\mathcal{R}_{d_1,d_2}(G)$, is the birigidity matroid of (G, p_1, p_2) for generic p_1, p_2. The special case when $d_1 = d_2$ has been studied in the context of the low rank matrix completion problem [96]. The case when $d_1 = 1$ or $d_2 = 1$ coincides with Whiteley's matroid from scene analysis [112].

A natural analogue of the gluing property holds in birigidity, and hence we can develop a bipartite analogue of abstract rigidity and formulate Graver-type questions in this bipartite setting. In the subsequent discussion, we will assume that all bipartite graphs $G = (V_1 \cup V_2, E)$ are given with a fixed bipartition (V_1, V_2) and will refer to V_1 and V_2 as the *left* vertex set and the *right* vertex set of G, respectively.

Suppose n_i, d_i are integers with $n_i \geq d_i \geq 1$ for $i = 1, 2$, and M is a matroid on $E(K_{n_1,n_2})$. Then we say that M is an *abstract* (d_1, d_2)*-birigidity matroid* if its closure operator cl_M satisfies the following two axioms.

(BG1) Suppose $E_i \subseteq V_{1,i} \times V_{2,i} \subseteq E(K_{n_1,n_2})$ for $i = 1, 2$, $|V_{1,1} \cap V_{1,2}| < d_1$, and $|V_{2,1} \cap V_{2,2}| < d_2$. Then $\mathrm{cl}_M(E_1 \cup E_2) \subseteq (V_{1,1} \times V_{2,1}) \cup (V_{1,2} \times V_{2,2})$.

(BG2) Suppose $E_i \subseteq V_{1,i} \times V_{2,i} \subseteq E(K_{n_1,n_2})$ with $\mathrm{cl}_M(E_i) = V_{1,i} \times V_{2,i}$ for $i = 1, 2$, $|V_{1,1} \cap V_{1,2}| \geq d_1$, and $|V_{2,1} \cap V_{2,2}| \geq d_2$. Then $\mathrm{cl}_M(E_1 \cup E_2) = (V_{1,1} \cup V_{1,2}) \times (V_{2,1} \cup V_{2,2})$.

The generic (d_1, d_2)-birigidity matroid is an example of an abstract (d_1, d_2)-birigidity matroid.

We can give equivalent characterisations of abstract birigidity matroids using bipartite analogues of properties (R1)-(R4) in Section 5.3.1. Consider the following properties of a matroid M on $E(K_{n_1,n_2})$, which all hold when $M = \mathcal{R}_{d_1,d_2}(K_{n_1,n_2})$ and $n_i \geq d_i \geq 1$ for $i = 1, 2$.

(BR1) The rank of M is $d_2 n_1 + d_1 n_2 - d_1 d_2$.

(BR2) Every copy of K_{d_1+1,d_2+1} is a circuit in M.

(BR3) Every copy of $K_{n_1-d_1+1,1}$ and K_{1,n_2-d_2+1} is a cocircuit in M.

(BR4) Suppose $G \subseteq K_{n_1,n_2}$, and v is either a left vertex of G of degree d_2, or a right vertex of G of degree d_1. If $G - v$ is independent in M, then G is M-independent.

Theorem 5.29 *Let n_1, n_2, d_1, d_2 be positive integers with $n_i \geq d_i$ and M be a matroid on $E(K_{n_1,n_2})$. Then the following statements are equivalent.*

- *M is an abstract (d_1, d_2)-birigidity matroid.*
- *M satisfies (BR1) and (BR2).*
- *M satisfies (BR1) and (BR3).*
- *M satisfies (BR1) and (BR4).*
- *M satisfies (BR2) and (BR3).*
- *M satisfies (BR2) and (BR4).*

The bipartite counterpart of Graver's conjecture is the following.

Conjecture 5.30 (Birigidity maximality conjecture) *The generic (d_1, d_2)-birigidity matroid of K_{n_1,n_2} is the unique maximal matroid in the family of abstract (d_1, d_2)-birigidity matroids of K_{n_1,n_2} for all $n_1 \geq d_1 \geq 1$ and $n_2 \geq d_2 \geq 1$.*

Unlike Graver's original conjecture for abstract rigidity, there are no known counterexamples to Conjecture 5.30. Currently, the only resolved cases are those where $d_1 = 1$ or $d_2 = 1$, in which case Whiteley's theorem [112] on scene analysis implies Conjecture 5.30, and those where $d_1 \geq n_1 - 3$ or $d_2 \geq n_2 - 3$, which follow from the characterisation of the dual matroids due to Brakensiek et al. [12].

The case when $d_1 = d_2 = 2$ is the smallest unsolved case. Bernstein [10] gave a noteworthy combinatorial characterization of independence in the generic $(2, 2)$-birigidity matroid. This characterization does not provide a co-NP certificate for independence, however, and giving a good characterization (in the sense of computational complexity) is wide open.

As for generic rigidity, generic birigidity has a natural dual counterpart. Brakensiek et al. [12] showed that the dual of the generic (d_1, d_2)-birigidity matroid of

K_{n_1,n_2} is a tensor product of U_{n_1,n_1-d_1} and U_{n_2,n_2-d_2}.[6] The following dual counterpart to Conjecture 5.30 is a special case of another question of Mason [80] for tensor products of matroids.

Conjecture 5.31 *There is a unique maximal element in the family of tensor products of two uniform matroids. Moreover, this unique maximal element is the dual of the generic birigidity matroid.*

A result by Brakensiek et al. [12] verifies Conjecture 5.31 in the case when one matroid is a uniform matroid of rank three, and the other is a uniform matroid of arbitrary rank. On the other hand, Las Vergnas [75] shows that Conjecture 5.31 is false for the larger family of products of uniform matroids. See also Bérczi et al. [8] for further recent progress.

Acknowledgements

Bill Jackson was supported by the MTA Distinguished Guest Scientist Fellowship Programme 2025. Tibor Jordán was supported by the National Research, Development and Innovation Office of Hungary, grant No. 152786, and the MTA-ELTE Momentum Matroid Optimization Research Group. Shin-ichi Tanigawa was supported by JSPS KAKENHI Grant Number 24K14835.

References

[1] T. Abott, *Generalizations of Kempe's universality theorem*, Master's thesis, Massachusetts Institute of Technology, 2008.

[2] Karim Adiprasito, *FAQ on the g-theorem and the hard Lefschetz theorem for face rings*, Rend. Mat. Appl. (7) **40** (2019), no. 2, 97–111.

[3] A. D. Alexandrov, *Convex polyhedra*, Russian ed., Springer Monographs in Mathematics, Springer-Verlag, Berlin, 2005, With comments and bibliography by V. A. Zalgaller and appendices by L. A. Shor and Yu. A. Volkov.

[4] Nicholas Anderson, *Matroid products in tropical geometry*, Res. Math. Sci. **11** (2024), no. 38, 28 pp.

[5] L. Asimow and B. Roth, *The rigidity of graphs*, Trans. Amer. Math. Soc. **245** (1978), 279–289.

[6] David Barnette, *The minimum number of vertices of a simple polytope*, Israel J. Math. **10** (1971), 121–125.

[7] ———, *A proof of the lower bound conjecture for convex polytopes*, Pacific J. Math. **46** (1973), 349–354.

[6] The family of tensor products of two matroids can be defined in a similar way to the family of symmetric powers of a single matroid, see [75].

[8] Kristóf Bérczi, Boglárka Gehér, András Imolay, László Lovász, Carles Padró, and Tamás Schwarcz, *Interaction between skew-representability, tensor products, extension properties, and rank inequalities*, 2025, available at: https://arxiv.org/abs/2507.10709.

[9] Alex R. Berg and Tibor Jordán, *A proof of Connelly's conjecture on 3-connected circuits of the rigidity matroid*, J. Combin. Theory Ser. B **88** (2003), no. 1, 77–97.

[10] Daniel Irving Bernstein, *Completion of tree metrics and rank-2 matrices*, Linear Algebra and its Applications **533** (2017), 1–13.

[11] E. D. Bolker and B. Roth, *When is a bipartite graph a rigid framework?*, Pacific J. Math. **90** (1980), no. 1, 27–44.

[12] Joshua Brakensiek, Manik Dhar, Jiyang Gao, Sivakanth Gopi, and Matt Larson, *Rigidity matroids and linear algebraic matroids with applications to matrix completion and tensor codes*, 2024, available at: https://arxiv.org/abs/2405.00778.

[13] Winfried Bruns and Jürgen Herzog, *Cohen-Macaulay rings*, Cambridge Studies in Advanced Mathematics, vol. 39, Cambridge University Press, Cambridge, 1993.

[14] Denys Bulavka, Eran Nevo, and Yuval Peled, *Volume rigidity and algebraic shifting*, J. Combin. Theory Ser. B **170** (2025), 189–202.

[15] A. L. Cauchy, *Recherche sur les polyédres - premier mémoire*, Journal de l'École Polytechnique **9** (1813), 66–86.

[16] Katie Clinch, Bill Jackson, and Shin-ichi Tanigawa, *Abstract 3-rigidity and bivariate C_2^1-splines I: Whiteley's maximality conjecture*, Discrete Anal. (2022), Paper No. 2, 50.

[17] ———, *Abstract 3-rigidity and bivariate C_2^1-splines II: Combinatorial characterization*, Discrete Anal. (2022), Paper No. 3, 32.

[18] Robert Connelly, *Questions, conjectures and remarks on globally rigid tensegrities*, preprint 2009.

[19] ———, *A counterexample to the rigidity conjecture for polyhedra*, Inst. Hautes Études Sci. Publ. Math. (1977), no. 47, 333–338.

[20] ———, *A flexible sphere*, Math. Intelligencer **1** (1978/79), no. 3, 130–131.

[21] ———, *Rigidity and energy*, Invent. Math. **66** (1982), no. 1, 11–33.

[22] ———, *On generic global rigidity*, Applied geometry and discrete mathematics, DIMACS Ser. Discrete Math. Theoret. Comput. Sci., vol. 4, Amer. Math. Soc., Providence, RI, 1991, pp. 147–155.

[23] ———, *Generic global rigidity*, Discrete Comput. Geom. **33** (2005), no. 4, 549–563.

[24] Robert Connelly and Simon D. Guest, *Frameworks, tensegrities, and symmetry*, Cambridge University Press, Cambridge, 2022.

[25] Robert Connelly, Tibor Jordán, and Walter Whiteley, *Generic global rigidity of body-bar frameworks*, J. Combin. Theory Ser. B **103** (2013), no. 6, 689–705.

[26] Robert Connelly and Walter Whiteley, *Global rigidity: the effect of coning*, Discrete Comput. Geom. **43** (2010), no. 4, 717–735.

[27] Collette R. Coullard and Lisa Hellerstein, *Independence and port oracles for matroids, with an application to computational learning theory*, Combinatorica **16** (1996), no. 2, 189–208.

[28] James Cruickshank, Bill Jackson, and Shin-ichi Tanigawa, *Vertex splitting, coincident realisations, and global rigidity of braced triangulations*, Discrete Comput. Geom. **69** (2023), no. 1, 192–208.

[29] ———, *Global rigidity of triangulated manifolds*, Adv. Math. **458** (2024), Paper No. 109953, 59.

[30] ———, *Rigidity of symmetric simplicial complexes and the lower bound theorem*, Forum Math. Sigma **13** (2025), Paper No. e4, 22.

[31] Max Dehn, *Über die starrheit konvexer polyeder*, Math. Ann. **77** (1916), 466–473.

[32] A. Dress, A. Drieding, and H. Haegi, *Classification of mobile molecules by category theory*, Symmetries and Properties of Non-Rigid Molecules: A Comprehensive Study (J. Maruana and J. Serre, eds.), Elsevier, Amsterdam, 1983, pp. 39–58.

[33] Olivier Durand de Gevigney, *On Frank's conjecture on k-connected orientations*, J. Combin. Theory Ser. B **141** (2020), 105–114.

[34] J. Edmonds, *Minimum partition of a matroid into independent sets*, J. Res. Nat. Bur. Standards **B69** (1965), 67–72.

[35] ———, *Submodular functions, matroids, and certain polyhedra*, Combinatorial Structures and their Applications, Gordon and Breach, New York, 1970, pp. 69–87.

[36] A. L. Fogelsanger, *The generic rigidity of minimal cycles*, Ph.D. thesis, Cornell University, 1988, available at:
http://www.armadillodanceproject.com/AF/Cornell/rigidity.htm.

[37] Samuel Frank and Jiayang Jiang, *New classes of counterexamples to Hendrickson's global rigidity conjecture*, Discrete Comput. Geom. **45** (2011), no. 3, 574–591.

[38] Dániel Garamvölgyi, Steven J. Gortler, and Tibor Jordán, *Globally rigid graphs are fully reconstructible*, Forum Math. Sigma **10** (2022), Paper No. e51, 27.

[39] Dániel Garamvölgyi, Tibor Jordán, and Csaba Király, *Count and cofactor matroids of highly connected graphs*, J. Combin. Theory Ser. B **166** (2024), 1–29.

[40] Dániel Garamvölgyi, Tibor Jordán, Csaba Király, and Soma Villányi, *Highly connected orientations from edge-disjoint rigid subgraphs*, Forum Math. Pi **13** (2025), Paper No. e11, 15.

[41] Herman Gluck, *Almost all simply connected closed surfaces are rigid*, Geometric Topology (Proc. Conf., Park City, Utah, 1974), Lecture Notes in Math., vol. 438, Springer, Berlin-New York, 1975, pp. 225–239.

[42] Jacob E. Goodman, Joseph O'Rourke, and Csaba D. Tóth (eds.), *Handbook of discrete and computational geometry*, 3rd ed., Discrete Mathematics and its Applications (Boca Raton), CRC Press, Boca Raton, FL, 2017.

[43] Steven J. Gortler, Alexander D. Healy, and Dylan P. Thurston, *Characterizing generic global rigidity*, Amer. J. Math. **132** (2010), no. 4, 897–939.

[44] Georg Grasegger, Hakan Guler, Bill Jackson, and Anthony Nixon, *Flexible circuits in the d-dimensional rigidity matroid*, Journal of Graph Theory **100** (2022), no. 2, 315–360.

[45] Jack Graver, *Rigidity matroids*, SIAM Journal on Discrete Mathematics **4** (1991), 355–368.

[46] Jack Graver, Brigitte Servatius, and Herman Servatius, *Combinatorial rigidity*, Graduate Studies in Mathematics, vol. 2, American Mathematical Society, Providence, RI, 1993.

[47] S. L. Hakimi, *On the degrees of the vertices of a directed graph*, J. Franklin Inst. **279** (1965), 290–308.

[48] Bruce Hendrickson, *Conditions for unique graph realizations*, SIAM J. Comput. **21** (1992), no. 1, 65–84.

[49] L. Henneberg, *Die graphische statik der starren systeme*, 1911, Leipzig.

[50] Jürgen Herzog and Takayuki Hibi, *Monomial ideals*, Graduate Texts in Mathematics, vol. 260, Springer-Verlag London, Ltd., London, 2011.

[51] Bill Jackson and Tibor Jordán, *Connected rigidity matroids and unique realizations of graphs*, J. Combin. Theory Ser. B **94** (2005), no. 1, 1–29.

[52] ______, *The Dress conjectures on rank in the 3-dimensional rigidity matroid*, Adv. in Appl. Math. **35** (2005), no. 4, 355–367.

[53] ______, *On the rank function of the 3-dimensional rigidity matroid*, Internat. J. Comput. Geom. Appl. **16** (2006), no. 5-6, 415–429.

[54] Bill Jackson, Tibor Jordán, and Zoltán Szabadka, *Globally linked pairs of vertices in equivalent realizations of graphs*, Discrete Comput. Geom. **35** (2006), no. 3, 493–512.

[55] ______, *Globally linked pairs of vertices in rigid frameworks*, Rigidity and symmetry, Fields Inst. Commun., vol. 70, Springer, New York, 2014, pp. 177–203.

[56] Bill Jackson and Shin-ichi Tanigawa, *Maximal matroids in weak order posets*, J. Combin. Theory Ser. B **165** (2024), 20–46.

[57] ______, *Symmetric tensor matroids, dual rigidity matroids, and the maximality conjecture*, 2025, available at: https://arxiv.org/abs/2503.14780.

[58] Tibor Jordán, *On the existence of k edge-disjoint 2-connected spanning subgraphs*, J. Combin. Theory Ser. B **95** (2005), no. 2, 257–262.

[59] ______, *Combinatorial rigidity: graphs and matroids in the theory of rigid frameworks*, Discrete geometric analysis, MSJ Mem., vol. 34, Math. Soc. Japan, Tokyo, 2016, pp. 33–112.

[60] ______, *The globally rigid complete bipartite graphs*, 2022, EGRES quick proofs QP- 2022-02 (2022), 3pp.

[61] ______, *Orientations of graphs with rooted connectivity requirements*, 13th Hungarian Japanese Symposium on Discrete Mathematics and Its Applications (Tokyo), May 2025.

[62] Tibor Jordán, Csaba Király, and Shin-ichi Tanigawa, *Generic global rigidity of body-hinge frameworks*, J. Combin. Theory Ser. B **117** (2016), 59–76.

[63] Tibor Jordán and Shin-ichi Tanigawa, *Global rigidity of triangulations with braces*, J. Combin. Theory Ser. B **136** (2019), 249–288.

[64] Tibor Jordán and Soma Villányi, *Globally linked pairs of vertices in generic frameworks*, Combinatorica **44** (2024), no. 4, 817–838.

[65] Tibor Jordán and Walter Whiteley, *Global rigidity*, in Handbook of Discrete and Computational Geometry (J.E. Goodman, J. O'Rourke, and C. D. Tóth, eds.), CRC Press, Boca Raton, FL, 2017, pp. 1661–1694.

[66] Gil Kalai, *Hyperconnectivity of graphs*, Graphs Combin. **1** (1985), no. 1, 65–79.

[67] ______, *Rigidity and the lower bound theorem. I*, Invent. Math. **88** (1987), no. 1, 125–151.

[68] ______, *Symmetric matroids*, J. Combin. Theory Ser. B **50** (1990), no. 1, 54–64.

[69] ______, *Algebraic shifting*, Computational commutative algebra and combinatorics (Osaka, 1999), Adv. Stud. Pure Math., vol. 33, Math. Soc. Japan, Tokyo, 2002, pp. 121–163.

[70] Gil Kalai, Eran Nevo, and Isabella Novik, *Bipartite rigidity*, Transactions of the American Mathematical Society **368** (2016), 5515–5545.

[71] Ken-ichi Kawarabayashi, Orlando Lee, Bruce Reed, and Paul Wollan, *A weaker version of Lovász path removal conjecture*, J. Combin. Theory Ser. B **98** (2008), no. 5, 972–979.

[72] Steven Klee, Eran Nevo, Isabella Novik, and Hailun Zheng, *A lower bound theorem for centrally symmetric simplicial polytopes*, Discrete Comput. Geom. **61** (2019), no. 3, 541–561.

[73] Steven Klee and Isabella Novik, *Lower bound theorems and a generalized lower bound conjecture for balanced simplicial complexes*, Mathematika **62** (2016), no. 2, 441–477.

[74] G. Laman, *On graphs and rigidity of plane skeletal structures*, J. Eng. Math. **4** (1970), 331–340.

[75] Michel Las Vergnas, *On products of matroids*, Discrete Mathematics **36** (1981), no. 1, 49–55.

[76] C. W. Lee, *P.L.-spheres, convex polytopes, and stress*, Discrete Comput. Geom. **15** (1996), no. 4, 389–421.

[77] Alan Lew, Eran Nevo, Yuval Peled, and Orit E. Raz, *Sharp threshold for rigidity of random graphs*, Bull. Lond. Math. Soc. **55** (2023), no. 1, 490–501.

[78] László Lovász, *Flats in matroids and geometric graphs*, Combinatorial Surveys (Proc. 6th British Combinatorial Conf., Royal Holloway Coll., Egham) (P. J. Cameron, ed.), Academic Press, 1977, pp. 45–86.

[79] László Lovász and Yechiam Yemini, *On generic rigidity in the plane*, SIAM J. Algebraic Discrete Methods **3** (1982), no. 1, 91–98.

[80] J. H. Mason, *Glueing matroids together: a study of Dilworth truncations and matroid analogues of exterior and symmetric powers*, Algebraic Methods in Graph Theory (L. Lovász and V. T. Sós, eds.), Colloquia Mathematica Societatis János Bolyai, vol. 25, North-Holland, 1981, pp. 519–561.

[81] J. C. Maxwell, *On the calculation of the equilibrium and stiffness of frames*, Philos. Mag. **27** (1864), 294–299.

[82] P. McMullen and D. W. Walkup, *A generalized lower-bound conjecture for simplicial polytopes*, Mathematika **18** (1971), 264–273.

[83] Ezra Miller and Bernd Sturmfels, *Combinatorial commutative algebra*, Graduate Texts in Mathematics, vol. 227, Springer-Verlag, New York, 2005.

[84] Bojan Mohar, Richard J. Nowakowski, and Douglas B. West, *Research problems from the 5th Slovenian Conference (Bled, 2003)*, Discrete Math. **307** (2007), no. 3-5, 650–658.

[85] C. St. J. A. Nash-Williams, *On orientations, connectivity and odd-vertex-pairings in finite graphs*, Canadian Journal of Mathematics **12** (1960), 555–567.

[86] ———, *Edge-disjoint spanning trees of finite graphs*, J. London Math. Soc. **36** (1961), 445–450.

[87] ———, *Decomposition of finite graphs into forests*, J. London Math. Soc. **39** (1964), 12.

[88] Eran Nevo, *Algebraic shifting and f-vector theory*, 2007, available at: https://arxiv.org/abs/0709.3265.

[89] V. H. Nguyen, *On abstract rigidity matroids*, SIAM Journal on Discrete Mathematics **24** (2010), 363–369.

[90] Ryoshun Oba, *Rigidity of balanced minimal cycle complexes*, SIAM Journal on Discrete Mathematics **39** (2025), no. 2, 681–697.

[91] James Oxley, *Matroid theory*, 2nd ed., Oxford Graduate Texts in Mathematics, vol. 21, Oxford University Press, Oxford, 2011.

[92] H. Pollaczek-Geiringer, *Über die gliederung ebener fachwerke*, Zeitschrift für Angewandte Mathematik und Mechanik (ZAMM) **7** (1927), 58–72.

[93] Luis Crespo Ruiz and Francisco Santos, *Bar-and-joint rigidity on the moment curve coincides with cofactor rigidity on a conic*, Combinatorial Theory **3** (2023), no. 1, 15.

[94] J. B. Saxe, *Embeddability of weighted graphs in k-space is strongly NP-hard*, Tech. report, Computer Science Department, Carnegie-Mellon University, Pittsburgh, PA, 1979.

[95] J. T. Schwartz, *Fast probabilistic algorithms for verification of polynomial identities*, J. Assoc. Comput. Mach. **27** (1980), no. 4, 701–717.

[96] A. Singer and M. Cucuringu, *Uniqueness of low-rank matrix completion by rigidity theory*, SIAM Journal on Matrix Analysis and Applications **31** (2010), 1621–1641.

[97] Richard P. Stanley, *Hilbert functions of graded algebras*, Advances in Mathematics **28** (1978), no. 1, 57–83.

[98] ———, *Combinatorics and commutative algebra*, 2nd ed., Progress in Mathematics, vol. 41, Birkhäuser Boston, 1996.

[99] Shin-ichi Tanigawa, *Sufficient conditions for the global rigidity of graphs*, J. Combin. Theory Ser. B **113** (2015), 123–140.

[100] Tiong-Seng Tay, *On generically dependent bar frameworks in space*, Structural Topology (1993), no. 20, 27–48, Dual French-English text.

[101] ———, *Lower-bound theorems for pseudomanifolds*, Discrete Comput. Geom. **13** (1995), no. 2, 203–216.

[102] ______, *Skeletal rigidity of p.l.-spheres*, Matroid theory (Seattle, WA, 1995), Contemp. Math., vol. 197, Amer. Math. Soc., Providence, RI, 1996, pp. 387–399.

[103] Tiong-Seng Tay, Neil White, and Walter Whiteley, *Skeletal rigidity of simplicial complexes. I*, European J. Combin. **16** (1995), no. 4, 381–403.

[104] ______, *Skeletal rigidity of simplicial complexes. II*, European J. Combin. **16** (1995), no. 5, 503–523.

[105] Tiong-Seng Tay and Walter Whiteley, *A homological interpretation of skeletal rigidity*, Adv. in Appl. Math. **25** (2000), no. 1, 102–151.

[106] Carsten Thomassen, *Configurations in graphs of large minimum degree, connectivity, or chromatic number*, Combinatorial Mathematics: Proceedings of the Third International Conference (New York, 1985), Ann. New York Acad. Sci., vol. 555, New York Acad. Sci., New York, 1989, pp. 402–412.

[107] ______, *Strongly 2-connected orientations of graphs*, J. Combin. Theory Ser. B **110** (2015), 67–78.

[108] W. T. Tutte, *On the problem of decomposing a graph into n connected factors*, J. London Math. Soc. **36** (1961), 221–230.

[109] Soma Villányi, *Every $d(d+1)$-connected graph is globally rigid in $\mathbf{R}^d$*, Journal of Combinatorial Theory, Series B **173** (2025), 1–13.

[110] D. J. A. Welsh, *Matroid theory*, LMS Monographs, vol. 8, Academic Press, London-New York, 1976.

[111] Walter Whiteley, *Infinitesimally rigid polyhedra. I. Statics of frameworks*, Trans. Amer. Math. Soc. **285** (1984), no. 2, 431–465.

[112] ______, *A matroid on hypergraphs, with applications in scene analysis and geometry*, Discrete and Computational Geometry **4** (1989), no. 1, 75–95.

[113] ______, *Vertex splitting in isostatic frameworks*, Structural Topology (1990), no. 16, 23–30, Dual French-English text.

[114] ______, *Matroids and rigid structures*, Matroid applications, Encyclopedia Math. Appl., vol. 40, Cambridge Univ. Press, Cambridge, 1992, pp. 1–53.

[115] ______, *Some matroids from discrete applied geometry*, Matroid theory (Seattle, WA, 1995), Contemp. Math., vol. 197, Amer. Math. Soc., Providence, RI, 1996, pp. 171–311.

[116] Richard Zippel, *Probabilistic algorithms for sparse polynomials*, Symbolic and algebraic computation (EUROSAM'79, Int. Symp., Marseille, 1979), Lecture Notes in Comput. Sci., vol. 72, Springer, Berlin-New York, 1979, pp. 216–226.

School of Mathematical and Statistical Sciences
University of Galway
University Road, Galway, H91 TK33
Ireland
james.cruickshank@universityofgalway.ie

School of Mathematical Sciences
Queen Mary University of London
Mile End Road, London, E1 4NS
United Kingdom
and
HUN-REN-ELTE Egerváry Research Group on Combinatorial Optimization
Department of Operations Research
ELTE Eötvös Loránd University
Pázmány Péter sétány 1/C, 1117 Budapest
Hungary
b.jackson@qmul.ac.uk

Department of Operations Research
ELTE Eötvös Loránd University
Pázmány sétány 1/C, 1117 Budapest
Hungary
tibor.jordan@ttk.elte.hu

Department of Mathematical Informatics
University of Tokyo
7-3-1 Hongo, Bunkyo-ku, Tokyo, 113-8656
Japan
tanigawa@mist.i.u-tokyo.ac.jp

Monadic Second-Order Logic for Hypergraphs and Matroids

Dillon Mayhew

Abstract

We consider the version of monadic second-order logic introduced in Hliněný's analogue of Courcelle's Theorem. This language establishes a natural connection between formal logics and the computational theory of hypergraphs. The theorems by Courcelle and Seese provide a framework for discussing this connection. Many of the ideas we use have their origins in the classical theorem by Büchi, Elgot, and Trakhtenbrot, showing that a set of strings forms a regular language if and only if it is defined by a monadic sentence.

1 Introduction

The topics in this survey lie in the intersection of three areas: structural discrete mathematics, the theory of computation, and logic and model theory. We are motivated to work in this intersection because we can gain insight into algorithms for graphs and other discrete structures by considering the power of certain logical languages. This connection has produced a great deal of beautiful mathematics, as exemplified by the theorems of Courcelle [13] and Seese [52]. The first of these is a meta-theorem providing polynomial-time recognition algorithms for graph properties defined by sentences in a particular logical language (as long as the input class has bounded structural complexity). Seese's Theorem describes when a minor-closed class of graphs has a decidable theory (meaning there is a finite procedure for deciding whether or not any given sentence in the logical language holds for every graph in the class).

The theorems by both Courcelle and Seese are focused on *monadic second-order logic*. This is a fragment of full second-order logic, in which we can quantify over elements and subsets of the domain, but not over any relation that is binary, ternary, or of any higher arity. Both theorems apply to monadic second-order languages for graphs. This survey will explore less-familiar territory: the monadic second-order logic of hypergraphs, especially matroids. Although the monadic logic of discrete objects has been well explored (for example, in the monograph by Courcelle and Engelfriet [15]), a different flavour emerges when we consider hypergraphs, as we shall discuss.

Let us start with a refresher of some fundamental definitions. A *hypergraph* is a pair $(E, \mathcal{I})$, where E is a finite set, and $\mathcal{I}$ is a family of subsets of E. Quite often, E is called the vertex-set of the hypergraph. In matroidal contexts, E is frequently the set of edges of some other graph. Referring to these edges as vertices would be disastrously confusing, so we adopt the convention of calling E the *ground set* of the hypergraph. The members of $\mathcal{I}$ are the *hyperedges*.

Matroids are axiomatic abstractions of various notions of independence, including linear independence. We will defer the formal definition until Section 2.2. For now, it suffices to note that a matroid is a hypergraph satisfying additional axioms. Frequently, we think of the hyperedges as being *independent sets*. This means the family of hyperedges will satisfy axioms that capture properties of linear independence. However, there is a famously large number of ways of viewing a matroid as a hypergraph. Just as a topology may be viewed as a collection of closed sets, open sets, or bases, a matroid may be considered to be a collection of independent sets, bases, or circuits, amongst many other possibilities.

Let us consider how we might develop a logical language that would allow us to make statements about matroids or other hypergraphs. We will present three possibilities, and argue that one is a little too weak, another a little too strong, and the third is about right. The first approach uses *first-order logic*. In this model, we have a supply of variables $z_1, z_2, z_3, \ldots$, which will be interpreted as elements of the ground set. We will allow ourselves a binary equality predicate, and for each non-negative integer k, we have a k-arity predicate I_k, which is interpreted as a subset of k-tuples of the ground set. In this approach, the k-element independent sets of the matroid are exactly the sets that can be ordered into a k-tuple that satisfies the I_k predicate. Since independent sets are unordered, the axiom schema for matroids will need to assert that any reordering of an independent subset will produce another independent subset. For example, we will need an axiom such as

$$I_3(z_1, z_2, z_3) \to I_3(z_1, z_3, z_2) \wedge I_3(z_2, z_1, z_3) \wedge I_3(z_2, z_3, z_1) \wedge I_3(z_3, z_1, z_2) \wedge I_3(z_3, z_2, z_1)$$

or something equivalent.

This first-order approach is the one used by Vámos, in his beautifully-titled paper *The missing axiom of matroid theory is lost forever* [59]. The advantage of the first-order system is that we have access to the tools of classical model theory, including compactness, which Vámos uses in a fundamental way. But the language is too weak for many natural purposes. For example, first-order logic is not able to say that a matroid has a given minor. Minors are central to the structural theory of matroids, so this is one reason we reject first-order logic as being too weak.

We move on to another approach. The adjective *monadic* means that we are allowed to quantify only over variables and unary relations. Quantifying over a unary relation is the same as quantifying over a subset of the domain. In full second-order logic, we would be able to quantify over all possible relations, of any possible arity. In the 2-*sorted* monadic second-order logic of hypergraphs, we have one sort of variables $z_1, z_2, z_3, \ldots$ representing elements of the ground set, and another sort $h_1, h_2, h_3, \ldots$ representing the hyperedges. We also have variables $Z_1, Z_2, Z_3, \ldots$ representing subsets of elements, and $H_1, H_2, H_3, \ldots$ representing subsets of hyperedges. We let ourselves quantify over all these variables. We give ourselves an equality predicate as well as binary predicates of the form $z_i \in Z_j$, $h_i \in H_j$, and $z_i \in h_j$, which are interpreted exactly as one would expect. We denote this logic by $\mathrm{CMSO}_2^{\mathrm{hyp}}$.

I suggest that $\mathrm{CMSO}_2^{\mathrm{hyp}}$ is stronger than we would like. In particular, it is strong enough that we should not expect an analogue of Courcelle's Theorem to hold. To be specific, $\mathrm{CMSO}_2^{\mathrm{hyp}}$ can make the statement that the ground set can be partitioned into hyperedges:

$$\exists H_1\ (\forall z_1\ \exists h_1\ (z_1 \in h_1 \wedge h_1 \in H_1 \wedge \forall h_2\ (z_1 \in h_2 \wedge h_2 \in H_1 \to h_1 = h_2))).$$

The problem of deciding whether a hypergraph contains a partition into hyperedges is the EXACT COVER problem, one of Karp's original 21 NP-complete problems [36]. Therefore, it is as computationally difficult as any problem that can be solved by a polynomial-time non-deterministic Turing machine. Consequently, we do not expect it can be solved by a (deterministic) polynomial-time algorithm. In the classical version of EXACT COVER, the hypergraph is described by a list of its hyperedges. I speculate that it remains NP-complete even when the hypergraph is described via a tree automaton with a bounded number of states (Conjecture 7.1). This would imply that an analogue of Courcelle's Theorem is unlikely to hold for hypergraphs using the logic $\mathrm{CMSO}_2^{\mathrm{hyp}}$. So, $\mathrm{CMSO}_2^{\mathrm{hyp}}$ is strong enough that we seem to lose some of the prime motivation for studying monadic second-order logic: the correspondence between the logical language and the theory of computation.

For this reason, we are going to concentrate on a third approach: the "Goldilocks" logic for hypergraphs, which we denote by $\mathrm{CMSO}_1^{\mathrm{hyp}}$. This is the logic introduced by Petr Hliněný when proving a version of Courcelle's Theorem for representable matroids (Theorem 3.2). In $\mathrm{CMSO}_1^{\mathrm{hyp}}$, we have variables $Z_1, Z_2, Z_3, \ldots$ which are interpreted as subsets of the ground set. We are able to quantify over these variables. We allow ourselves a binary predicate $\subseteq$ to let us speak about the relation between subsets, and we have a unary predicate $\mathsf{hyp}(\cdot)$, which will be satisfied exactly when the input variable is interpreted as a hyperedge. Note that in $\mathrm{CMSO}_1^{\mathrm{hyp}}$, we have dropped variables representing elements of the ground set, and we use only subset variables. We lose no expressive power with this change, because it is easy to state when a variable represents the empty subset or a singleton subset.

If a class of hypergraphs has bounded structural complexity (as defined in Section 2.6), then any subclass that is defined in $\mathrm{CMSO}_1^{\mathrm{hyp}}$ can be recognised by a tree automaton (Lemma 2.5). This means that analogues of Courcelle's Theorem do hold in this new regime, as we discuss in Section 3. The major difference between $\mathrm{CMSO}_1^{\mathrm{hyp}}$ and more classical variants of monadic second-order logic, such as $\mathrm{CMSO}_2^{\mathrm{hyp}}$, is the presence of predicates that do not have fixed arity. In most cases, the *signature* of a monadic second-order logic will specify, amongst other things, the relational predicates used in the language. Importantly, each such predicate is interpreted as a set of tuples of the ground set, and each predicate has a specified *arity*, which is simply the number of elements in those tuples. In $\mathrm{CMSO}_1^{\mathrm{hyp}}$, we do not want predicates operating on a fixed number of elements of the ground set. That would require us to use an infinite sequence of predicates to talk about hyperedges of unbounded size, just as in the first-order logic of matroids. So instead the predicate $\mathsf{hyp}(\cdot)$ operates on variables that represent subsets of arbitrary size. (Quite often, traditional monadic second-logic allows predicates operating on subset variables, when we want to speak of the cardinality of a subset modulo some integer. The presence of these predicates distinguishes monadic second-order logic and *counting monadic second-order logic*. So, in some ways, $\mathrm{CMSO}_1^{\mathrm{hyp}}$ is just extending this idea by introducing another such predicate.) Having predicates that operate on subset variables gives the monadic second-order logic of hypergraphs and matroids a different flavour to the fixed-arity logic used in the context of graphs. We shall talk more about this difference in Section 4 while discussing Seese's Conjecture.

1.1 The Büchi-Elgot-Trakhtenbrot Theorem

We are exploring the interaction between logical languages and questions from complexity theory and computability theory. Many of the important ideas in this interaction can be found in a classical theorem, which was proved independently by Büchi [11], Elgot [18], and Trakhtenbrot [56] (the BET Theorem). By analysing this theorem, we give ourselves a framework for understanding much of the work in the survey. To state the BET Theorem, we need to discuss strings over finite alphabets and a formal language which can express statements about those strings. For any positive integer n, we use $[n]$ to denote $\{1, 2, \ldots, n\}$. If n is zero, then $[n]$ is the empty set. From this point onwards, Σ stands for a finite set called an *alphabet*.

Definition 1.1 (Strings and languages) A (finite) *string* (over Σ) is a function $\mathbf{w}\colon [n] \to \Sigma$, for some non-negative integer n. The elements of $[n]$ are called the *positions* of the string. The set of strings over Σ is denoted by Σ^*. Any subset of Σ^* is a *language*.

Next, we develop a version of monadic second-order logic for strings. The variables we use are $Z_1, Z_2, Z_3, \ldots$, which are interpreted as subsets of positions. The *atomic formulas* include $Z_i \subseteq Z_j$, which is interpreted in the obvious way, as well as:

- $\mathsf{term}(Z_i)$, which is interpreted as saying that Z_i represents a *terminal* set of positions, meaning a set of the form $\{i, i+1, \ldots, n\}$ for a word $\mathbf{w}\colon [n] \to \Sigma$.
- $\mathsf{char}_\alpha(Z_i)$, where α is an element of the alphabet Σ. This predicate is interpreted as saying that Z_i contains a single position and the character in this position is equal to α.

Now, we consider the collection of formulas that we can construct using the atomic formulas, the usual logical connectives $\wedge$, $\vee$, $\neg$, $\rightarrow$, $\leftrightarrow$, and the quantifiers $\forall$ and $\exists$. This collection is denoted by MSO^Σ. A variable is *bound* in a formula if it is quantified over, and otherwise it is *free*. A formula with no free variables is a *sentence*. We write $Z_i = Z_j$ to mean $(Z_i \subseteq Z_j) \wedge (Z_j \subseteq Z_i)$ and $Z_i \not\subseteq Z_j$ to stand for $\neg(Z_i \subseteq Z_j)$. Let $\mathsf{empty}(Z_i)$ be the formula $\forall Z_j\ (Z_j \subseteq Z_i \rightarrow Z_j = Z_i)$. Note that $\mathsf{empty}(Z_i)$ is satisfied exactly when Z_i is interpreted as the empty set. Next let $\mathsf{sing}(Z_i)$ be $\neg\mathsf{empty}(Z_i) \wedge \forall Z_j\ (Z_j \subseteq Z_i \rightarrow (\mathsf{empty}(Z_j) \vee Z_j = Z_i))$. Thus, $\mathsf{sing}(Z_i)$ is satisfied exactly when Z_i is interpreted as a singleton set.

Let $\mathcal{L} \subseteq \Sigma^*$ be a language. We say that $\mathcal{L}$ is MSO^Σ-*definable* if there is some sentence that is satisfied by exactly the strings in $\mathcal{L}$.

Example 1.2 Let Σ be the alphabet $\{A, B\}$. We consider the language defined by the MSO^Σ-sentence $\exists Z_1\ (\mathsf{term}(Z_1) \wedge \forall Z_2\ (\mathsf{sing}(Z_2) \rightarrow (\mathsf{char}_B(Z_2) \leftrightarrow Z_2 \subseteq Z_1)))$. This language is the subset of Σ^* consisting of all strings, where no A symbol comes after a B symbol.

Definition 1.3 (Finite-state automata) Let Σ be an alphabet. A finite-state automaton (FSA) (with alphabet Σ) is a tuple $(D, \mathsf{start}, \mathsf{acc})$, where D is a directed graph, start is a vertex of D, and acc is a set of vertices of D. Each arc of D is labelled with a single element of Σ. The vertices of D are known as *states*, and we use Q to denote the set of states. Generally, "state" refers to a possible configuration of the memory in a computing machine. We call start the *initial state* and refer to acc as the set of *accepting states*. Note that we allow self-directed loops and parallel arcs.

Say that $\mathbf{D} = (D, \mathsf{start}, \mathsf{acc})$ is an FSA and that $\mathbf{w}\colon [n] \to \Sigma$ is a string. A *run* of $\mathbf{D}$ on $\mathbf{w}$ is a function $r\colon \{0\} \cup [n] \to Q$ such that $r(0) = \mathsf{start}$ and for every $i \in [n]$, if $\mathbf{w}(i) = \alpha$, then there is an arc from $r(i-1)$ to $r(i)$ that is labelled with α. If $r(n)$ is in acc, then the run *accepts* $\mathbf{w}$. We say that $\mathbf{D}$ *accepts* $\mathbf{w}$ if there exists an accepting run.

Let $\mathcal{L} \subseteq \Sigma^*$ be a language. We say that $\mathcal{L}$ is *regular* if there exists an FSA that accepts exactly the strings in $\mathcal{L}$. In this case we say that the FSA *recognises* the language.

Example 1.4 Let Σ be the alphabet $\{A, B\}$. Let D be the directed graph shown in Figure 1. Let acc be the set of shaded vertices. A string is accepted by $(D, \mathsf{start}, \mathsf{acc})$ if and only if there is no A character that appears after a B character.

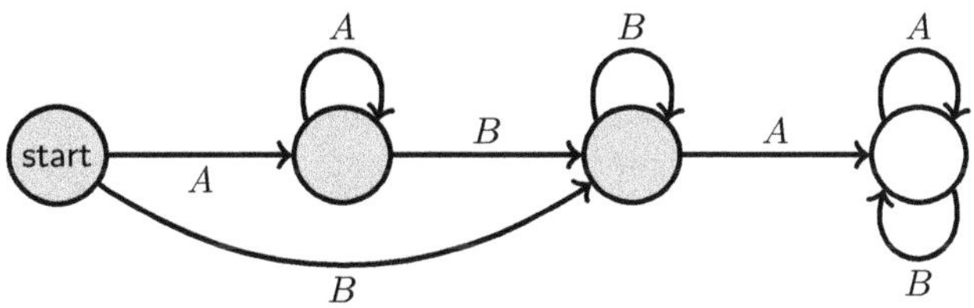

Figure 1: A finite-state automaton

It is no coincidence that the language defined in Example 1.2 is the same as the language recognised by the FSA in Example 1.4. The Büchi-Elgot-Trakhtenbrot Theorem tells us we are always able to construct these parallel definitions of regular languages. In fact, the languages defined by the monadic language MSO^{Σ} are exactly the languages recognised by finite-state automata.

Theorem 1.5 (BET Theorem) *Let Σ be a finite alphabet and let $\mathcal{L} \subseteq \Sigma^*$ be a language. Then $\mathcal{L}$ is regular if and only if it is MSO^{Σ}-definable.*

The proof of the BET Theorem contains many of the seminal ideas linking the monadic languages of discrete structures to the theory of computing machines. We will spend some time sketching the proof, since the key ideas inform so much of the work in this survey. The easier direction involves proving that a regular language is MSO^{Σ}-definable. We assume that $\mathbf{D} = (D, \mathsf{start}, \mathsf{acc})$ is an FSA that recognises the language $\mathcal{L}$. We construct an MSO^{Σ}-sentence $\varphi_{\mathbf{D}}$ that defines $\mathcal{L}$. The idea is that $\varphi_{\mathbf{D}}$ will assert the existence of an accepting run of $\mathbf{D}$ on a string. Let $Q = \{q_1, q_2, \ldots, q_s\}$ be the set of states of D. We start the construction by asserting that there is a partition $\{P_1, P_2, \ldots, P_s\}$ of the set of positions. We are essentially labelling each position in the string with one of the states $q_1, q_2, \ldots, q_s$, and this is equivalent to creating a function $r\colon \{0\} \cup [n] \to Q$ for any string with n positions. The job of $\varphi_{\mathbf{D}}$ is to enforce that this function is an accepting run of $\mathbf{D}$ on the string. We must assert that r takes the final position to an accepting state, and $\varphi_{\mathbf{D}}$ does this by saying that any terminal singleton set belongs to $\cup_{q_i \in \mathsf{acc}} P_i$. Also, φ needs to ensure that every transition from one state to another obeys the arc-labellings of D. Imagine that the arcs of D that depart from q_a with the label $\alpha \in \Sigma$ have $q_{k_1}, q_{k_2}, \ldots, q_{k_s}$ as their destinations. Then $\varphi_{\mathbf{D}}$ will have a clause saying that if Z_i and Z_j are singleton sets of positions and there is exactly one terminal set that contains Z_j but not Z_i, and $\mathsf{char}_\alpha(Z_i)$ holds, and Z_i is in P_a, then Z_j is in one of $P_{k_1}, P_{k_2}, \ldots, P_{k_s}$. We repeat this construction for every choice of state q_a and label α. We also need to assert that the state assigned to the first position is appropriate. We do this with a clause saying that if the singleton set Z_i is in exactly one terminal set and $\mathsf{char}_\alpha(Z_i)$ holds, then Z_i is in one of $P_{k_1}, P_{k_2}, \ldots, P_{k_s}$, where $q_{k_1}, q_{k_2}, \ldots, q_{k_s}$ are the destinations of arcs departing start with the label α. With this, we have sketched the construction of $\varphi_{\mathbf{D}}$. It is not difficult to check that a string satisfies $\varphi_{\mathbf{D}}$ if and only if it is accepted by $\mathbf{D}$. So we have completed the "only if" direction of the BET Theorem.

The "if" direction is more convoluted, since it involves using the power of non-determinism to simulate the existential quantifiers in an MSO^Σ-sentence. Let $\mathbf{D} = (D, \mathsf{start}, \mathsf{acc})$ be an FSA. Then $\mathbf{D}$ is *deterministic* if, for every state q and every label $\alpha \in \Sigma$, there is exactly one arc departing from q that is labelled with α. Otherwise the automaton is *non-deterministic*. If the automaton is deterministic, then there is exactly one run of the automaton on any string. In a non-deterministic FSA, there may be multiple runs (or zero runs) on a given string. Given that the power of non-determinism vs. determinism in general computing machines leads to the notorious P vs. NP problem, it is still a little surprising that non-determinism gives finite-state automata no additional power, as was demonstrated by Rabin and Scott in 1959 [48].

Lemma 1.6 *Let* $\mathbf{D} = (D, \mathsf{start}, \mathsf{acc})$ *be an FSA. There is a deterministic FSA that accepts exactly the same strings as* $\mathbf{D}$.

Sketch proof. We replace the non-deterministic machine $\mathbf{D}$ with a deterministic automaton $\mathbf{D}_{\mathrm{det}}$. The price we must pay is an exponential blow-up in the number of states. If Q is the set of states of D, then $\mathbf{D}_{\mathrm{det}}$ will have the power set of Q as its set of states. When $A, B \subseteq Q$ are sets of states, there is an arc labelled with $\alpha \in \Sigma$ from A to B in $\mathbf{D}_{\mathrm{det}}$ if and only if B is the set of states q such that there exists $q' \in A$ and an arc of D from q' to q that is labelled by α. That is, the unique arc departing from A labelled by α has as its destination the set of states that can be reached from a state in A using an arc labelled by α. This makes it clear that $\mathbf{D}_{\mathrm{det}}$ is indeed deterministic. The starting state of $\mathbf{D}_{\mathrm{det}}$ is $\{\mathsf{start}\}$, and the accepting states are the subsets of Q that have non-empty intersection with acc. It is very easy to see that if there is an accepting run of $\mathbf{D}$ on $\mathbf{w} \in \Sigma^*$, then there is an accepting run of $\mathbf{D}_{\mathrm{det}}$. There is a fairly mechanical inductive proof which shows that an accepting run of $\mathbf{D}_{\mathrm{det}}$ implies the existence of an accepting run of $\mathbf{D}$. Thus $\mathbf{D}$ and $\mathbf{D}_{\mathrm{det}}$ accept exactly the same strings. □

To complete the proof of the BET Theorem, we must show that any language defined by a sentence φ in MSO^Σ can be recognised by an automaton. As is often the case in model theory, the proof is inductive on the number of steps needed to construct the sentence φ. The issue with this strategy is that the process of building a sentence involves formulas that feature free variables, and which therefore are not sentences. So, our inductive strategy requires us to prove that a language defined by any formula (possibly with free variables) can be recognised by an automaton. The problem is that it makes no sense to ask whether a formula with free variables is satisfied by a string, just as it makes no sense to ask whether the equation $x = y$ is correct without specifying what these variables stand for. So we introduce a function which will give meaning to the free variables by assigning values to them. Let φ be some formula in MSO^{Σ} and let F be its set of free variables. Let $\mathbf{w}\colon [n] \to \Sigma$ be a string. An *interpretation* is a function $\theta\colon F \to 2^{[n]}$. Now, we say that φ is satisfied by the pair $(\mathbf{w}, \theta)$ if it is true when every variable in F is interpreted as its image under θ.

Having specified what it means for a formula to be satisfied, we must explain how an automaton can process a string along with an interpretation of free variables. We do this by stacking the string on top of a sequence of strings from $\{\checkmark, \times\}^*$. Each of these additional strings defines a subset of $[n]$ — the subset of positions that contain a ✓ symbol. We now make this more formal. As always, Σ is a finite alphabet. Assume that n is a non-negative integer and that $\mathbf{w}\colon [n] \to \Sigma$ is a string. Let s be a non-negative integer and let $P_1, P_2, \ldots, P_s$ be subsets of $[n]$. We define the string

$$S(\mathbf{w}; P_1, P_2, \ldots, P_s)\colon [n] \to \Sigma \times \{\checkmark, \times\}^s.$$

For any position $i \in [n]$, the image of i under $S(\mathbf{w}; P_1, P_2, \ldots, P_s)$ is the tuple $(\mathbf{w}(i), P_1(i), P_2(i), \ldots, P_s(i))$, where $P_j(i)$ is ✓ if i is in P_j, and ✗ otherwise. A finite-state automaton with alphabet $\Sigma \times \{\checkmark, \times\}^s$ is said to be a *Σ-FSA* with *arity* s. The next result provides the final step in the proof of the BET Theorem. It uses a "boot-strapping" approach that starts with the atomic formulas and builds automata that can recognise any arbitrary formula.

Lemma 1.7 *Let Σ be a finite alphabet. Let φ be a formula in* MSO^{Σ} *and let $Z_{i_1}, Z_{i_2}, \ldots, Z_{i_s}$ be the free variables of φ. There exists* $\mathbf{D}$*, a Σ-FSA with arity s, where the following property is satisfied: for any string $\mathbf{w}\colon [n] \to \Sigma$ and any choice of subsets $P_1, P_2, \ldots, P_s \subseteq [n]$, we have that* $\mathbf{D}$ *accepts $S(\mathbf{w}; P_1, P_2, \ldots, P_s)$ if and only if φ is satisfied by $(\mathbf{w}, \theta)$, where θ is the interpretation that takes each Z_{i_j} to P_j.*

Sketch proof. We again apply induction on the number of steps required to build φ from atomic formulas. This means there are three base cases, corresponding to the three possible forms of atomic formula. In these three cases, φ is one of the formulas $Z_{i_1} \subseteq Z_{i_2}$, or $\mathsf{term}(Z_{i_1})$, or $\mathsf{char}_{\alpha}(Z_{i_1})$. In the first case, $\mathbf{D}$ will be an FSA with alphabet $\Sigma \times \{\checkmark, \times\}^2$. We need $\mathbf{D}$ to accept if and only if it never encounters a triple of the form $(\alpha, \times, \checkmark)$, because this will correspond to an element of $[n]$ that is in P_2 but not in P_1. We leave it as an exercise to construct $\mathbf{D}$ in the other cases.

For the inductive step of the argument, we make our lives simpler by noting that any formula is logically equivalent to one constructed using only negation $\neg$, conjunction $\wedge$, and the existential quantifier $\exists$. This leaves us with three cases to which we apply the inductive strategy. To start with, we assume that φ is a formula of the form $\neg\psi$. Note that the free variables of φ are exactly the free variables of ψ. Because ψ is constructed using fewer steps than $\varphi = \neg\psi$, the inductive hypothesis implies there exists a Σ-FSA $\mathbf{D}'$ that accepts $S(\mathbf{w}; P_1, P_2, \ldots, P_s)$ if and only if ψ is satisfied by $\mathbf{w}$ and the interpretation θ. Lemma 1.6 shows that we may as well assume $\mathbf{D}'$ is deterministic. Now the rest of this case is easy: we construct $\mathbf{D}$ from $\mathbf{D}'$ by making the accepting states non-accepting, and vice versa. This means that $\mathbf{D}$ will accept precisely when $\mathbf{D}'$ does not accept, which is exactly the behaviour that we want. Note that this argument relies on the determinism of $\mathbf{D}'$ in a critical way, because we are assuming there is a unique state that $\mathbf{D}'$ can arrive at as it finishes its computation on $S(\mathbf{w}; P_1, P_2, \dots, P_s)$.

If φ is equal to $\psi_1 \wedge \psi_2$, then we let $\mathbf{D}_1$ and $\mathbf{D}_2$ be the automata provided to us by the inductive hypothesis, where $\mathbf{D}_i$ decides the formula ψ_i. There are some technicalities in this case, since a free variable of ψ_1 may not appear in ψ_2. Thus, the arities of $\mathbf{D}_1$ and $\mathbf{D}_2$ may be different. But we will elide these difficulties, since the idea of the construction is fairly simple: we construct the FSA $\mathbf{D}$ as the product of $\mathbf{D}_1$ and $\mathbf{D}_2$. By this we mean that the set of states in $\mathbf{D}$ is the Cartesian product of the set of states of $\mathbf{D}_1$ and $\mathbf{D}_2$. The starting state of $\mathbf{D}$ is the ordered pair containing the two starting states of $\mathbf{D}_1$ and $\mathbf{D}_2$. An ordered pair of states in $\mathbf{D}$ is accepting in $\mathbf{D}$ if and only if the two states are accepting in $\mathbf{D}_1$ and $\mathbf{D}_2$. There is an arc from (q_1, q_2) to (q_1', q_2') with the label α if and only if each $\mathbf{D}_i$ has an arc labelled by α from q_i to q_i'.

The final case of the proof is where φ is the formula $\exists Z_i \psi$, where Z_i is a free variable in ψ. This is the case where we use non-determinism to mimic the power of the existential quantifier. By induction, there is an FSA $\mathbf{D}'$ that recognises the formula ψ. Say that the arity of $\mathbf{D}'$ is $s+1$. Thus, every arc-label of $\mathbf{D}'$ has the form $(\alpha, c_1, c_2, \dots, c_{s+1})$, where each c_j belongs to $\{\checkmark, \text{✗}\}$. For the sake of simplicity, we assume without any loss of generality that Z_i is the last of the free variables in ψ, so that it corresponds to the character c_{s+1}. The states of the FSA $\mathbf{D}$ are the same as those of $\mathbf{D}'$. The arc-labels of $\mathbf{D}$ are exactly the tuples we obtain by deleting the final character of the arc-labels in $\mathbf{D}'$. Thus the arity of $\mathbf{D}$ is one less than that of $\mathbf{D}'$, as we would expect. Let q and q' be states in $\mathbf{D}'$. There is an arc labelled $(\alpha, c_1, c_2, \dots, c_s)$ from q to q' in $\mathbf{D}$ if and only if $\mathbf{D}'$ has an arc from q to q' labelled either $(\alpha, c_1, c_2, \dots, c_s, \checkmark)$ or $(\alpha, c_1, c_2, \dots, c_s, \text{✗})$. Note that $\mathbf{D}$ will typically not be deterministic, even though $\mathbf{D}'$ may be. A simple inductive argument shows that $\mathbf{D}$ accepts $S(\mathbf{w}, P_1, P_2, \dots, P_s)$ if and only if there is some subset P_{s+1} of positions such that $\mathbf{D}'$ accepts $S(\mathbf{w}, P_1, P_2, \dots, P_{s+1})$, and this is exactly the behaviour we were seeking. □

We obtain the proof of the BET Theorem from Lemma 1.7 by restricting to the case where φ is a sentence (which corresponds to the case with $s = 0$). In this case, the lemma provides an FSA $\mathbf{D}$ with alphabet Σ which accepts $S(\mathbf{w}) = \mathbf{w}$ if and only if φ is satisfied by $\mathbf{w}$. This shows that the language defined by φ is recognisable by an FSA, and completes the proof of the "if" direction.

The BET Theorem is relevant to this survey for multiple reasons. The proof of the theorem provides a framework that helps us understand the proof of other related theorems, including Courcelle's Theorem and its matroid-theoretical descendants, as we see in Section 3. Let us think about the computational difficulty of recognising whether a given string belongs to a specified language. This problem can be arbitrarily hard and may not be solvable by any Turing machine. But once we know that a language is defined by a monadic sentence, then the complexity of testing membership in the language becomes almost trivial. The BET Theorem tells us that any such language is exactly the language recognised by some FSA, and the whole point of an FSA is that it never backtracks. It therefore runs in time that is linearly bounded by the length of the string. So if a language is monadically-defined, then there is an algorithm which will test whether an input string belongs to the language and this algorithm runs in linear time. This phenomenon underlies all Courcelle-type theorems. Courcelle shows that any class of graphs with bounded structural complexity that is defined by a monadic sentence can be recognised by an automaton. This leads to an efficient algorithm for testing membership in the class. The main difficulty in upgrading the BET Theorem to Courcelle's Theorem is that graphs do not come equipped with a linear ordering in the same way that strings do. Instead we consider classes of graphs with tree-like structure (that is, classes with bounded *tree-width*), which can therefore be processed by automata that operate on trees, rather than strings.

The BET Theorem also illustrates a theme that has dominated recent research into monadic languages and discrete structures: recognisability vs. definability. The first of these terms refers to the power of a computing machine to test membership of a class and the second refers to the expressive power of a logical language to define a class. The BET Theorem shows that in at least one context, these two properties are equivalent: a language of strings is definable by a monadic sentence if and only if it is recognisable by an automaton. In other contexts, it is much more difficult to show that recognisability is equivalent to definability. It is not too hard to show that when a class of graphs with bounded tree-width is definable by a monadic sentence, it is recognisable by a tree automaton. The other direction is the subject of a conjecture by Courcelle [14, Conjecture 1] which was open for decades until it was proved in a beautiful paper by Bojańczyk and Pilipczuk [7].

Considering languages of strings can also inform our thinking around decidable theories. To show that a language of strings has a decidable monadic second-order theory, we must show that there is a computing machine which can take any sentence from MSO^{Σ} as input, and decide in finite time whether or not every string in the language satisfies the sentence. Clearly, this is equivalent to deciding if there is a string in the language that is a counterexample to the sentence. The crucial idea here is that given any finite-state automaton, there is a finite procedure for deciding if there is a string that will be accepted by the automaton. This follows from the *pumping lemma* [17, Proposition 5.3.1], which provides an upper bound on the length of a smallest-possible string that is accepted by the automaton. Say that the language $\mathcal{L}$ is regular, so that it is recognised by an automaton. The BET Theorem says there is a sentence $\varphi_{\mathcal{L}}$ that is satisfied exactly by the strings in $\mathcal{L}$. Now assume we want to test whether the arbitrary monadic sentence ψ is satisfied by all the strings in $\mathcal{L}$. We construct the automaton that accepts exactly the strings satisfying $\varphi_{\mathcal{L}} \wedge (\neg\psi)$, using the boot-strapping techniques of Lemma 1.7. The pumping lemma provides us with an explicit upper bound on the length of a smallest-possible string satisfied by this automaton. All we need do is test the strings up to that length. If any of these strings is accepted then we have found a string satisfying $\varphi_{\mathcal{L}} \wedge (\neg\psi)$, which therefore belongs to $\mathcal{L}$ but does not satisfy ψ. If none of the strings is accepted, then ψ has no counterexample in $\mathcal{L}$, so it is a theorem for this class.

We will revisit all these uses of the BET theorem throughout the survey.

2 Fundamentals

In this section, we review some fundamental definitions and ideas.

2.1 Graphs and hypergraphs

A *graph*, G, consists of a set of *vertices* (denoted by $V(G)$), a set of *edges* (denoted by $E(G)$), and an *incidence function*, which takes each edge to a set containing either one or two vertices. Note that this definition allows graphs to contain *loops* (edges that are incident with a single vertex) and *parallel edges* (pairs of distinct non-loop edges which are incident with the same vertices). If X is a subset of $E(G)$, then $G[X]$ is a graph with X as its edge-set. The vertices of $G[X]$ are the vertices of G that are incident with at least one edge in X.

Recall from the introduction that a *hypergraph* (also called a *set-system*) is a pair $(E, \mathcal{I})$, where E is a finite set, and $\mathcal{I}$ is a family of subsets of E. We refer to E as the *ground set* of the hypergraph and to the sets in $\mathcal{I}$ as *hyperedges*.

2.2 Matroids

Matroids are axiomatic abstractions of a notion of (in)dependence that arises naturally in linear algebra, graph theory, and the theory of field extensions, amongst other places. Birkhoff [2, 3], Mac Lane [40], Steinitz [54], and Van der Waerden [60] were among the prominent mathematicians who knew that these disparate areas of mathematics exhibited similar behaviour. The neglected mathematician Takeo Nakasawa gave a set of axioms to capture this similarity [44], but it was Hassler Whitney who has become known as the founder of matroid theory. The axioms below are essentially the same as his [61]. Any facts stated here without proof can be found in Oxley's *Matroid Theory* [47], which is the standard text for a thorough introduction to the subject.

Definition 2.1 A *matroid* is a hypergraph $(E, \mathcal{I})$ satisfying the following axioms.

(I1) $\mathcal{I}$ is non-empty,

(I2) every subset of a hyperedge is a hyperedge,

(I3) if I and J are hyperedges and $|I| < |J|$, then there exists an element $e \in J \backslash I$ such that $I \cup \{e\}$ is a hyperedge.

The set E is known as the *ground set* of the matroid, and the members of $\mathcal{I}$ are the *independent sets*.

A subset of the ground set that is not independent is *dependent*. A *circuit* is a dependent set that has no dependent proper subset. A *basis* is an independent set that has no independent proper superset. Any set that contains a basis is *spanning*. A set that is disjoint with at least one basis is *coindependent*. The bases of a matroid are equicardinal, and the shared cardinality is the *rank* of that matroid.

Assume that $M = (E, \mathcal{I})$ is a matroid. The coindependent sets of M are themselves the independent sets of a matroid with E as its ground set. This matroid is the *dual* of M, denoted by M^*. If X is a subset of E, then $r_M(X)$ stands for the *rank* of X in M, which is the maximum cardinality of an independent subset of X. Thus, the rank of M is $r_M(E)$, which we abbreviate to $r(M)$.

Let X and Y be disjoint subsets of E such that X is independent in M. Then

$$M/X\backslash Y = (E\backslash(X \cup Y), \{I \subseteq E\backslash(X \cup Y) \colon I \cup X \in \mathcal{I}\})$$

is a matroid, and any matroid arising from M in this way is a *minor* of M. We say $M/X\backslash Y$ is obtained by *contracting* X and *deleting* Y. A minor of M that is not equal to M is a *proper minor*.

If $\mathcal{M}$ is a class of matroids and any minor of a matroid in $\mathcal{M}$ is also in $\mathcal{M}$, then $\mathcal{M}$ is *minor-closed*. If $\mathcal{M}$ is a minor-closed class of matroids, then an *excluded minor* for $\mathcal{M}$ is a matroid N such that N does not belong to $\mathcal{M}$, but any proper minor of N belongs to $\mathcal{M}$. If M is a matroid, then M is in $\mathcal{M}$ if and only if it does not have a minor isomorphic to an excluded minor for $\mathcal{M}$. An *antichain* is a set $\mathcal{A}$ of pairwise non-isomorphic matroids such that whenever M and N are distinct members of $\mathcal{A}$, no minor of M is isomorphic to N. The excluded minors for a minor-closed class form an antichain. A class of matroids is *well-quasi-ordered* if it contains no infinite antichain.

Matroids can be seen as abstractions of matrices (explaining their name). To see why this is the case, let A be a matrix with entries over the field $\mathbb{F}$. If A has r rows, then every column is a vector from $\mathbb{F}^r$. We let E be the set of these vectors and we let $\mathcal{I}$ be the family of linearly independent subsets of E. Then $(E, \mathcal{I})$ is a matroid, and any such matroid is $\mathbb{F}$-*representable*.

Figure 2 shows the *Fano matroid*, which has rank 3. (You may recognise it as the projective plane PG(2, 2).) The ground set of the Fano matroid is the set of points in the diagram. Because the matroid has rank 3, any subset of size four or more is dependent. All subsets of size two or less are independent. A set of size three is independent if and only if it is not contained in one of the lines of the diagram. In this diagram, the Fano matroid is represented over $\mathbb{F}_2$, the field of two elements, so each point is labelled with a vector from $\mathbb{F}_2^3$. We see that when a matroid is representable it means that its points can be given coordinates in a way that captures the dependencies of the matroid.

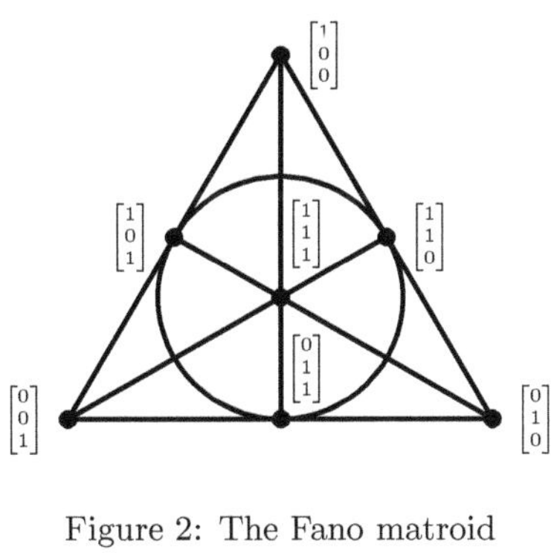

Figure 2: The Fano matroid

For any field $\mathbb{F}$, the class of $\mathbb{F}$-representable matroids is minor-closed, which means that we can hope to characterise the class by explicitly listing the excluded minors. Such a characterisation has been completed only when $\mathbb{F}$ has size two [57], three [4, 53], or four [24]. In these cases, the numbers of excluded minors are one, four, and seven respectively. The number of excluded minors when $\mathbb{F}$ is the field of five elements is known to be more than 2000 [9], so an explicit excluded-minor characterisation in this next case may be out of reach for some time.

The famous Robertson-Seymour project on graph minors shows that any minor-closed class of graphs has a finite number of excluded minors [50]. This is not necessarily the case for minor-closed classes of matroids: for example, when $\mathbb{F}$ is an infinite field, there are infinitely many excluded minors for the class of $\mathbb{F}$-representable matroids. (Oxley attributes a proof to Geelen and Whittle [47, Theorem 6.5.17].) Rota published the following influential conjecture in 1971 [51].

Conjecture 2.2 (Rota's Conjecture) *Let $\mathbb{F}$ be a finite field. There are only finitely many excluded minors for the class of $\mathbb{F}$-representable matroids.*

A proof of Rota's Conjecture has been announced by Geelen, Gerards, and Whittle [29] as the culmination of a massive project investigating the structure of minor-closed classes of $\mathbb{F}$-representable matroids [25, 27]. These structural results are inspired by the project of Robertson and Seymour that illuminated the structure of minor-closed classes of graphs.

2.3 Biased graphs

Let G be a graph. A *theta subgraph* of G comprises two distinct vertices, u and v, and three paths that join u to v, where u and v are the only vertices that are in more than one of these paths. Thus, any theta subgraph contains three cycles. A *linear class* is a set $\mathcal{B}$ of cycles of G such that no theta subgraph contains exactly two cycles in $\mathcal{B}$. In this case, we say $(G, \mathcal{B})$ is a *biased graph*. The cycles in $\mathcal{B}$ are said to be *balanced*. If $(G, \mathcal{B})$ is a biased graph then $F(G, \mathcal{B})$ is the corresponding *frame matroid*. The ground set of this matroid is $E(G)$ and the independent sets are the subsets $X \subseteq E(G)$ such that $G[X]$ contains no balanced cycles and no more than one cycle per connected component. Biased graphs and their related matroids were introduced by Zaslavsky [63, 64]. If every cycle is balanced, the independent sets of the frame matroid are exactly the forests of G, so we have reduced to the case in which $F(G, \mathcal{B})$ is the *graphic matroid* $M(G)$.

Say that $\Omega = (G, \mathcal{B})$ is a biased graph. Let A and B be disjoint subsets of edges such that $G[A]$ is acyclic. Let $\mathcal{P}$ be the set of connected components of $G[A]$. The graph $G/A\backslash B$ has $\mathcal{P}$ as its vertex-set. The edges of $G/A\backslash B$ are the edges of $E(G)\backslash(A \cup B)$ and an edge $e \in E(G)\backslash(A \cup B)$ is incident with a component $P \in \mathcal{P}$ if and only if P contains a vertex v such that v and e are incident in G. Let C be a cycle of $G/A\backslash B$. There is a unique cycle C' of G such that $C'\backslash C \subseteq A$. We declare C to belong to the class $\mathcal{B}/A\backslash B$ if and only if C' is in $\mathcal{B}$. Now $\Omega/A\backslash B = (G/A\backslash B, \mathcal{B}/A\backslash B)$ is a biased graph, and any biased graph that arises in this way is a *minor* of Ω. This definition is constructed in such a way that $F(\Omega/A\backslash B) = F(\Omega)/A\backslash B$ [64, Theorem 2.5]. The definitions of minor-closed classes, excluded minors, antichains, and well-quasi-ordered classes in the context of biased graphs are analogous to the corresponding definitions for matroids.

Let G be a graph and let Γ be a group (written multiplicatively). Let L be the set of tuples $(e, u, v) \in E(G) \times V(G) \times V(G)$, where $\{u, v\}$ is the set of vertices incident with e. A function $\gamma\colon L \to \Gamma$ is a Γ-*gaining* of G if $\gamma(e, u, v) = \gamma(e, v, u)^{-1}$ whenever e is an edge joining distinct vertices u and v. If $W = v_0, e_0, v_1, \ldots, v_{n-1}, e_{n-1}, v_n$ is a walk in G, then we define $\gamma(W)$ to be $\gamma(e_0, v_0, v_1) \cdots \gamma(e_{n-1}, v_{n-1}, v_n)$ and refer to this product as the *gain* of the walk. The gain of a closed walk may depend on the choice of the starting point and the orientation of travel around the walk, but any two gains obtained by different choices will be conjugates in Γ. This means that the class of cycles with gain equal to the identity id_Γ is well defined. We denote this class by $\mathcal{B}_\gamma$. It is easy to see that $\mathcal{B}_\gamma$ is a linear class. If Ω is a biased graph and $\Omega = (G, \mathcal{B}_\gamma)$ for some Γ-gaining γ, then Ω is Γ-*gainable*. A biased graph that is Γ-gainable for at least one group Γ is *gainable*. If γ is a Γ-gaining of G, then $F(G, \mathcal{B}_\gamma)$ is a Γ-*gain-graphic* matroid.

Figure 3 gives an illustration of a gain-graphic matroid. The group Γ in this case is the Klein four-group, isomorphic to the direct product $\mathbb{Z}_2 \otimes \mathbb{Z}_2$. Let id be the identity element and let a, b, and c be the non-identity elements, so that $a+b+c = \mathrm{id}$. On the left side of the diagram, we show a Γ-gained graph. Each edge is labelled with an integer between 1 and 9, as well as an element of Γ. When an edge e is directed from the vertex u to v, and is labelled with $\alpha \in \Gamma$ this indicates that α is the image of (e, u, v) under the gaining. On the right side of the diagram we see a geometric representation of the corresponding Γ-gain-graphic matroid, which has the ground set $\{1, 2, \ldots, 9\}$. Since this matroid has rank three, this diagram represents independent subsets in exactly the same way as Figure 2. We note that $\{1, 2, 3\}$ is the set of loops in the graph, and is a basis of the matroid. Every other element of the matroid is positioned on a line between two of these basis elements, and its location on that line is essentially determined by its gain label. In this way, we can think of the points in a gain-graphic matroid as being given coordinates by elements of a group, in the same way that the points in a representable matroid are given coordinates from a field. So representable and gain-graphic matroids are analogues — both are coordinatised by algebraic structures.

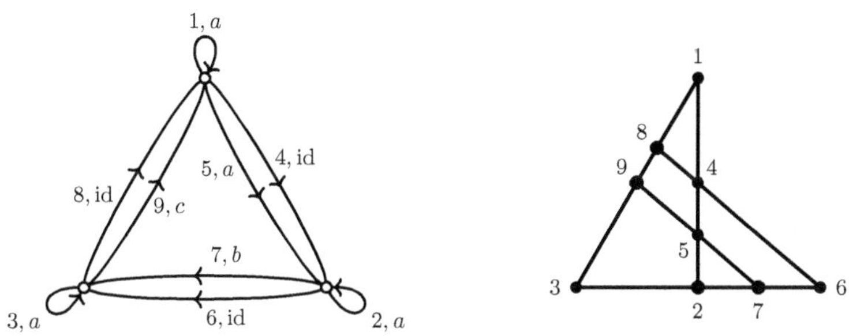

Figure 3: A graph with a gaining and the corresponding gain-graphic matroid

Classes of representable matroids are central in the history of matroid theory. There is a symmetry between classes of representable matroids and classes of gain-graphic matroids. This symmetry can be seen in a theorem by Kahn and Kung [35]

and in the work of Geelen, Gerards, and Whittle [28] that describes the structure of classes of finite-field-representable matroids. These results show that the structure of finite-field-representable classes and finite-group-gain-graphic classes are intrinsically linked. Despite the symmetry of the two types of classes, gain-graphic matroids have not received as much attention as representable matroids. A major theme of this survey will be looking for gain-graphic analogues of theorems that apply to representable matroids.

2.4 Monadic logic

The defining characteristic of monadic second-order logic is that quantifications take place over both elements and subsets of the domain. This distinguishes it from first-order logic (where we may quantify only over elements of the domain) and full second-order logic (where we may quantify over relations of any arity). In this section we will give formal definitions of the syntax and semantics for some variants of monadic second-order logic in the context of discrete structures.

All of the languages we define are recursively constructed from *atomic formulas*. The form that the atomic formulas take depends on the particular language. If φ is a formula, then $\mathrm{Var}(\varphi)$ and $\mathrm{Free}(\varphi)$, respectively, are the sets of *variables* and *free variables* that appear in φ. A free variable is one that has not been quantified over. In addition, we let $\mathrm{Bound}(\varphi) = \mathrm{Var}(\varphi) \setminus \mathrm{Free}(\varphi)$ stand for the set of *bound variables*. Every formula is a sequence of symbols, and the formulas of the language are exactly the sequences that are constructed using a finite application of the following rules.

- Every atomic formula is a formula.
- If φ is a formula, then so is $(\neg\varphi)$. $\mathrm{Var}(\neg\varphi)$ and $\mathrm{Free}(\neg\varphi)$ are respectively equal to $\mathrm{Var}(\varphi)$ and $\mathrm{Free}(\varphi)$.
- If φ_1 and φ_2 are formulas such that $\mathrm{Free}(\varphi_1) \cap \mathrm{Bound}(\varphi_2) = \emptyset = \mathrm{Bound}(\varphi_1) \cap \mathrm{Free}(\varphi_2)$, then $(\varphi_1 \wedge \varphi_2)$ is a formula. $\mathrm{Var}(\varphi_1 \wedge \varphi_2)$ and $\mathrm{Free}(\varphi_1 \wedge \varphi_2)$ are respectively equal to $\mathrm{Var}(\varphi_1) \cup \mathrm{Var}(\varphi_2)$ and $\mathrm{Free}(\varphi_1) \cup \mathrm{Free}(\varphi_2)$.
- If φ is a formula and Z_i is in $\mathrm{Free}(\varphi)$, then $(\exists Z_i \varphi)$ is a formula. $\mathrm{Var}(\exists Z_i \varphi)$ is equal to $\mathrm{Var}(\varphi)$ and $\mathrm{Free}(\exists Z_i \varphi) = \mathrm{Free}(\varphi) \setminus \{Z_i\}$.

We omit parentheses freely from formulas in order to improve readability. We also freely use notations such as $\vee$, $\rightarrow$, $\leftrightarrow$, and $\forall$ with their usual logical meanings, since each of these can be defined in terms $\neg$, $\wedge$, and $\exists$. A formula with no free variables is a *sentence*.

We cannot speak directly about the cardinality of sets in these languages. However, we often allow ourselves atomic formulas that say when the cardinality is congruent to a particular residue modulo some integer q. The set of formulas that are constructed without using any modular cardinality predicates form *monadic second-order logic* (MSO), and the set of all formulas (including those that use modular counting) is *counting monadic second-order logic* (CMSO).

As an example, the logic MSO^{Σ} we deployed in Section 1.1 has atomic formulas of the form $Z_i \subseteq Z_j$, $\mathsf{term}(Z_i)$, and $\mathsf{char}_{\alpha}(Z_i)$ where α is in the alphabet Σ. If φ is equal to $Z_i \subseteq Z_j$ then $\mathrm{Var}(\varphi) = \mathrm{Free}(\varphi) = \{Z_i, Z_j\}$. If $\varphi = \mathsf{char}_{\alpha}(Z_i)$ or $\mathsf{term}(Z_i)$, then $\mathrm{Var}(\varphi) = \mathrm{Free}(\varphi) = \{Z_i\}$.

2.4.1 Logic for graphs. The logic $\mathrm{CMSO}_1^{\mathrm{G}}$ has variables $v_1, v_2, v_3, \ldots$ representing vertices, and $V_1, V_2, V_3, \ldots$ representing subsets of vertices. The atomic formulas have the form: (i) $v_i = v_j$, (ii) $v_i \in V_j$, (iii) $\mathsf{adj}(v_i, v_j)$, and (iv) $|V_i| \equiv p \bmod q$, where p and q are integers satisfying $0 \leq p < q$.

In all of these cases, the set of variables in an atomic formula is exactly as one would expect and every variable is free in an atomic formula. Note that $\mathrm{MSO}_1^{\mathrm{G}}$ is the set of formulas constructed without using any formulas of type (iv). The interpretation of $\mathsf{adj}(v_i, v_j)$ is that vertices v_i and v_j are *adjacent* (joined by an edge).

The logic $\mathrm{CMSO}_2^{\mathrm{G}}$ has variables $v_1, v_2, v_3, \ldots$ representing vertices, $e_1, e_2, e_3, \ldots$ representing edges, and variables $V_1, V_2, V_3, \ldots$ and $E_1, E_2, E_3, \ldots$ which respectively represent subsets of vertices and edges. The atomic formulas have the form: (i) $v_i = v_j$ or $e_i = e_j$, (ii) $v_i \in V_j$ or $e_i \in E_j$, (iii) $\mathsf{inc}(v_i, e_j)$, and (iv) $|V_i| \equiv p \bmod q$ or $|E_j| \equiv p \bmod q$, where p and q are integers satisfying $0 \leq p < q$.

We interpret $\mathsf{inc}(v_i, e_j)$ to mean that e_j is incident with v_i. The language $\mathrm{CMSO}_2^{\mathrm{G}}$ is strictly more powerful than $\mathrm{CMSO}_1^{\mathrm{G}}$. For example, the former language can define the class of Hamiltonian graphs and the latter cannot [15, Proposition 5.19]. It is common to denote these languages by CMSO_1 and CMSO_2, or CMS_1 and CMS_2.

2.4.2 Logic for hypergraphs. The logics we have defined up to this point are well established in the literature. Now we move to languages that have not been explored so extensively. Let $\mathcal{P}$ be a finite set of unary predicates. The language $\mathrm{CMSO}(\mathcal{P})$ has $Z_1, Z_2, Z_3, \ldots$ as its variables, and the following atomic formulas: (i) $Z_i \subseteq Z_j$, (ii) $P(Z_i)$, for any predicate $P \in \mathcal{P}$, and (iii) $|Z_i| \equiv p \bmod q$, where p and q are integers satisfying $0 \leq p < q$.

In the list below, we describe some variants of monadic second-order logic for hypergraphs.

(i) $\mathrm{CMSO}_1^{\mathrm{hyp}} = \mathrm{CMSO}(\{\mathsf{hyp}\})$. We will use this as the general monadic second-order logic for hypergraphs.

(ii) $\mathrm{CMSO}_1^{\mathrm{ind}} = \mathrm{CMSO}(\{\mathsf{ind}\})$. We will use this logic in the specific case that the hyperedges are intended to be the independent sets of a matroids. This is the logic introduced by Hliněný [31, 32]. It has been denoted by various notations in the past, including MSOL, CMS_M, and CMS_0. Obviously, we could just as well use logics such as $\mathrm{CMSO}(\{\mathsf{circuit}\})$ or $\mathrm{CMSO}(\{\mathsf{basis}\})$ in the cases where the hyperedges are circuits or bases of matroids. Any of these logics will have equivalent expressive power.

(iii) $\mathrm{CMSO}_1^{\mathrm{star}} = \mathrm{CMSO}(\{\mathsf{star}\})$. In this logic, the variables will represent sets of edges, and the subsets satisfying star will be the *vertex stars* of a graph (the sets of edges that are incident with a vertex).

(iv) $\mathrm{CMSO}_1^{\mathrm{bias}} = \mathrm{CMSO}(\{\mathsf{star}, \mathsf{bal}\})$. This is the logic we use to discuss biased graphs. The sets satisfying bal will be the edge-sets of balanced cycles.

The subscript 1 is a reminder that these logics are 1-*sorted* logics, meaning that there is only one type of variable. This distinguishes the logic from the 2-*sorted* logic $\mathrm{CMSO}_2^{\mathrm{hyp}}$ that we discussed in the introduction.

Now that we have formally defined the syntax of monadic logic for hypergraphs, we will move to the semantics: that is, we define what it means for a formula to be satisfied by a discrete structure. Let $\mathcal{P}$ be a collection of unary predicates. We say that a *$\mathcal{P}$-structure* is a finite set E, along with a collection of subsets $\mathcal{I}_P \subseteq 2^E$ for each predicate $P \in \mathcal{P}$. Let φ be a formula in CMSO($\mathcal{P}$) and let $\theta\colon \mathrm{Free}(\varphi) \to 2^E$ be a function taking free variables to subsets of E. We refer to θ as an *interpretation*. We will define what it means for φ to be *satisfied* by $(E, \{\mathcal{I}_P\}_{P\in\mathcal{P}})$ under the interpretation θ. If φ is the atomic formula $Z_i \subseteq Z_j$, then it is satisfied if $\theta(Z_i)$ is a subset of $\theta(Z_j)$. If φ is $P(Z_i)$, for some P in $\mathcal{P}$, then φ is satisfied if $\theta(Z_i)$ is a member of $\mathcal{I}_P$. And finally, if φ is $|Z_i| \equiv p \bmod q$, then φ is satisfied if the cardinality of $\theta(Z_i)$ is congruent to p modulo q.

Now we define what satisfaction means when φ is not atomic. If φ is $\neg\psi$ for some formula ψ, then φ is satisfied exactly when ψ is *not* satisfied by $(E, \{\mathcal{I}_P\}_{P\in\mathcal{P}})$ under θ. Next assume that φ is a conjunction $\psi_1 \wedge \psi_2$. For $i = 1, 2$, let θ_i be the restriction of θ to $\mathrm{Free}(\psi_i)$. Then φ is satisfied exactly when ψ_i is satisfied by $(E, \{\mathcal{I}_P\}_{P\in\mathcal{P}})$ under θ_i, for both $i = 1$ and $i = 2$. Next, if φ is $\exists Z_i \psi$, then φ is satisfied if there is some subset $X \subseteq E$ such that ψ is satisfied by $(E, \{\mathcal{I}_P\}_{P\in\mathcal{P}})$ under the interpretation obtained from θ by adding the ordered pair (Z_i, X). This completes our definition of the semantics of CMSO($\mathcal{P}$).

We illustrate the power of CMSO($\mathcal{P}$) by showing how we can define the structures that most interest us. First we define some useful formulas. The formula $\mathsf{union}(Z_i, Z_j, Z_k)$ is $\forall Z_l\ (\mathsf{sing}(Z_l) \to (Z_l \subseteq Z_k \leftrightarrow (Z_l \subseteq Z_i \vee Z_l \subseteq Z_j)))$, which will be satisfied exactly when Z_k is interpreted as the union of Z_i and Z_j. If P is a predicate in $\mathcal{P}$, then $\mathsf{max}_P(Z_i)$ stands for $P(Z_i) \wedge \forall Z_j\ ((Z_i \subseteq Z_j \wedge P(Z_j)) \to Z_i = Z_j)$. Thus $\mathsf{max}_P(Z_i)$ is satisfied when Z_i is interpreted as a member of $\mathcal{I}_P$ that has no proper superset in $\mathcal{I}_P$.

A matroid (axiomatised via independent sets) is a pair $(E, \mathcal{I}_{\mathsf{ind}})$ satisfying the following $\mathrm{MSO}_1^{\mathrm{ind}}$-sentences.

(I1) $\exists Z_1\ \mathsf{ind}(Z_1)$

(I2) $\forall Z_1\ \forall Z_2\ ((Z_1 \subseteq Z_2 \wedge \mathsf{ind}(Z_2)) \to \mathsf{ind}(Z_1))$

(I3′) $\forall Z_1\ \forall Z_2\ (\mathsf{ind}(Z_1) \wedge \mathrm{max}_{\mathsf{ind}}(Z_2) \wedge \neg\, \mathrm{max}_{\mathsf{ind}}(Z_1)) \to$
$\exists Z_3\ ((\mathsf{sing}(Z_3) \wedge Z_3 \subseteq Z_2 \wedge Z_3 \not\subseteq Z_1) \wedge \forall Z_4\ (\mathsf{union}(Z_1, Z_3, Z_4) \to \mathsf{ind}(Z_4)))$

Because we are not able to say that one set is larger than another in $\mathrm{MSO}_1^{\mathrm{ind}}$, we have replaced **(I3)** from Section 2.2. The new statement **(I3′)** says that if I and J are independent sets, where J is maximal and I is not, then there is an element $e \in J \backslash I$ such that $I \cup \{e\}$ is independent. These axiom schemes are equivalent.

We can think of a graph as being a hypergraph $(E, \mathcal{I}_{\mathsf{star}})$ that satisfies the $\mathrm{MSO}_1^{\mathrm{star}}$-sentences

- $\forall Z_1\ (\mathsf{sing}(Z_1) \to \exists Z_2\ (\mathsf{star}(Z_2) \wedge Z_1 \subseteq Z_2))$
- $\forall Z_1\ \forall Z_2\ \forall Z_3\ \forall Z_4\ (\mathsf{sing}(Z_1) \wedge \mathsf{star}(Z_2) \wedge \mathsf{star}(Z_3) \wedge \mathsf{star}(Z_4) \wedge$
$Z_1 \subseteq Z_2 \wedge Z_1 \subseteq Z_3 \wedge Z_1 \subseteq Z_4) \to (Z_2 = Z_3 \vee Z_2 = Z_4 \vee Z_3 = Z_4)$

With these statements we are saying that every edge needs to be in either one or two vertex stars. Now we can think of the sets satisfying star as being vertices, and the elements of the ground set as being edges that joins those vertices, where incidence of an edge and a vertex is just element inclusion. It is interesting that this is the logical language for graphs that arises naturally in some matroid contexts, given that it de-emphasises vertices and emphasises edges. This recalls Tutte's comment: "If a theorem about graphs can be expressed in terms of edges and circuits only, it probably exemplifies a more general theorem about matroids" [58, pp. 497].

We can also use $\mathrm{MSO}_1^{\mathrm{bias}}$ to define biased graphs. From [21, Proposition 3.6] we have a $\mathrm{MSO}_1^{\mathrm{bias}}$-formula cycle with one free variable, which is satisfied when that variable is sent to the set of edges in a cycle. From this, we construct a formula theta with a free variable, which is satisfied when the variable is sent to the edges of a theta subgraph: a set X of edges is a theta subgraph precisely when it contains exactly three distinct cycles and any proper subset of X contains at most one cycle. To define a bias graph, we start with the $\mathrm{MSO}_1^{\mathrm{star}}$-formulas above, and we add a sentence saying that if Z_1, Z_2, and Z_3 are subsets satisfying cycle, where their union satisfies theta, then it is not the case that exactly two of Z_1, Z_2, and Z_3 satisfy bal.

Hliněný observed that for any matroid N, there is an $\mathrm{MSO}_1^{\mathrm{ind}}$-sentence that is satisfied exactly by the matroids that have a minor isomorphic to N [32, Lemma 5.1]. We sketch a proof of an analogous fact for biased graphs, using [21, Proposition 3.5], which gives us a formula satisfied by connected subsets of edges.

Proposition 2.3 *Let Ω be a biased graph. There is an $\mathrm{MSO}_1^{\mathrm{bias}}$-sentence that is satisfied exactly by the biased graphs that have a minor isomorphic to Ω.*

Sketch proof. We assume $\Omega = (G, \mathcal{B})$ has vertices $v_1, \ldots, v_n$ and edges $e_1, \ldots, e_m$. We build a sentence that says there exist connected subsets $F_1, \ldots, F_n$ of edges such that none of these sets contains a cycle, and no vertex star contains edges from more than one of these sets. This establishes the n trees that will correspond to the vertices of a Ω-minor. Next we assert that there exist singleton sets $E_1, \ldots, E_m$ of edges such that the edge in E_i is incident with a vertex in $G[F_j]$ if and only if e_i is incident with v_j. The final step involves a separate clause for each cycle C of Ω. There will exist a unique cycle that contains the edge-set $\cup_{e_i \in C} E_i$, and this cycle must be balanced exactly when C is balanced in Ω. □

From Hliněný's observation, we see that any minor-closed class of matroids with finitely many excluded minors can be defined by an $\mathrm{MSO}_1^{\mathrm{ind}}$-sentence. The class of *lattice-path matroids* provides an example of a natural minor-closed class that is $\mathrm{MSO}_1^{\mathrm{ind}}$-definable [34, Theorem 5.7], despite the fact that it has infinitely many excluded minors [8].

2.5 Tree automata

The BET Theorem implies there is a fast algorithm for testing membership in any language defined by a monadic sentence. This is because running a finite-state automaton on a string takes linear time. Courcelle's Theorem and its descendants rely on the same idea. However, graphs and hypergraphs do not typically have a linear ordering in the same way that a string does. So, it is more natural to use automata that process trees. In this section, we explain how hypergraphs can be described by tree-shaped data structures.

We describe a rooted binary tree where we distinguish between left and right children. Let T be a *tree*, meaning a connected graph without cycles. We distinguish a *root vertex*, denoted by root. Any vertex with degree one is a *leaf*. The set of leaves of T is denoted by $L(T)$. We let $I(T)$ be the set of non-leaf vertices, which we say are *internal*. We require that every non-root internal vertex has degree three and that the root has degree two. Every vertex $v \in I(T)$ has two neighbours that are not contained in the path from v to root. We say that these two neighbours are the *children* of v. We distinguish the *left* and *right* children of v using a bijection c_v from the set of children of v to the set $\{L, R\}$.

Let Q be a finite set of *states* and let $\mathsf{acc} \subseteq Q$ be the set of *accepting states*. Let s be a non-negative integer, which we will call the *arity*. To each internal vertex v, we associate a *transition relation* τ_v. If both the children of v are leaves, then τ_v is a subset of $\{✓, ✗\}^s \times \{✓, ✗\}^s \times Q$. If neither child is a leaf, then τ_v is a subset of Q^3. If the left child of v is a leaf and the right child is not, then τ_v is contained in $\{✓, ✗\}^s \times Q^2$, and in the final case, where the right child is a leaf but the left child is not, then τ_v is a subset of $Q \times \{✓, ✗\}^s \times Q$. (We use the convention that $\{✓, ✗\}^0 = \{\epsilon\}$.) We refer to the entire system

$$\mathbf{T} = (T, \mathsf{root}, \{c_v\}_{v \in I(T)}, s, Q, \mathsf{acc}, \{\tau_v\}_{v \in I(T)})$$

as a *tree automaton*.

Let S be a function from $L(T)$ to $\{✓, ✗\}^s$. A *run* of $\mathbf{T}$ on S is a function $r \colon I(T) \to Q$ such that for any vertex $v \in I(T)$ with left child v_L and right child v_R, the following conditions hold: (i) if v_L and v_R are leaves, then $(S(v_L), S(v_R), r(v))$ is in τ_v, (ii) if neither v_L nor v_R is a leaf, then $(r(v_L), r(v_R), r(v))$ is in τ_v, (iii) if v_L is a leaf but v_R is not, then $(S(v_L), r(v_R), r(v))$ is in τ_v, and (iv) if v_R is a leaf but v_L is not, then $(r(v_L), S(v_R), r(v))$ is in τ_v. A run r is *accepting* if $r(\mathsf{root})$ is in acc. If there exists an accepting run on S, then we say that $\mathbf{T}$ *accepts* S. We define $\mathbf{T}$ to be *deterministic* if for every choice of the function S, there is a unique run of $\mathbf{T}$ on S.

Any function $S \colon L(T) \to \{✓, ✗\}$ picks out a subset of $L(T)$, which we denote by $L(S)$. To be precise, $l \in L(T)$ belongs to $L(S)$ if and only if $S(l) = ✓$. More generally, if s is a positive integer and S is a function from $L(T)$ to $\{✓, ✗\}^s$, then for any integer $j \in \{1, \ldots, s\}$, we set $L_j(S) \subseteq L(T)$ to be the set such that l is in $L_j(S)$ if and only if the coordinate in position j of $S(l)$ is ✓. We identify $\{✓, ✗\}$ with $\{✓, ✗\}^1$, so $L(S)$ is the same as $L_1(S)$.

Let $\mathbf{T}$ be a tree automaton with arity 1. Then we can think of $\mathbf{T}$ as accepting some of the subsets of $L(T)$. We let $\mathcal{I}(\mathbf{T})$ be

$$\{L(S) \colon S \text{ is a function from } L(T) \text{ to } \{✓, ✗\},\ \mathbf{T} \text{ accepts } S\}.$$

A tree automaton $\mathbf{T} = (T, \mathsf{root}, \{c_v\}_{v \in I(T)}, 1, Q, \mathsf{acc}, \{\tau_v\}_{v \in I(T)})$ with arity 1 corresponds to the hypergraph $(L(T), \mathcal{I}(\mathbf{T}))$. This will be our primary method for describing a hypergraph for the purposes of computation. It is a fairly easy exercise to see that any hypergraph can be described as $(L(T), \mathcal{I}(\mathbf{T}))$ for some choice of automaton $\mathbf{T}$. The question is, how many states are required for a tree automaton that will describe the hypergraph? This becomes a measure of its structural complexity. In the next section, we will discuss the fact that a class of hypergraphs can be described by tree automata with a bounded number of states if and only if the class has bounded *decomposition-width*.

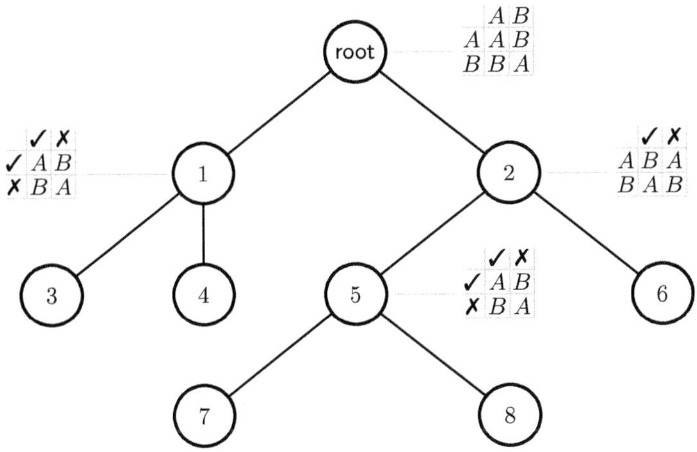

Figure 4: A tree automaton

Example 2.4 Figure 4 illustrates a tree automaton $\mathbf{T}$. In this case the arity s is 1, and Q is the set $\{A, B\}$. Let us define acc to be $\{A\}$. Each internal vertex v is labelled with a 2×2 table. The transition relation τ_v is exactly the set of triples (α, β, γ) where γ appears in the table in the row labelled by α and the column labelled by β. Note that $\mathbf{T}$ is deterministic. Let $S\colon L(T) \to \{✓, ✗\}$ be the function such that $L(S) = \{3, 6, 7\}$. Then $\mathbf{T}$ has a unique run r on S and r acts as follows: $r\colon 5 \mapsto B$ $\quad r\colon 2 \mapsto A$ $\quad r\colon 1 \mapsto B$ $\quad r\colon \mathsf{root} \mapsto B$. So $\mathbf{T}$ does not accept S. In fact, we can easily see that $\mathcal{I}(\mathbf{T})$ is exactly the collection of subsets of $L(T)$ that have even cardinality.

The next result is proved using the same boot-strapping approach as Lemma 1.7 (see [22, Lemma 4.7]). In the language of Section 6, it shows that definability implies recognisability for hypergraphs of bounded decomposition-width.

Lemma 2.5 *Let φ be any formula in* $\mathrm{CMSO}_1^{\mathrm{hyp}}$ *and let $Z_{i_1}, Z_{i_2}, \ldots, Z_{i_s}$ be the free variables of φ. Let $\mathbf{T} = (T, \mathsf{root}, \{c_v\}_{v \in I(T)}, 1, Q, \mathsf{acc}, \{\tau_v\}_{v \in I(T)})$ be a tree automaton and let M be the hypergraph $(L(T), \mathcal{I}(\mathbf{T}))$. There exists a tree automaton $\mathbf{T}_\varphi = (T, \mathsf{root}, \{c_v\}_{v \in I(T)}, s, Q_\varphi, \mathsf{acc}_\varphi, \{\rho_v\}_{v \in I(T)})$ such that for any function $S\colon L(T) \to \{✓, ✗\}^s$, we have that $\mathbf{T}_\varphi$ accepts S if and only if φ is satisfied by M under the interpretation that takes each Z_{i_j} to $L_j(S)$.*

2.6 Decomposition-width and branch-width

Numerical *width parameters* are fundamental in modern structural discrete mathematics. Algorithmic meta-theorems such as Courcelle's Theorem often require that the input class has bounded complexity, meaning that some width parameter is bounded for objects in the class. In the graph context, *tree-width* and *clique-width* are commonly used. For matroids, the most frequently used width parameter is *branch-width*, which we now define. Let T be a tree such that every vertex has degree one or three and let $L(T)$ be the set of leaves of T. Let $M = (E, \mathcal{I})$ be any hypergraph and let π be a bijection from E to $L(T)$ — we say that (T, π) is a *tree decomposition* of M. Assume that e is an edge joining vertices u and v in T. Then e partitions E into sets U_e and V_e, where U_e is the set of elements $x \in E$ such that the path from $\pi(x)$ to u does not contain v. Assuming that M is a matroid, for any subset $X \subseteq E$, the *width* of e is the numerical parameter $r_M(U_e)+r_M(V_e)-r(M)+1$. The branch-width of M is the smallest value k such that M has a tree decomposition where every edge has width at most k. The branch-width of a matroid M is bounded above by $r(M) + 1$.

Decomposition-width uses the same framework, but it applies to all hypergraphs and does not require a matroidal rank function. We change the definition of the width of an edge. Say that J is any subset of E. We define an equivalence relation on subsets of J. Say that $X, X' \subseteq J$ are equivalent if for any subset $Y \subseteq E \backslash J$, either both $X \cup Y$ and $X' \cup Y$ are in $\mathcal{I}$, or neither one is. The width of the edge e in T is the maximum number of equivalence classes when J is equal to either U_e or V_e. Now the decomposition-width of M is the smallest number k such that there is a tree decomposition of M where every edge has width at most k. The terminology of decomposition-width was first used by Král [38] and Strozecki [55]. The definition I have given here differs from theirs, but is equivalent in the sense that when $\mathcal{M}$ is a class of hypergraphs, the maximum decomposition-width of hypergraphs in $\mathcal{M}$ is finite under both definitions, or under neither. The same comment applies to the parameter *hyper-rankwidth*, introduced by Bojańczyk [6].

If a class of matroids has bounded decomposition-width, then it has bounded branch-width, but the converse does not hold [22, Corollary 2.8]. Theorem 1.4 in [22] shows that a class $\mathcal{M}$ of hypergraphs has bounded decomposition-width if and only if there is an integer k such that every $M \in \mathcal{M}$ can be described as $(L(T), \mathcal{I}(\mathbf{T}))$ for some arity-1 tree automaton $\mathbf{T}$ with at most k states. So this becomes our notion of a class of hypergraphs with bounded structural complexity — a class that can be described via tree automata with a bounded number of states.

For the purposes of applying tree automata to test membership of monadically-defined classes, we would like to be able to produce a tree-automata description of a matroid, given some description via a matrix or biased graph. Assuming that the input is restricted to a class with bounded branch-width, Theorem 6.5 in [22] provides us with an algorithm for doing so. The foundation of this algorithm is work by Oum and Seymour [46, Corollary 7.2] that produces tree decompositions for matroids, relative to branch-width. Their algorithm either produces a tree decomposition where every edge has width at most $3\lambda + 1$, or it gives a certificate that the matroid has branch-width greater than λ.

2.7 Transductions

Transductions are one of the most useful tools in the monadic theory of discrete structures. Multiple definitions of transductions are found in the literature [6, 7, 15]. All of them are technical, but express the same intuition. Rather than diving into another formal definition, I will attempt to explain this intuition with examples. The formal definition I have in mind is tailored to hypergraphs, and can be found in [34].

We think of a transduction as a relation which takes as input a class of structures relative to some language, and produces as output some other class (relative to a possibly different language). Such a relation allows us to use the language of the input class to make statements about objects in the output class. In our conception, we will have a formula dom (for domain), and every satisfying assignment of this formula will correspond to an output of the transduction. The elements of the ground set in the output hypergraph will be the subsets that satisfy a further formula, univ.

We give an example from matroid theory that takes each matroid to its dual. The input and output languages will be $\mathrm{CMSO}_1^{\mathrm{ind}}$. Let dom be a sentence which is satisfied by $M = (E, \mathcal{I})$ when $\mathcal{I}$ is the collection of independent sets of a matroid. Let univ be the formula sing, which is satisfied by singleton sets. Let coindep be a formula with one free variable that is satisfied when that variable is interpreted as a coindependent set. The trio dom, univ, coindep comprises the transduction. In this case, there will be exactly one output for every input that is a matroid. The ground set of the output will be $\{\{e\} \colon e \in E\}$, which we can identify with E. The hyperedges of the output will be the coindependent sets of M, which is to say that the output will be isomorphic to the dual matroid M^*.

For another example, we consider the transduction taking each matroid to its connected components. Let $M = (E, \mathcal{I})$ be a matroid. A *separator* of M is a set X such that $r_M(X)+r_M(E\backslash X) = r(M)$. A non-empty separator that does not properly contain a non-empty separator is a *connected component*. In this example we let dom be a formula with a free variable Z_i such that dom is satisfied when $M = (E, \mathcal{I})$ is a matroid and Z_i is interpreted as a connected component. The formula univ has a free variable, and will be satisfied when this variable is interpreted as a singleton subset of Z_i. So the ground set of the output will be $\{\{e\} \colon e \in Z_i\}$, which we can identify with Z_i. Let indep be the formula that is satisfied when its free variable is interpreted as an independent set. The transduction comprising dom, univ, indep takes each matroid to its connected components.

We can use a transduction to simplify the problem of finding a sentence that defines a particular property by using the *Backwards Translation Theorem*. Say that a hypergraph property is defined by a monadic sentence φ. The Backwards Translation Theorem says that for any transduction, there exists a sentence that will be satisfied by the hypergraph M exactly when the image of M under the transduction contains at least one hypergraph satisfying φ. (Alternatively, when every hypergraph in the image satisfies φ.) So for example, say that we wish to show that a matroid property is monadically-definable. Assume this property holds for a matroid whenever all of its connected components have the property. We have noted that there is a transduction taking each matroid to its connected components, so the Backwards Translation Theorem means that we need only show that the property is monadically-definable for connected matroids. The theorem has become an essential tool in the investigation of recognisability and definability, as we will discuss in Section 6. Versions of the backwards translation theorem can be found in [5, Proposition 20], [7, Lemma B.1], [15, Theorem 1.40], and [34].

3 Complexity

Tree automata provide the engine that powers Courcelle's Theorem and its descendants. These theorems give us *fixed-parameter tractable* algorithms. An algorithm of this type receives input with an associated numerical parameter, λ, and runs in time $O(f(\lambda)n^c)$, where $f(\lambda)$ is a value that depends only on λ. Here n is the size of the input and c is a positive constant. A fixed-parameter tractable algorithm is more desirable than one with running time $O(n^{f(\lambda)})$, because it will typically be useful for a wider range of parameter values. See [16] for an introduction to parameterized complexity.

Theorem 3.1 (Courcelle's Theorem) *Let φ be a $\mathrm{CMSO}_2^{\mathrm{G}}$-sentence. There is a fixed-parameter tractable algorithm for testing φ in graphs with respect to the parameter of tree-width.*

Hliněný introduced the matroid logic $\mathrm{CMSO}_1^{\mathrm{ind}}$ when proving the following result, an analogue of Courcelle's Theorem for finite-field-representable matroids [31, Corollary 5.2].

Theorem 3.2 *Let φ be a $\mathrm{CMSO}_1^{\mathrm{ind}}$-sentence. There is a fixed-parameter tractable algorithm for testing φ with respect to the parameter of branch-width, when the input consists of $\mathbb{F}$-representable matroids (and $\mathbb{F}$ is a finite field).*

Král [38], Strozecki [55], and Funk, Mayhew, and Newman [22] all present abstract versions of Hliněný's Theorem, meaning that they extend the result to matroid classes that are not necessarily representable. All of these theorems rely on statements similar to Lemma 2.5 to show that monadically definable properties can be recognised by tree automata. Results in [23] shows that Hliněný's Theorem applies to other classes that arise naturally in matroid theory: in particular, the class of Γ-gain-graphic matroids, when Γ is a finite group.

We can certify that the hypotheses of Theorem 3.2 are tight by relaxing them and showing that testing some monadic property becomes NP-hard. This is exactly what Hliněný accomplishes in [30, Theorem 5.6], where he shows that it is NP-complete to test whether a matroid contains a particular fixed minor when the input consists of matroids represented over the rational numbers (even when the input is restricted to matroids with branch-width three). Thus, we cannot relax the finiteness of $\mathbb{F}$ in Theorem 3.2. A technique from [41] essentially develops a transduction from the graph language $\mathrm{CMSO}_2^{\mathrm{G}}$ to the language $\mathrm{CMSO}_1^{\mathrm{ind}}$ applied to rank-three matroids. Because every such matroid has branch-width at most four, this transduction allows us to show that we cannot relax the condition of bounded branch-width in Theorem 3.2. It should be easy enough to show that similar results apply in the case of Γ-gain-graphic matroids (Conjecture 7.3).

4 Decidability

In this section, we talk about classes with *decidable theories*. We discussed this idea in Section 1.1. Say that $\mathcal{A}$ is a logical language which can make statements about a class of structure. This class has a decidable $\mathcal{A}$-theory if there is a theorem-testing machine: that is, a Turing machine which will take as input any sentence in the language and halt in finite time while deciding whether or not the sentence is satisfied by every structure in the class. The idea of a theorem-testing machine belongs squarely to Hilbert's philosophical program of the first part of the twentieth century. This program was devastated by the discoveries of Gödel, Tarski, Turing, Post, and Church in the 1930s. Because of those discoveries, we now know that classes with decidable theories should be regarded as exceptional: decidability is strong evidence that a class is tractable and well-behaved. The next theorem beautifully characterises the boundary between monadic decidability and undecidability for minor-closed classes of graphs [52].

Theorem 4.1 (Seese's Theorem) *Let $\mathcal{G}$ be a minor-closed class of graphs. The* $\mathrm{CMSO}_2^{\mathrm{G}}$*-theory of $\mathcal{G}$ is decidable if and only if $\mathcal{G}$ has bounded tree-width.*

Seese's Theorem is an extraordinary piece of work. I find the manner in which it is proved even more amazing than the statement of the theorem. The proof depends upon the structural results of Robertson and Seymour, in particular the Grid Theorem [49]. This theorem tells us that a minor-closed class of graphs is either tree-like (in the sense that it has bounded tree-width), or contains arbitrarily large square grids as minors. In the first case, the minor-closed family is recognised by a tree automaton, and we can prove the "if" direction by using an analogue of the pumping lemma for tree automata. Hliněný and Seese [33] attribute this direction to Courcelle, and separately to Arnborg, Lagergren, Seese. In the second case the class contains arbitrarily large grids, and these are exactly the structures we need to encode the evolution of a computation being carried out by a Turing machine. (Seese does this indirectly by instead encoding a tiling problem.) We think of each row of the grid as encoding the state of the machine tape at one step of its computation. This encoding of computation means that the monadic-theory of the class cannot possibly be decidable, or else we would be able to solve the Halting Problem in finite time. So the structures provided to us by Robertson and Seymour's theory are exactly the structures that correspond to the two strategies we need to prove decidability or undecidability. I cannot see any trivial reason for this correspondence to be so precise — it is one of the most striking mathematical facts I know.

It would be lovely to find an analogue of Seese's Theorem for general minor-closed classes of matroids (Question 7.6). I believe this will be extremely hard, maybe impossible. However, Hliněný and Seese [33, Theorem 6.2] have proved a version for $\mathbb{F}$-representable matroids when $\mathbb{F}$ is finite. This theorem is stated in terms of branch-width rather than tree-width. For the "only if" direction of their theorem, they rely on a matroidal version of the Grid Theorem due to Geelen, Gerards, Whittle [26].

The finiteness of the field $\mathbb{F}$ is necessary for the theorem by Hliněný and Seese. Assume that $\mathbb{F}$ is infinite. The antichain of rank-3 matroids presented in [47, Example 14.1.2] is contained in the class of $\mathbb{F}$-representable matroids. If we take any subset of this antichain and close it under minors, we obtain a class with bounded branch-width. There will therefore be uncountably many such minor-closed classes. Since there are only countably many Turing machines, it follows easily that uncountably many of these classes have undecidable theories.

We note that the analogue of Seese's Theorem for general minor-closed classes of matroids does not hold when we replace tree-width with either branch-width or decomposition-width. We just saw how to construct uncountably many minor-closed classes that have undecidable monadic theories, despite having bounded branch-width. These classes also have bounded decomposition-width. So neither bounded branch-width nor bounded decomposition-width is enough to guarantee a decidable monadic theory in general minor-closed classes of matroids. Nor is it true that a class with a decidable monadic theory must have bounded branch-width or decomposition-width. The counterexample here is the class of *uniform* matroids. These are the matroids where every circuit is spanning. It is easy to see from [22, Corollary 2.8] that the class of uniform matroids has unbounded decomposition-width, and it also has unbounded branch-width. Despite this, the class of uniform matroids has a decidable $\mathrm{CMSO}_1^{\mathrm{ind}}$-theory (personal communication from Nathan Bowler). So, currently, I do not know of a statement that would plausibly serve as an analogue of Seese's Theorem for general matroid classes, and I am somewhat sceptical that such an analogue exists.

The decidability of the $\mathrm{CMSO}_1^{\mathrm{ind}}$-theory for uniform matroids illustrates an important way in which monadic logic for hypergraphs differs from monadic logic with fixed-arity predicates. Seese has conjectured that in the latter type of logic, when a class of structures has a decidable monadic theory, there is a transduction from this class to the class of trees ([52], see also [33, Conjecture 4.12]). This conjecture remains open. There is nothing tree-like about the class of uniform matroids, despite the fact that the $\mathrm{CMSO}_1^{\mathrm{ind}}$-theory of uniform matroids is decidable. So there is no obvious analogue of Seese's Conjecture for matroids, and this is one reason that it will be difficult to characterise the minor-closed classes of matroids with decidable monadic theories.

Now, we will look at analogues for gain-graphic matroids of the work we have just discussed. Let Γ be a finite group, and let $\mathcal{M}$ be a minor-closed class of Γ-gain-graphic matroids. We would like to know that the $\mathrm{CMSO}_1^{\mathrm{ind}}$-theory of $\mathcal{M}$ is decidable if and only if $\mathcal{M}$ has bounded branch-width. I certainly believe this to be the case (Conjecture 7.4). However, it is not possible to assemble either direction of the statement from results currently in the literature. Proving the "if" direction could be accomplished by applying a tree-automaton version of the pumping lemma to classes of Γ-gain-graphic matroids with bounded branch-width. I am sure that a minor-closed class of frame matroids with unbounded branch-width contains arbitrarily large grids among its underlying graphs (but I not sure that this proof has been written down). I believe that any class with arbitrarily large grids will have an undecidable monadic theory (Conjecture 7.5). This feels quite provable, and should be enough to establish the "only if" direction of the gain-graphic analogue of the theorem by Hliněný and Seese.

We close this section by noting a result from [22, Theorem 7.4]. The proof relies on the tree automaton version of the pumping lemma.

Lemma 4.2 *Let $\mathcal{M}$ be a $\mathrm{CMSO}_1^{\mathrm{hyp}}$-definable class of hypergraphs. If $\mathcal{M}$ has bounded decomposition-width, then the $\mathrm{CMSO}_1^{\mathrm{hyp}}$-theory of $\mathcal{M}$ is decidable.*

5 Definability

The results in Sections 3 and 4 tell us that if a class of hypergraphs is definable in $\mathrm{CMSO}_1^{\mathrm{hyp}}$, we gain access to powerful tools from the theory of computation. So, we are well motivated to ask which classes are definable in this way.

We have already discussed Hliněný's observation that a minor-closed class of matroids is $\mathrm{MSO}_1^{\mathrm{ind}}$-definable if it has a finite number of excluded minors. The resolution of Rota's Conjecture (Conjecture 2.2) therefore means that the class of $\mathbb{F}$-representable matroids is $\mathrm{MSO}_1^{\mathrm{ind}}$-definable when $\mathbb{F}$ is a finite field. The other direction of the following theorem is due to Mayhew, Newman, and Whittle [42].

Theorem 5.1 *Let $\mathbb{F}$ be a field. The class of $\mathbb{F}$-representable matroids is MSO_1-definable if and only if $\mathbb{F}$ is finite.*

We will explain how we prove the "only if" direction of this theorem: that is, how we prove non-definability of a class of hypergraphs. The idea is similar to the characterisation of regular languages developed by Myhill [43] and Nerode [45]. Their characterisation is built on an equivalence relation that uses the concatenation operation on strings. For our analogue, we need a method to "concatenate" hypergraphs. The techniques in [42] apply to particular matroidal methods of concatenation. These techniques are generalised by Funk, Matthews, and Mayhew [20] to methods for "gluing" together two structures in such a way that the output is a hypergraph.

Definition 5.2 Let U, V, and C be finite sets such that $U \cap V = \emptyset$. Let c be a function from 2^U to C, and let d be a function from $2^V \times C$ to $\{\checkmark, \times\}$. We define $(U, c) \boxplus (V, d)$ to be the hypergraph

$$(U \cup V, \{X \cup Y \colon X \subseteq U,\ Y \subseteq V,\ d(Y, c(X)) = \checkmark\}).$$

Definition 5.3 Let $\mathcal{M}$ be a class of hypergraphs and let C be a finite set. When U_1 and U_2 are finite sets and c_i is a function from 2^{U_i} to C for each i, we write $(U_1, c_1) \sim_{\mathcal{M},C} (U_2, c_2)$ to mean that

$$(U_1, c_1) \boxplus (V, d) \in \mathcal{M} \leftrightarrow (U_2, c_2) \boxplus (V, d) \in \mathcal{M}$$

for every finite set V such that $(U_1 \cup U_2) \cap V = \emptyset$ and every function d from $2^V \times C$ to $\{\checkmark, \times\}$.

It is easy to see that the relation $\sim_{\mathcal{M},C}$ is an equivalence relation. The next result [20] provides a hypergraph analogue of the Myhill-Nerode lemma.

Lemma 5.4 *Let $\mathcal{M}$ be a $\mathrm{CMSO}_1^{\mathrm{hyp}}$-definable class of hypergraphs. Let C be a finite set. The equivalence relation $\sim_{\mathcal{M},C}$ has only finitely many equivalence classes.*

We show how to apply the lemma in the next result. An *exact cover* of the hypergraph $(E, \mathcal{I})$ is a subfamily $\mathcal{I}' \subseteq \mathcal{I}$ such that every element of E belongs to exactly one member of $\mathcal{I}'$.

Proposition 5.5 *There is no sentence in $\mathrm{CMSO}_1^{\mathrm{hyp}}$ that defines the class of hypergraphs with exact covers.*

Proof Let $\mathcal{M}$ be the class of hypergraphs with exact covers. For each positive integer s, let U_s and V_s be finite sets with cardinality s. We assume $U_s \cap V_t = \emptyset$ for each pair s and t. Let C be the set $\{1, 2\}$, and for each s, let $c_s : 2^{U_s} \to C$ be the function that takes every singleton subset of U_s to 1 and every other subset to 2. Let $d_s : 2^{V_s} \times C \to \{\checkmark, \text{✗}\}$ be the function that takes the pair (Y, i) to ✓ when $Y \subseteq V_s$ is a singleton set and $i = 1$. Every other pair in $2^{V_s} \times C$ is taken to ✗. We can see that $(U_s, c_s) \boxplus (V_t, c_t)$ is the complete bipartite graph $K_{s,t}$, where the two sides of the bipartition are U_s and V_t. Now, this hypergraph has an exact cover if and only if $s = t$ (because $K_{s,t}$ has a perfect matching if and only if $s = t$). So, $\sim_{\mathcal{M},C}$ has one equivalence class for each positive integer s. Lemma 5.4 implies the result. □

We mention some other techniques which have been used to prove non-definability results. Ehrenfeucht–Fraïssé (EF) games are often used to show that a property cannot be defined in first-order logic. We give a very informal description of these games. They feature two antagonists, Splitter and Duplicator. Splitter attempts to distinguish between two structures by making a sequence of choices. Each choice interprets a variable as an element of one of the structures. Duplicator attempts to show that for any choice made by Splitter, it is possible to interpret that variable in the other structure, in such a way that at the end of the game the two interpretations appear identical. If we can prove that Duplicator has a winning strategy, then we have proved a non-definability result.

It appears to be extremely difficult to use EF games to show that a property is not definable in monadic second-order logic, because in the monadic context, Splitter is able to interpret variables not only as elements of the structure, but also as subsets. This provides Splitter with a huge amount of additional power, and makes it much more difficult to prove that Duplicator can always defeat Splitter. Libkin gives a proof that the class of hypergraphs with even-cardinality ground sets is not $\mathrm{CMSO}_1^{\mathrm{hyp}}$-definable [39, Proposition 7.12], and it seems unlikely that we will be able to use EF games to prove non-definability results much more complicated than this.

Van Bevern et al. [1, Corollary 3.16] develop a Myhill-Nerode method for hypergraphs which is somewhat similar to Lemma 5.4. The important difference is that their "gluing" operations produce hypergraphs where only a bounded number of hyperedges can contain elements from both terms in the operation. This contrasts with the operation in Definition 5.2, which allows arbitrarily many such hyperedges. (For example, in the proof of Proposition 5.5, the gluing operation produces $K_{s,t}$, which has st edges crossing between the sets U_s and V_t.) Many natural matroid operations will have an unbounded number of independent sets that cross the join, which makes Lemma 5.4 more useful in the matroid context.

Kotek and Makowsky use the method of *connection matrices* to prove non-monadic-definability of graph parameters [37]. Using their language, the operation $\boxtimes$ that we use in Proposition 5.5 is not *smooth*, so we are not able to use their techniques for that result, although the use of connection matrices to prove non-definability for some variants of 2-sorted hypergraph logic remains an intriguing possibility.

Theorem 5.1 provides us with a clean dichotomy for the monadic definability of the class of $\mathbb{F}$-representable matroids: if $\mathbb{F}$ is finite the class is definable, and otherwise it is not. It is natural to ask the parallel question for classes of gain-graphic matroids. For what groups Γ is the class of Γ-gain-graphic matroids $\mathrm{MSO}_1^{\mathrm{ind}}$-definable? Daryl Funk and I are currently engaged in a lengthy project to investigate the different ways in which biased graphs can represent frame matroids. Using the language of this survey, this work will show that there is a $\mathrm{CMSO}_1^{\mathrm{ind}}$-sentence φ and a transduction from $\mathrm{CMSO}_1^{\mathrm{ind}}$ to $\mathrm{CMSO}_1^{\mathrm{bias}}$ which will take any frame matroid *not* satisfying φ to the biased graphs that represent it. The sentence φ characterises certain pathological families. Work in preparation by myself, Ben Shạhar, and Funk shows that if Γ is a finite group, then the Γ-gainable biased graphs are $\mathrm{CMSO}_1^{\mathrm{bias}}$-definable. Combining these pieces of work and applying the Backwards Translation Theorem shows that the class of Γ-gain-graphic matroids is $\mathrm{CMSO}_1^{\mathrm{ind}}$-definable when Γ is finite. So, one direction of the dichotomy continues to hold in the context of gain-graphic matroids.

The other direction is false! There are infinite groups Γ such that the class of Γ-gain-graphic matroids is $\mathrm{CMSO}_1^{\mathrm{ind}}$-definable. The same group of authors will show that if F is a finite group and $\Gamma = \otimes_{i\geq 1} F$ is an infinite direct product, then the class of Γ-gain-graphic matroids is $\mathrm{CMSO}_1^{\mathrm{ind}}$-definable. So, Conjecture 1.5 in [21] is false.

A group Γ is *uniformly locally-finite* if there is a function g_Γ such that any subgroup of Γ generated by at most k elements has size at most $g_\Gamma(k)$. Funk, Matthews, and Mayhew use Lemma 5.4 to prove that if Γ is an infinite group and Γ is *not* uniformly locally-finite, then the class of Γ-gain-graphic matroids is *not* $\mathrm{CMSO}_1^{\mathrm{ind}}$-definable. Despite this progress in both the positive and negative directions, it is currently not clear which infinite groups Γ have the property that the class of Γ-gain-graphic matroids is $\mathrm{CMSO}_1^{\mathrm{ind}}$-definable. Nonetheless, I remain hopeful that we will ultimately be able to resolve this problem (Question 7.7).

6 Recognisability vs. definability

When we say a class of structures is *recognisable*, we mean there is some computing machine that accepts structures in the class and rejects structures not in the class. A class is *definable* if it can be characterised by a sentence in some formal logical language. The connection between recognisability and definability is mysterious and has attracted much attention. The canonical result that relates these properties is the BET Theorem, which says that a language of strings is recognised by a finite-state automaton if and only if it can be defined by an MSO^{Σ}-sentence.

Proving that definability implies recognisability is often fairly easy: we must show that for each sentence, there is some machine that recognises the structures satisfying that sentence. This is what we do in Lemmas 1.7 and 2.5. The other direction is often hard. The issue with proving that recognisability implies definability is that there is no obvious way for an automaton to process structures such as graphs, hypergraphs, and matroids, and therefore it is not clear what it means for a class of graphs to be recognisable. On the other hand, a tree decomposition can be processed by an automaton, and in fact, it is not difficult to show that recognisability implies definability for classes of tree decompositions: the proof involves a sentence that asserts the existence of an accepting run, just as in the proof of the BET Theorem. Now transductions come to the rescue. Let $\mathcal{G}$ be a class of graphs with bounded tree-width. Bojańczyk and Pilipczuk [7] prove that there is a transduction from the monadic language of $\mathcal{G}$ to the monadic language of tree decompositions taking each graph in $\mathcal{G}$ to possible tree decompositions. The Backwards Translation Theorem, applied to this transduction, shows that if the tree decompositions of graphs in $\mathcal{G}$ are recognised by an automaton, then $\mathcal{G}$ is definable. This motivates us to define recognisability of $\mathcal{G}$ to be synonymous with recognisability of its tree decompositions. Under this definition, Bojańczyk and Pilipczuk [7] have proved that recognisability implies definability for minor-closed class of graphs with bounded tree-width, and have thereby settled Courcelle's Conjecture.

I want to conjecture that recognisability implies definability for classes of hypergraphs with bounded decomposition-width. I continue to identify bounded-decomposition-width classes of hypergraphs with the corresponding classes of tree automata. This means that, for me, recognisability of a class of hypergraphs can be defined in terms of a tree automaton that processes tree automata. The following definition is technical, but it attempts to capture the intuition that tree automata can themselves be processed by tree automata.

Let Q be a finite set of states, and let s be a non-negative integer. Define $\mathcal{R}(Q,s)$ to be the union of the power sets of $\{\checkmark,\text{✗}\}^s \times \{\checkmark,\text{✗}\}^s \times Q$, Q^3, $\{\checkmark,\text{✗}\}^s \times Q^2$, and $Q \times \{\checkmark,\text{✗}\}^s \times Q$. Thus, $\mathcal{R}(Q,s)$ is the set of all possible transition relations. Let Q and Q_1 be finite sets of states, and let $\mu\colon Q \to Q_1$ be an injective function. If $\tau_v \in \mathcal{R}(Q,s)$ is a transition relation, then $\mu(\tau_v) \in \mathcal{R}(Q_1,s)$ is the transition relation obtained by applying μ to every coordinate of every triple in τ_v (we assume that μ acts as the identity on $\{\checkmark,\text{✗}\}^s$). The idea here is that we can use the function μ to rename the states of an automaton.

Let Q_1 and Q_2 be finite sets such that $|Q_1| = k$, and let acc_2 be a subset of Q_2. The set Q_1 is a "canonical" set of k states, and Q_2 is the set of states of the automaton that processes tree automata. We also have a function $\xi\colon \mathcal{R}(Q_1,1) \to \mathcal{R}(Q_2,0)$. If $\mathbf{T} = (T, \mathsf{root}, \{c_v\}_{v\in I(T)}, 1, Q, \mathsf{acc}, \{\tau_v\}_{v\in I(T)})$ is a tree automaton with $|Q| \le k$, we say that $(Q_1, Q_2, \mathsf{acc}_2, \xi)$ *recognises* $\mathbf{T}$ if there is an injection $\mu\colon Q \to Q_1$ such that $(T, \mathsf{root}, \{c_v\}_{v\in I(T)}, 0, Q_2, \mathsf{acc}_2, \{\xi(\mu(\tau_v))\}_{v\in I(T)})$ is a tree automaton with an accepting run.

Example 6.1 Let $Q_1 = \{0,1\}$ and let $Q_2 = \{Y, N\}$, where $\mathsf{acc}_2 = \{Y\}$. We define ξ so that it takes $\{(\checkmark,\checkmark,0),(\checkmark,\text{✗},1),(\text{✗},\checkmark,1),(\text{✗},\text{✗},0)\}$ to the relation $\{(\epsilon,\epsilon,Y)\}$, and every other subset of $\{\checkmark,\text{✗}\} \times \{\checkmark,\text{✗}\} \times Q_1$ to $\{(\epsilon,\epsilon,N)\}$. (Remember that $\{\checkmark,\text{✗}\}^0 = \{\epsilon\}$.) Similarly, we set ξ to take $\{(1,1,0),(1,0,1),(0,1,1),(0,0,0)\}$ to the relation $\{(Y,Y,Y),(Y,N,N),(N,Y,N),(N,N,N)\}$ and every other subset of Q_1^3 to $\{(Y,Y,N),(Y,N,N),(N,Y,N),(N,N,N)\}$. We will leave it as an exercise to complete the definition of $\xi\colon \mathcal{R}(Q_1,1) \to \mathcal{R}(Q_2,0)$ in such a way that $(Q_1,Q_2,\mathsf{acc}_2,\xi)$ recognises the arity-1 automaton $\mathbf{T}$ if and only if it is deterministic, has at most two states, and accepts exactly the leaf subsets that have even cardinality.

Note that the class of hypergraphs where the hyperedges are exactly the subsets of even cardinality is $\mathrm{CMSO}_1^{\mathrm{hyp}}$-definable, using the modular arithmetic predicates. So, in this example, we have a recognisable class that is also definable. The following statement conjectures that recognisability implies definability for general classes of hypergraphs with bounded decomposition-width. Work by Campbell et al. provides progress on the conjecture for finite-field-representable matroids [12].

Conjecture 6.2 *Let $\mathcal{M}$ be a class of hypergraphs with decomposition-width at most k. Assume there is a 4-tuple $(Q_1, Q_2, \mathsf{acc}_2, \xi)$, where Q_1 and Q_2 are finite sets with $|Q_1| = k$, acc_2 is a subset of Q_2, ξ is a function from $\mathcal{R}(Q_1, 1)$ to $\mathcal{R}(Q_2, 0)$, and the following holds for every hypergraph M with decomposition-width at most k:*

(i) *if M is in $\mathcal{M}$, there is a tree automaton $\mathbf{T}$ with at most k states such that $M = (L(\mathbf{T}, \mathcal{I}(\mathbf{T}))$ and $(Q_1, Q_2, \mathsf{acc}_2, \xi)$ recognises $\mathbf{T}$,*

(ii) *if M is not in $\mathcal{M}$ and $\mathbf{T}$ is a tree automaton with at most k states such that $M = (L(\mathbf{T}, \mathcal{I}(\mathbf{T}))$, then $(Q_1, Q_2, \mathsf{acc}_2, \xi)$ does not recognise $\mathbf{T}$.*

Then $\mathcal{M}$ is $\mathrm{CMSO}_1^{\mathrm{hyp}}$*-definable.*

7 Open problems

In this section, I summarise conjectures and questions arising from the survey.

7.1 Complexity

The NP-complete EXACT COVER problem has as its input a hypergraph $(E, \mathcal{I})$. It asks if there is a subcollection $\mathcal{I}' \subseteq \mathcal{I}$ such that each element of E is contained in exactly one member of $\mathcal{I}'$. I conjecture that this problem remains NP-complete even for hypergraphs of bounded decomposition-width. Recall that, for the purposes of describing a hypergraph as input to an algorithm, we think of a tree automaton $\mathbf{T}$ as being a description of $(L(\mathbf{T}), \mathcal{I}(\mathbf{T}))$.

Conjecture 7.1 *There exists an integer k, such that* EXACT COVER *is* NP*-complete, even when the input is restricted to hypergraphs described by tree automata with at most k states.*

Hypergraphs with exact covers are $\mathrm{CMSO}_2^{\mathrm{hyp}}$-definable, so Conjecture 7.1 would give us reason to believe that we cannot expect an analogue of Courcelle's Conjecture for the $\mathrm{CMSO}_2^{\mathrm{hyp}}$ language.

The DECOMPOSITION WIDTH problem has as its input an integer k and a hypergraph $M = (E, \mathcal{I})$, described via a list of the members of $\mathcal{I}$. The question is whether M is equal to $(L(\mathbf{T}), \mathcal{I}(\mathbf{T}))$ for some tree automaton $\mathbf{T}$ with at most k states. It is not even clear that the problem is in NP. Given a tree automaton $\mathbf{T}$, we could verify in polynomial-time that $\mathcal{I} \subseteq \mathcal{I}(\mathbf{T})$, but there is no obvious way to certify the reverse containment.

Conjecture 7.2 DECOMPOSITION-WIDTH *is* NP*-hard.*

The next conjecture should be quite simple to resolve.

Conjecture 7.3 *It is* NP*-hard to test* $\mathrm{CMSO}_1^{\mathrm{ind}}$*-definable properties for Γ-gain-graphic matroids when Γ is infinite, even when the input is restricted to a class with bounded branch-width. It is* NP*-hard to test* $\mathrm{CMSO}_1^{\mathrm{ind}}$*-definable properties for Γ-gain-graphic matroids when the input has unbounded branch-width, even when Γ is finite.*

7.2 Decidability

Conjecture 7.4 *Let Γ be a finite group, and let $\mathcal{M}$ be a minor-closed class of Γ-gain-graphic matroids. The* $\mathrm{CMSO}_1^{\mathrm{ind}}$*-theory of $\mathcal{M}$ is decidable if and only if $\mathcal{M}$ has bounded branch-width.*

Let Q_n stand for the square grid graph with n^2 vertices. The next conjecture is probably not too difficult to prove, and would help to prove the "only if" direction of Conjecture 7.4.

Conjecture 7.5 *Let $\mathcal{M}$ be a minor-closed class of frame matroids. Assume that for every positive integer n, there is a linear class of cycles $\mathcal{B}_n$ in Q_n such that $F(Q_n, \mathcal{B}_n)$ is isomorphic to a matroid in $\mathcal{M}$. The* $\mathrm{CMSO}_1^{\mathrm{ind}}$*-theory of $\mathcal{M}$ is undecidable.*

The following open question concerns a general analogue of Seese's Theorem for matroids. I am not optimistic that it will have a clean answer.

Question 7.6 *Which minor-closed classes of matroids have decidable* $\mathrm{CMSO}_1^{\mathrm{ind}}$*-theories?*

7.3 Definability

I would dearly like an answer to the following question.

Question 7.7 *What are the infinite groups Γ such that the class of Γ-gain-graphic matroids is* $\mathrm{CMSO}_1^{\mathrm{ind}}$*-definable?*

This is closely related to the question: what are the infinite groups Γ such that the class of Γ-gainable biased graphs is $\mathrm{CMSO}_1^{\mathrm{bias}}$-definable?

The class of representable matroids is not $\mathrm{CMSO}_1^{\mathrm{ind}}$-definable [20]. By analogy, we make the following conjecture.

Conjecture 7.8 *The class of gainable biased graphs is not* $\mathrm{CMSO}_1^{\mathrm{bias}}$*-definable.*

The next conjecture lists a handful of well-studied classes of matroids. I do not believe any of them can be defined in $\mathrm{CMSO}_1^{\mathrm{ind}}$, but, at the same time, I cannot see how to prove this fact using Lemma 5.4.

Conjecture 7.9 *The following classes of matroids are not* $\mathrm{CMSO}_1^{\mathrm{ind}}$*-definable:* (i) *orientable matroids,* (ii) *transversal matroids,* (iii) *matroids that are transversal and cotransversal,* (iv) *gammoids,* (v) *positroids.*

7.4 Excluded minors and well-quasi-ordering

The work by myself and Funk to build a transduction from frame matroids to their biased graph representations should allow us to attack the following variant of Rota's Conjecture.

Conjecture 7.10 *Let Γ be a finite group. There are only finitely many excluded minors for the class of Γ-gain-graphic matroids.*

The hypothesis of finiteness for Γ is necessary for Conjecture 7.10, because if Γ is infinite, then there are infinitely many excluded minors [19, Theorem 2.19].

The class of Γ-gainable biased graphs is minor-closed, so we can ask about its excluded minors in the minor order of biased graphs. Zaslavsky found that there is a single excluded minor when Γ is the group $\mathbb{Z}_2$ [62, Theorem 6], and Brettell, Campbell, Funk, and Mayhew have found the three excluded minors for the case when Γ is $\mathbb{Z}_3$ [10]. Funk shows there are infinitely many excluded minors when Γ is infinite [19, Theorem 2.12]. We make a different analogue of Rota's Conjecture.

Conjecture 7.11 *Let Γ be a finite group. There are only finitely many excluded minors for the class of Γ-gainable biased graphs.*

Despite their similarity, Conjectures 7.10 and 7.11 are probably independent, because a Γ-gain-graphic matroid may be represented by multiple different Γ-gainable biased graphs. For the same reason, the next two conjectures will probably be independent. The first appears as Conjecture 1.4 in [21]. Both conjectures fail if Γ is infinite (Theorem 2.4 and Corollary 2.31 in [19]).

Conjecture 7.12 *Let Γ be a finite group. The class of Γ-gain-graphic matroids is well-quasi-ordered under minors.*

Conjecture 7.13 *Let Γ be a finite group. The class of Γ-gainable biased graphs is well-quasi-ordered under minors.*

Acknowledgements

My thinking on the topics in this survey have been influenced by conversations with many people, including Sapir Ben Shahar, Nathan Bowler, Nick Brettell, Rutger Campbell, Rod Downey, Daryl Funk, Noam Greenberg, Eun Jung Kim, Noleen Koehler, Dugald Macpherson, Janos Makowsky, Angus Matthews, Mike Newman, and Geoff Whittle. Rutger Campbell also provided detailed comments on an earlier draft. I thank them all, as well as an anonymous reviewer for their thoughtful suggestions.

References

[1] René van Bevern, Rodney G. Downey, Michael R. Fellows, Serge Gaspers, and Frances A. Rosamond, *Myhill-Nerode methods for hypergraphs*, Algorithmica **73** (2015), no. 4, 696–729.

[2] Garrett Birkhoff, *Abstract Linear Dependence and Lattices*, Amer. J. Math. **57** (1935), no. 4, 800–804.

[3] ———, *Combinatorial relations in projective geometries*, Ann. of Math. (2) **36** (1935), no. 3, 743–748.

[4] Robert E. Bixby, *On Reid's characterization of the ternary matroids*, J. Combin. Theory Ser. B **26** (1979), no. 2, 174–204.

[5] A. Blumensath and B. Courcelle, *Recognizability, hypergraph operations, and logical types*, Inform. and Comput. **204** (2006), no. 6, 853–919.

[6] Mikołaj Bojańczyk, *The category of MSO transductions*, 2023, https://arxiv.org/abs/2305.18039.

[7] Mikołaj Bojańczyk and Michał Pilipczuk, *Definability equals recognizability for graphs of bounded treewidth*, Proceedings of the 31st Annual ACM-IEEE Symposium on Logic in Computer Science (LICS 2016), ACM, New York, 2016, p. 10.

[8] Joseph E. Bonin, *Lattice path matroids: the excluded minors*, J. Combin. Theory Ser. B **100** (2010), no. 6, 585–599.

[9] Nick Brettell, *The excluded minors for* GF(5)*-representable matroids on ten elements*, Adv. in Appl. Math. **166** (2025), 102864.

[10] Nick Brettell, Rutger Campbell, Daryl Funk, and Dillon Mayhew, *The excluded minors for* $\mathbb{Z}_3$*-gainable and regular biased graphs*, in preparation.

[11] J. Richard Büchi, *Weak second-order arithmetic and finite automata*, Z. Math. Logik Grundlagen Math. **6** (1960), 66–92.

[12] Rutger Campbell, Bruno Guillon, Mamadou Moustapha Kanté, Eun Jung Kim, and Sang-il Oum, *Recognisability equals definability for finitely representable matroids of bounded path-width*, 2025 40th Annual ACM/IEEE Symposium on Logic in Computer Science (LICS), 2025, pp. 678–690.

[13] Bruno Courcelle, *The monadic second-order logic of graphs. I. Recognizable sets of finite graphs*, Inform. and Comput. **85** (1990), no. 1, 12–75.

[14] ———, *The monadic second-order logic of graphs. V. On closing the gap between definability and recognizability*, Symposium on Mathematical Foundations of Computer Science '89 (Porabka-Kozubnik, 1989), vol. 80, 1991, pp. 153–202.

[15] Bruno Courcelle and Joost Engelfriet, *Graph structure and monadic second-order logic*, Encyclopedia of Mathematics and its Applications, vol. 138, Cambridge University Press, Cambridge, 2012, A language-theoretic approach, With a foreword by Maurice Nivat.

[16] R. G. Downey and M. R. Fellows, *Parameterized complexity*, Monographs in Computer Science, Springer-Verlag, New York, 1999.

[17] Heinz-Dieter Ebbinghaus and Jörg Flum, *Finite model theory*, Perspectives in Mathematical Logic, Springer-Verlag, Berlin, 1995.

[18] Calvin C. Elgot, *Decision problems of finite automata design and related arithmetics*, Trans. Amer. Math. Soc. **98** (1961), 21–51.

[19] Daryl Funk, *On excluded minors and biased graph representations of frame matroids*, Ph.D. thesis, Simon Fraser University, 2015.

[20] Daryl Funk, Angus Matthews, and Dillon Mayhew, *Monadic non-definability and gain-graphic matroids*, 2025, https://arxiv.org/abs/2510.00139.

[21] Daryl Funk, Dillon Mayhew, and Mike Newman, *Defining bicircular matroids in monadic logic*, Q. J. Math. **73** (2022), no. 1, 65–92.

[22] ———, *Tree automata and pigeonhole classes of matroids: I*, Algorithmica **84** (2022), no. 7, 1795–1834.

[23] ———, *Tree automata and pigeonhole classes of matroids: II*, Electron. J. Combin. **30** (2023), no. 3, Paper No. 3.6, 34.

[24] J. F. Geelen, A. M. H. Gerards, and A. Kapoor, *The excluded minors for* GF(4)-*representable matroids*, J. Combin. Theory Ser. B **79** (2000), no. 2, 247–299.

[25] Jim Geelen, Bert Gerards, and Geoff Whittle, *Towards a structure theory for matrices and matroids*, International Congress of Mathematicians. Vol. III, Eur. Math. Soc., Zürich, 2006, pp. 827–842.

[26] ———, *Excluding a planar graph from* GF(q)*-representable matroids*, J. Combin. Theory Ser. B **97** (2007), no. 6, 971–998.

[27] ———, *Towards a matroid-minor structure theory*, Combinatorics, complexity, and chance, Oxford Lecture Ser. Math. Appl., vol. 34, Oxford Univ. Press, Oxford, 2007, pp. 72–82.

[28] ———, *Structure in minor-closed classes of matroids*, Surveys in combinatorics 2013, London Math. Soc. Lecture Note Ser., vol. 409, Cambridge Univ. Press, Cambridge, 2013, pp. 327–362.

[29] ———, *Solving Rota's conjecture*, Notices Amer. Math. Soc. **61** (2014), no. 7, 736–743.

[30] Petr Hliněný, *On matroid representability and minor problems*, Mathematical Foundations of Computer Science 2006 (Berlin, Heidelberg) (Rastislav Královič and Paweł Urzyczyn, eds.), Springer Berlin Heidelberg, 2006, pp. 505–516.

[31] Petr Hliněný, *Branch-width, parse trees, and monadic second-order logic for matroids (extended abstract)*, STACS 2003, Lecture Notes in Comput. Sci., vol. 2607, Springer, Berlin, 2003, pp. 319–330.

[32] ———, *On matroid properties definable in the MSO logic*, Mathematical foundations of computer science 2003, Lecture Notes in Comput. Sci., vol. 2747, Springer, Berlin, 2003, pp. 470–479.

[33] Petr Hliněný and Detlef Seese, *Trees, grids, and MSO decidability: from graphs to matroids*, Theoret. Comput. Sci. **351** (2006), no. 3, 372–393.

[34] Susan Jowett, Dillon Mayhew, Songbao Mo, and Christopher Tuffley, *Monadic transductions and definable classes of matroids*, 2024, https://arxiv.org/abs/2401.12969.

[35] J. Kahn and J. P. S. Kung, *Varieties of combinatorial geometries*, Trans. Amer. Math. Soc. **271** (1982), no. 2, 485–499.

[36] Richard M. Karp, *Reducibility among combinatorial problems*, Complexity of computer computations (Proc. Sympos., IBM Thomas J. Watson Res. Center, Yorktown Heights, N.Y., 1972), The IBM Research Symposia Series, Plenum, New York-London, 1972, pp. 85–103.

[37] Tomer Kotek and Johann A. Makowsky, *Connection matrices and the definability of graph parameters*, Log. Methods Comput. Sci. **10** (2014), no. 4, 4:1, 33.

[38] Daniel Král, *Decomposition width of matroids*, Discrete Appl. Math. **160** (2012), no. 6, 913–923.

[39] Leonid Libkin, *Elements of finite model theory*, Texts in Theoretical Computer Science. An EATCS Series, Springer-Verlag, Berlin, 2004.

[40] Saunders Mac Lane, *Some Interpretations of Abstract Linear Dependence in Terms of Projective Geometry*, Amer. J. Math. **58** (1936), no. 1, 236–240.

[41] Dillon Mayhew, *Matroid complexity and nonsuccinct descriptions*, SIAM J. Discrete Math. **22** (2008), no. 2, 455–466.

[42] Dillon Mayhew, Mike Newman, and Geoff Whittle, *Yes, the 'missing axiom' of matroid theory is lost forever*, Trans. Amer. Math. Soc. **370** (2018), no. 8, 5907–5929.

[43] John Myhill, *Finite automata and the representation of events*, WADD Technical Report **57** (1957), 112–137.

[44] Takeo Nakasawa, *Zur Axiomatik der linearen Abhängigkeit. I [Sci. Rep. Tokyo Bunrika Daigaku Sect. A* **2** *(1935), no. 43, 129–149; Zbl 0012.22001]*, A lost mathematician, Takeo Nakasawa, Birkhäuser, Basel, 2009, pp. 68–88.

[45] A. Nerode, *Linear automaton transformations*, Proc. Amer. Math. Soc. **9** (1958), 541–544.

[46] Sang-il Oum and Paul Seymour, *Approximating clique-width and branch-width*, J. Combin. Theory Ser. B **96** (2006), no. 4, 514–528.

[47] James Oxley, *Matroid theory*, second ed., Oxford Graduate Texts in Mathematics, vol. 21, Oxford University Press, Oxford, 2011.

[48] M. O. Rabin and D. Scott, *Finite automata and their decision problems*, IBM J. Res. Develop. **3** (1959), 114–125.

[49] Neil Robertson and P. D. Seymour, *Graph minors. V. Excluding a planar graph*, J. Combin. Theory Ser. B **41** (1986), no. 1, 92–114.

[50] ———, *Graph minors. XX. Wagner's conjecture*, J. Combin. Theory Ser. B **92** (2004), no. 2, 325–357.

[51] Gian-Carlo Rota, *Combinatorial theory, old and new*, Actes du Congrès International des Mathématiciens (Nice, 1970), Tome 3, Gauthier-Villars Éditeur, Paris, 1971, pp. 229–233.

[52] D. Seese, *The structure of the models of decidable monadic theories of graphs*, Ann. Pure Appl. Logic **53** (1991), no. 2, 169–195.

[53] P. D. Seymour, *Matroid representation over* GF(3), J. Combin. Theory Ser. B **26** (1979), no. 2, 159–173.

[54] Ernst Steinitz, *Algebraische Theorie der Körper*, J. Reine Angew. Math. **137** (1910), 167–309.

[55] Yann Strozecki, *Monadic second-order model-checking on decomposable matroids*, Discrete Appl. Math. **159** (2011), no. 10, 1022–1039.

[56] B. A. Trahtenbrot, *Finite automata and the logic of one-place predicates*, Sibirsk. Mat. Ž. **3** (1962), 103–131.

[57] W. T. Tutte, *A homotopy theorem for matroids. I, II*, Trans. Amer. Math. Soc. **88** (1958), 144–174.

[58] William Thomas Tutte, *Selected papers of W. T. Tutte. Vol. II*, Charles Babbage Research Centre, Winnipeg, MB, 1979.

[59] P. Vámos, *The missing axiom of matroid theory is lost forever*, J. London Math. Soc. (2) **18** (1978), no. 3, 403–408.

[60] Bartel Leendert van der Waerden, *Moderne Algebra*, J. Springer, Berlin, 1940.

[61] Hassler Whitney, *On the Abstract Properties of Linear Dependence*, Amer. J. Math. **57** (1935), no. 3, 509–533.

[62] Thomas Zaslavsky, *Characterizations of signed graphs*, J. Graph Theory **5** (1981), no. 4, 401–406.

[63] ———, *Biased graphs. I. Bias, balance, and gains*, J. Combin. Theory Ser. B **47** (1989), no. 1, 32–52.

[64] ———, *Biased graphs. II. The three matroids*, J. Combin. Theory Ser. B **51** (1991), no. 1, 46–72.

School of Computer Science
University of Leeds
Woodhouse, Leeds LS2 9JT
United Kingdom
d.mayhew@leeds.ac.uk

Domino Tilings Beyond 2D

Caroline J. Klivans and Nicolau Saldanha

Abstract

There is a rich history of domino tilings in two dimensions. Through a variety of techniques, we can answer questions such as: how many tilings are there of a given region, or what does the space of all tilings look like? These questions and their answers become significantly more difficult in dimension three and above. Despite this curse of dimensionality, there have been exciting recent advances in the theory. Here we briefly review foundational results of two-dimensional domino tilings and highlight where challenges arise when generalizing to higher dimensions. We then survey results in higher dimensional domino tilings, focusing on the question of connectivity of the space of tilings, and the open problems that still remain.

1 Introduction

There is a rich literature on two dimensional domino tilings. The theory has been studied from a variety of perspectives, including combinatorics, probability, and statistical physics. In two dimensions, we consider quadriculated regions in the plane: a *quadriculated* region is a union of unit squares, see Section 2. We consider tilings or coverings of such regions, where our tiles are 2×1 rectangles, consisting therefore of two unit squares glued together along an edge. Equivalently, we seek a perfect matching in planar subgraphs of the $\mathbf{Z}^2$ lattice. Fundamental questions in the area include: Which regions can be tiled? Given a tileable region, how many tilings are there? Is there a natural way to move between one tiling and another? These and other questions have been investigated and answered in many ways using various interesting techniques.

In three or more dimensions, we consider cubiculated regions of $\mathbf{R}^d$: a *cubiculated* region is a union of unit cubes, see Section 3. Our tiles are *dominoes*, consisting of two unit cubes glued together along a common face. Equivalently, we seek a perfect matching in subgraphs of $\mathbf{Z}^d$. The seemingly subtle move to higher dimensions or equivalently higher degree lattices proves to add great complexity. While most of the questions from two dimensions translate naturally, many of the techniques used to answer them are no longer sufficient. In higher dimensions, we have very few answers to any of the questions above. Nonetheless, much interesting progress has been made recently, especially in the development of new tools and techniques. Here we survey the recent advancements in higher dimensional tilings, providing many partial answers and stating the open problems that remain.

We start by briefly reviewing two dimensions and then introduce three dimensional domino tiling. In three dimensions, it is already more challenging to simply have visual representations of tilings. We provide examples and include data for small regions in order to appreciate the size and complexity of tilings that appears already in these small cases. The article highlights the question of connectedness of the space of tilings; i.e., given a region, can we move from one tiling to another via a sequence of local moves? A topological statistic called the *twist* proves to be quite relevant for this question and others. We see the twist throughout the survey and again include computations in order to demonstrate the difficulty of the higher dimensional theory. We will also touch on the topics of random tilings, enumeration, and moving beyond three dimensions.

The reader will notice that we give a lot of space to our own work. This bias is mostly caused by our own interests and expertise. There is some very interesting work about which we did not write in great detail, but we hope that the material here might lead an interested reader to investigate more into the area of higher dimensional tiling.

2 Domino tilings in two dimensions

A *quadriculated region* $\mathcal{R}$ is a finite union of unit squares with disjoint interiors and fitting together as the squares $[i, i+1] \times [j, j+1]$ in the plane. A *domino* is the union along an edge of two adjacent unit squares of $\mathcal{R}$. A *domino tiling* is a set of dominoes with disjoint interiors which covers the region $\mathcal{R}$.

Equivalently, we may construct a dual graph $\mathcal{R}^*$. Vertices of $\mathcal{R}^*$ are centers of unit squares of $\mathcal{R}$, and edges of $\mathcal{R}^*$ join centers of unit squares of $\mathcal{R}$ which are adjacent along an edge of $\mathcal{R}$. Thus, $\mathcal{R}^*$ is a bipartite graph with several nice properties. A domino tiling then corresponds to a perfect matching of $\mathcal{R}^*$; this is also called a *dimer* cover.

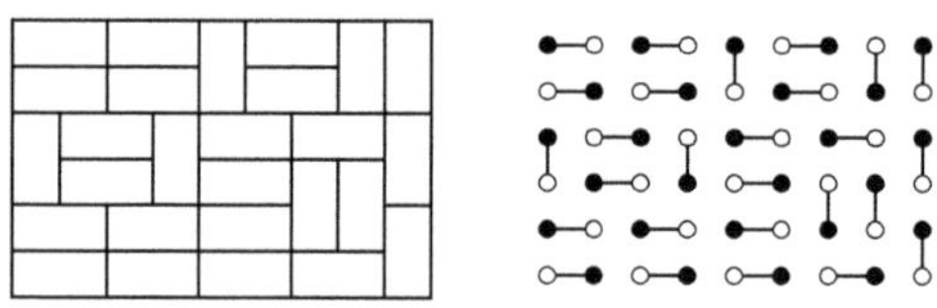

Figure 1: A two dimensional tiling and the corresponding dimer cover of the dual graph.

The theory of domino tilings in two dimensions spans many disciplines and draws on various techniques including from combinatorics, probability, and statistical physics. We provide here only a very quick overview of this vast subject and refer the reader to the surveys [17, 28].

A natural first question is enumerative.

Question 2.1 *Consider a quadriculated region $\mathcal{R}$. How many domino tilings of $\mathcal{R}$ are there?*

This question was addressed in Kasteleyn [16] and Temperley and Fisher [29]. Their methods rely on linear algebra: Pfaffians, determinants, and the transfer matrix-method. More explicitly, given a region, one constructs its adjacency matrix A (with entries in $\{0, 1\}$). There is a correspondence between tilings of $\mathcal{R}$ and monomials in the expansion of $\det A$, but there is also unwanted cancelling. By changing signs of entries in a clever way, a matrix K is obtained (with entries in $\{0, \pm 1\}$), for which $\det K$ gives the number of tilings of $\mathcal{R}$.

For some special regions, the methods yield useful formulas. For instance, the number $N(m, n)$ of domino tilings of the $m \times n$ rectangle is given by the somewhat surprising expression:

$$N(m,n) = \prod_{j=1}^{\lceil \frac{m}{2} \rceil} \prod_{k=1}^{\lceil \frac{n}{2} \rceil} \left(4\cos^2 \frac{\pi j}{(m+1)} + 4\cos^2 \frac{\pi k}{n+1} \right).$$

This formula can be used to obtain both estimates of $N(m,n)$ and arithmetical properties of this number. Other regions would be studied later, such as the Aztec diamond, shown in Figure 2.

Question 2.2 *What does a random tiling look like?*

This is a deep question, and it has received a lot of attention. We only mention a few notable contributions.

The *Aztec diamond* is the region shown in Figure 2, see [24]. The Arctic circle theorem by Jockush, Propp, and Shor [15], and the work by Cohn, Elkies, and Propp [7] show that there are (at least) two different kinds of behavior for tiles in such regions. Inside of a circle inscribed in the diamond, the tiling is disordered (as a liquid or a gas); outside of the same circle, the tiling is frozen into a solid wall of dominoes, all with the same orientation.

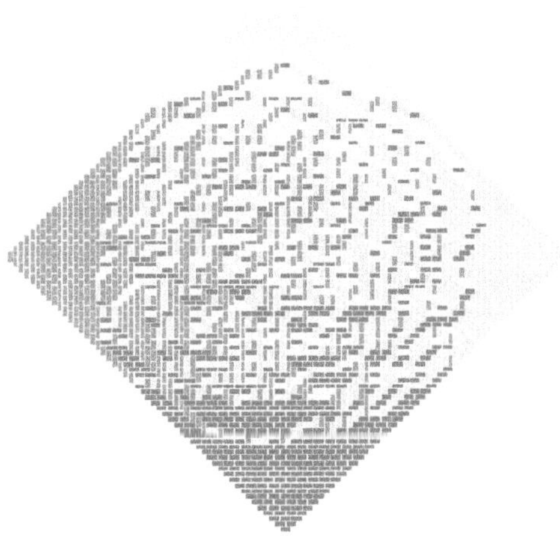

Figure 2: The Aztec diamond: a random tiling evinces the Arctic circle, see [24].

This beautiful result inspired a lot of research. For instance, a similar behavior occurs in other regions, as discussed in [18] by Kenyon, Okounkov and Sheffield.

Another natural question is the following.

Question 2.3 *Given a region $\mathcal{R}$, can we move from one tiling to another? Is the space of tilings connected via local moves?*

We first define the simplest kind of local move, a *flip*. In order to perform a flip, remove two adjacent parallel dominoes (forming a 2×2 block) and place them back in the only other possible manner (rotated by $\pi/2$), see Figure 3.

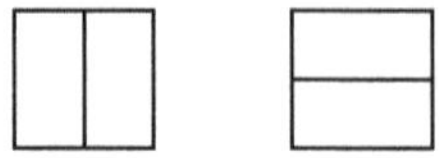

Figure 3: A flip move.

A fundamental result here is the following, shown by Thurtson [30].

Theorem 2.4 *In two dimensions, any two tilings of a simply connected region are flip connected.*

Similar results hold for regions which are not simply connected, such as tori or planar regions with holes (see [25]). Topological properties must then be taken into account, and the statement for quadriculated tori is slightly more complicated, with tilings which admit no flip.

The above questions and others have inspired much beautiful mathematics. In this article, we will pose the same questions for domino tilings in higher dimensions. Generally speaking, we have far fewer answers in higher dimensions, but will discuss recent progress in various directions.

3 Basics of three-dimensional domino tilings

Our survey will primarily focus on domino tilings in three dimensions. The regions of consideration are *cubiculated*: cubical polyhedral complexes $\mathcal{R}$, usually embedded in $\mathbf{R}^3$, safely thought of as collections of unit cubes which intersect along an entire face, edge or corner, or not at all. The dual graph $\mathcal{G}(\mathcal{R}^*)$ of $\mathcal{R}$ is formed by taking a vertex for each cube in $\mathcal{R}$ and an edge for any pair of cubes sharing a side. The graph $\mathcal{G}(\mathcal{R}^*)$ is a subgraph of the $\mathbf{Z}^3$ lattice and induces a bipartite two coloring of the cubes in $\mathcal{R}$ – color a cube white if the sum of the coordinates of the corresponding vertex is even and black otherwise.

We will assume certain niceness conditions on our regions. For starters, we require that $\mathcal{R}$ is *balanced*, i.e., the number of white cubes is equal to the number of black cubes, otherwise a domino tiling would not be possible. Nice regions include boxes and tori. Later, we will need other subtler conditions, for instance, that $\mathcal{R}$ is a topological manifold (i.e., the boundary of $\mathcal{R}$ should be a surface).

A three-dimensional domino is a $2 \times 1 \times 1$ union of two adjacent cubes in $\mathcal{R}$. A three-dimensional tiling of $\mathcal{R}$ is a collection of dominoes which exactly cover $\mathcal{R}$. Two dominoes may only intersect along faces of their boundary, and the union of all dominoes must equal $\mathcal{R}$. A domino in $\mathcal{R}$ corresponds to an edge in $\mathcal{G}(\mathcal{R}^*)$. A domino tiling of $\mathcal{R}$ corresponds to a perfect matching of $\mathcal{G}(\mathcal{R}^*)$, sometimes called a dimer cover of $\mathcal{G}(\mathcal{R}^*)$, see Figure 4.

As we can see already in this small example, it is hard to visualize domino tilings in three dimensions. To better visualize such tilings, we use a system of floors, as illustrated in Figure 5. The region $\mathcal{R}$ is broken into slices or floors by fixing values of the z-direction. The left picture of Figure 5 shows the same tiling that we saw in Figure 4, with floors parallel to the xy plane. Domino tiles that sit properly in a single plane are drawn within that plane. Vertical domino tiles that span two planes are drawn as two corresponding squares, one black, one white. The image in the

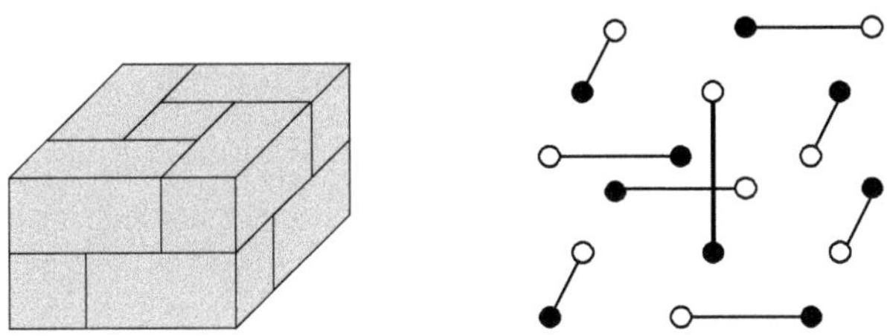

Figure 4: Three dimensional dominoes are $2 \times 1 \times 1$ bricks of two adjacent cubes. On the left, a three dimensional tiling of the $3 \times 3 \times 2$ box. On the right, the corresponding dimer cover of the dual graph.

middle of Figure 5 is the domino tiling of $\mathcal{R}$ with all vertical dominoes. The third tiling is obtained by reflection.

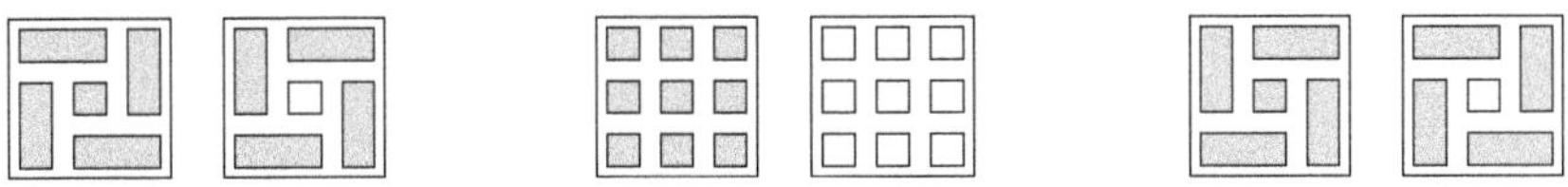

Figure 5: Representations of tilings of the $3 \times 3 \times 2$ box by floors.

3.1 Questions

As we saw above, for a simply connected region in two dimensions, any two tilings are connected by a series of flip moves. The flip move is natural in three dimensions as well. A *flip* removes two adjacent parallel dominoes and places them back rotated within a $2 \times 2 \times 1$ block.

We are led to the question: Given a region in three dimensions, are any two tilings connected by a series of flip moves? Or equivalently, in three dimensions, is the space of tilings connected by flips? Unlike two dimensions, the answer is no. Of course, the two dimensional case required the region to be simply connected. In three dimensions, even for nice regions, the answer is still no. For example, the first and last tilings in Figure 5 both have no possible flip moves.

The $3 \times 3 \times 2$ box is a small region, but it is not the size of the box that is causing the obstruction. There are also examples of tilings of larger boxes that similarly have no flip moves. For even N, if L and M are multiples of 4, it is easy to adapt the example in Figure 6 to obtain tilings of the $L \times M \times N$ box which admit no flips. For large boxes, it would be interesting to know more about the set of tilings which admit no flips.

Question 3.1 *Which regions have tilings which admit no flip? Which boxes? How many such tilings are there?*

In fact, the flip connected components of nice regions can be quite complicated. In Table 1, we show the number and size of the flip connected components for the $3 \times 3 \times 2$ box and the $4 \times 4 \times 4$ box.

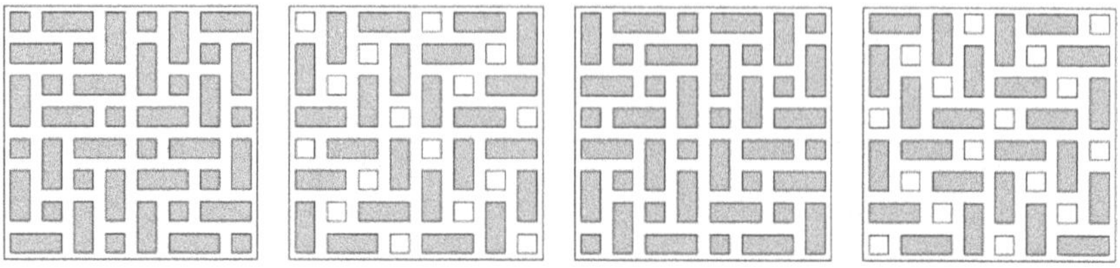

Figure 6: A tiling of the $8 \times 8 \times 4$ box with no possible flip moves.

$3 \times 3 \times 2$ box	$4 \times 4 \times 4$ box
Number of tilings: 229	Number of tilings: $5,051,532,105$
Connected components: 3	Connected components: 93
Sizes:	Sizes:
227	4,412,646,453
1	2 × 310,185,960
1	2 × 8,237,514
	2 × 718,308
	2 × 283,044
	6 × 2,576
	24 × 618
	24 × 236
	6 × 4
	24 × 1

Table 1: The number of flip connected components of small boxes.

For the $3 \times 3 \times 2$ box, the two components of size one are shown in Figure 5, first and third tiling. There are additional 227 tilings of this size box, all of which are connected by a series of flip moves. For the $4 \times 4 \times 4$ box, components are listed by size and multiplicity. For example, there are 6 components of size $2,576$ and 24 components of size 1, i.e., tilings with no flip moves. All of this leads to the following motivating question:

Question 3.2 *For three dimensional regions, how and when can we move from one tiling to another via local moves?*

Although Question 3.2 is not fully resolved, there has been significant progress recently, which we outline in Section 4.

The data in Table 1 also reveals that there are a very large number of tilings of even quite small regions. This leads to another natural question.

Question 3.3 *For a fixed region $\mathcal{R}$ in three (or more) dimensions, how many tilings are there of $\mathcal{R}$?*

Question 3.3 is also not resolved; we discuss the enumerative perspective of higher dimensional domino tiling in Section 8.

4 Connectivity by local moves

As we have seen above, unlike in the two-dimensional case, the collection of three dimensional tilings of a fixed region is not generally connected by flip moves. The flip move is, in essence, two-dimensional. For three dimensional regions, a new move is needed.

4.1 A three dimensional move

Definition 4.1 A *trit* move replaces 3 dominoes, one parallel to each axis inside a $2 \times 2 \times 2$ box, see Figure 7.

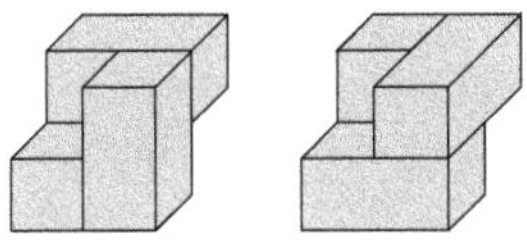

Figure 7: A positive trit move replaces the 3 dominoes seen on the left with the three dominoes on the right. A negative trit moves replaces the dominoes on the right with those on the left.

The trit move is fundamentally three dimensional in nature. It alters a domino in each of the three axis directions. Moreover, it cannot be reproduced by a sequence of flip moves. Instead, it can often be used as a local move for a tiling with no flip moves. For example, the first and third tilings of Figure 5 have no flip moves but can be connected via a sequence of trit and flip moves. The driving open question in this area is whether or not these two local moves, flips and trits, suffice to connect three dimensional regions.

Question 4.2 *Let $\mathcal{R}$ be a fixed connected and simply connected three-dimensional region. Is the set of all tilings of $\mathcal{R}$ connected by flips and trits?*

In general, the answer is no. Figure 8 shows a counterexample: no flip or trit is possible. As we shall see, if the region is not simply connected, then the flux (to be defined below) is an invariant under both flips and trits, so the answer is no in this case.

Figure 8: Two tilings of a very special region.

However, the problem may be that these regions are not "nice" enough. In two dimensions, regions need to be simply connected to guarantee that you can flip between all possible tilings. In three dimensions, it is not clear what the correct notion of a nice region is. The cylinders mentioned in the previous paragraph are contractible, so the notion cannot be of a purely topological nature. Whatever the notion, we assume it includes boxes. Even for these nicest of regions, the question of connectedness remains open.

Question 4.3 *Are all tilings of the* $[0, L] \times [0, M] \times [0, N]$ *box connected by flips and trits?*

4.2 A topological approach

Our first approach to studying such questions is topological, using homology, knot theory, and a cubical approximation of surfaces.

For this approach, we define two topological invariants. The *flux* of a tiling can be thought of as a "flow across surfaces" and takes on values in $H_1(\mathcal{R}; \mathbf{Z})$. The *twist* of a tiling can be thought of as how much a tiling is "twisted by trits". The twist is an integer if the flux is equal to 0; otherwise, the twist is in $\mathbf{Z}/m\mathbf{Z}$, where the value of m depends on the flux. We will define these invariants more precisely shortly.

For this approach, we also need the notion of *refinement*. A *refinement of a region* decomposes each cube into $5 \times 5 \times 5$ smaller cubes. A *refinement of a tiling* decomposes each domino into $5 \times 5 \times 5$ smaller dominoes, each parallel to the original. Refinement gives us a little extra space to perform local moves. First, we note the following.

Proposition 4.4 *If* t_0 *and* t_1 *are two tilings of a region connected by flips (resp., flips and trits), then their refinements are also connected by flips (resp., flips and trits).*

The converse is false, examples exist in the $4 \times 4 \times 2$ or $4 \times 4 \times 4$ boxes.

Let $\mathcal{R}$ be a three dimensional region, and t_0, t_1 be two tilings of $\mathcal{R}$. The *difference of tilings* is defined as follows:

$$t_1 - t_0 := \text{union of tiles (with the orientation of } t_0 \text{ reversed)}.$$

The difference of two tilings yields a system of cycles in the dual graph $\mathcal{G}(\mathcal{R}^*)$, where we ignore trivial 2-cycles formed when a tile is the same in both tilings, see Figure 9. We provide a more formal definition using homology theory. In a complex, such as $\mathcal{R}^*$, $C_0(\mathcal{R}^*; \mathbf{Z})$ is the additive abelian group formally generated by the vertices of $\mathcal{R}^*$. Similarly, $C_1(\mathcal{R}^*; \mathbf{Z})$ and $C_2(\mathcal{R}^*; \mathbf{Z})$ are generated by edges and faces, respectively. There are boundary maps $\partial : C_k \to C_{k-1}$ with $\partial^2 = 0$. The kernel and image of ∂ are $Z_k \subseteq C_k$ and $B_{k-1} \subseteq Z_{k-1} \subseteq C_{k-1}$, respectively. Finally, the homology H_k is the quotient Z_k/B_k.

Back to tilings, we interpret tilings t_0, t_1 as elements of $C_1(\mathcal{R}^*; \mathbf{Z})$ (each tiling is a sum of oriented edges) with a common boundary $\partial t_0 = \partial t_1 \in C_0(\mathcal{R}^*; \mathbf{Z})$ (the sum of all white vertices minus the sum of all black vertices). The difference is then $t_1 - t_0 \in Z_1(\mathcal{R}^*; \mathbf{Z})$: the homological version of the sum (or union) of a collection of disjoint cycles in the complex $\mathcal{R}^*$.

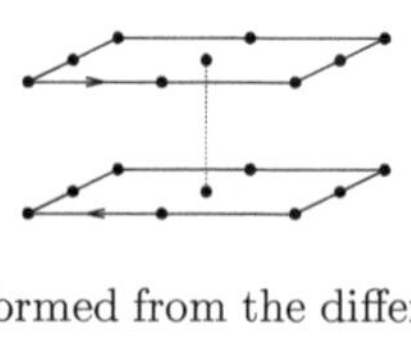

Figure 9: The system of cycles formed from the difference of the first and last tilings seen in Figure 5.

We are now ready to define the flux, the first of our two topological invariants. Given a region $\mathcal{R}$, fix a base tiling $t_\oplus$. For the box, we often choose the tiling with all vertical tiles for $t_\oplus$. The flux of a tiling t is the homology class of the difference between t and the base tiling.

Definition 4.5 For a region $\mathcal{R}$ with base tiling $t_\oplus$, the flux of a tiling t is

$$\mathrm{Flux}(t) := [t - t_\oplus] \in H_1(\mathcal{R}^*; \mathbf{Z}).$$

In order to define the second topological invariant, we borrow a notion from knot theory. In knot theory, a *Seifert surface* is a connected and oriented compact surface whose boundary is a given knot or link. It is a classical theorem that any link has a Seifert surface associated to it. In our setting, the difficulty is that we are working with discrete spaces and not smooth curves. Our boundaries are differences of tilings, and the surfaces we want must live in $\mathcal{R}^*$.

Definition 4.6 A (discrete) Seifert surface for a pair t_0, t_1 is a connected embedded oriented topological surface S with boundary $t_0 - t_1$.

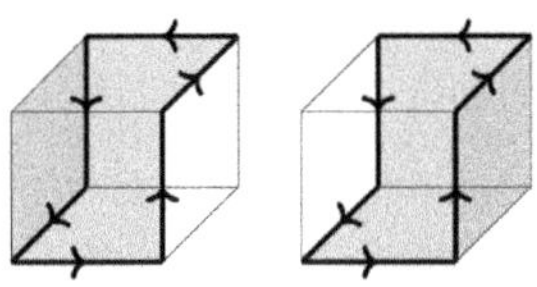

Figure 10: Two Seifert surfaces for the difference of the two tilings of a trit move.

Figure 10 shows two discrete Seifert surfaces for the same boundary. The boundary is the difference of two tilings that differ in a single trit move. In the paper [12], the authors show the following key technical Lemma.

Lemma 4.7 *If* $\mathrm{Flux}(t_0) = \mathrm{Flux}(t_1)$, *then, for sufficiently large* k, *there exists a discrete Seifert surface for the pair after* k *refinements.*

Informally, we think of refinements as giving a little more space to move; formally, we need refinements because of Lemma 4.7.

The *twist* of a tiling will prove to be a fundamental statistic for tilings, not just for the connectedness result below, see Section 6. We first present a general definition (following [12]): while general, this definition has the drawback of relying on a significant amount of topology. We shall later present an alternative definition for boxes. Given a region, fix a base tiling. For a tiling t, take a Seifert surface S for the pair t and the base tiling. At each vertex of $\mathcal{G}(\mathcal{R})$, we record a 0 or ± 1, depending on how the tiles of t intersect with the surface S. There is an additional sign recorded, denoted by $c(v)$ below, based on the color of the vertex ($c(v) = +1$ for black vertices, $c(v) = -1$ for white). Finally, we take the sum of all values across all vertices.

We define, for a tiling t, a Seifert surface S and a vertex v:

$$\phi(t;S) = \sum_v \varphi(v;t;S), \qquad \varphi(v;t;S) = \mathrm{c}(v) \cdot \begin{cases} +1, & \text{end above } S \\ 0, & \text{end on } S \\ -1, & \text{end below } S. \end{cases}$$

Definition 4.8 Fix a base tiling $t_\oplus$. The *twist* of a tiling t is

$$\mathrm{Twist}(t) := \phi(t; t - t_\oplus).$$

This definition assumes that, given t and $t_\oplus$, the value of ϕ is independent on the choice of the Seifert surface S.

The above definition of twist has the advantage of being general, but has the disadvantage of relying on topological tools. We provide an alternative definition. It is not intuitive and not very general, but it allows us to compute the twist for examples of tilings of small boxes. It is elementary, self-contained, and can also easily be used for programming the computation of the twist of tilings of larger boxes.

Consider a tiling of a box drawn as in Figures 5 and 6. For each column of each floor, list all possible pairs of an isolated square (denoting a vertical domino) and half of a horizontal domino. Each such pair contributes additively with ± 1, where the sign depends on the following 4 bits: whether the square is painted or not (thus indicating whether the vertical domino also uses the previous or the following floor); who is above (the square or the horizontal domino); parity of the distance between square and domino; whether the domino also uses the column to the right or to the left. The rule is illustrated in Figure 11, where we show examples of the 16 cases. Add all such contributions for all floors, all columns and all pairs: divide the result by 4 to obtain the twist (an integer!).

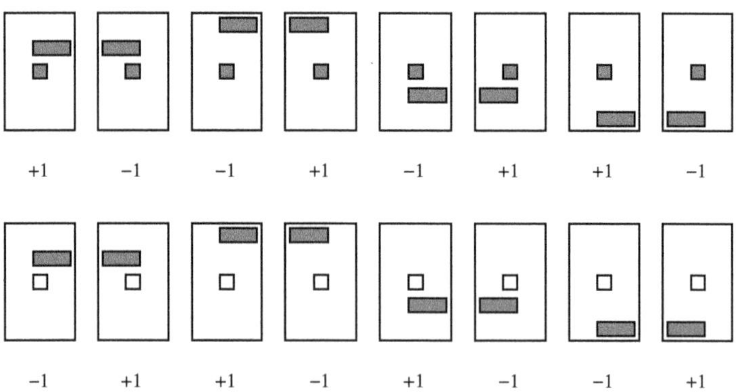

Figure 11: Signs of contributions to the twist.

We turn back to Figure 5 for a few small examples. Consider first the left tiling. There are two contributing pairs on the second column of the first floor and another two on the second column of the second floor. By comparing with Figure 11, we see that all four contributions are negative, and the twist is therefore $-4/4 = -1$. For the central tiling, there are no pairs, and the twist is 0. For the right tiling, there are again four contributing pairs: since they are all positive, the twist is $4/4 = +1$. Figure 6 gives us a larger example. There are four contributing pairs on the first column of the first floor, three negative and one positive. There are eight contributing pairs on the second column of the first floor, another eight on the third column and four more on the fourth column. This long list of signed pairs will eventually tell us that the twist equals 0.

The flux and twist behave nicely with respect to our local moves. If t_0 is connected to t_1 by a sequence of flip and trit moves:

$$t_0 \rightsquigarrow \text{trit} \rightsquigarrow t_1,$$

then the flux remains the same:

$$\text{Flux}(t_0) = \text{Flux}(t_1).$$

If the sequence includes exactly one trit, then the twist changes by exactly 1:

$$\text{Twist}(t_0) = \text{Twist}(t_1) \pm 1;$$

the sign is defined by the orientation of the trit. If the equivalence between the two definitions of twist is assumed, the fact that a flip does not change the value of the twist follows by taking a unit square as a Seifert surface. Similarly, for a trit, the two natural Seifert surfaces are shown in Figure 10: each of them consists of three squares and has a single interior vertex. It is then easy to compute the difference between the two values of the twist.

The main theorem of [12, Theorem 1] shows how tilings are connected by flips and trits with respect to flux and twist.

Theorem 4.9 *For two tilings t_0 and t_1 of a region $\mathcal{R}$:*

1. *There exists a sequence of flips and trits connecting refinements of t_0 and t_1* if and only if $\text{Flux}(t_0) = \text{Flux}(t_1)$.
2. *There exists a sequence of flips connecting refinements of t_0 and t_1* if and only if $\text{Flux}(t_0) = \text{Flux}(t_1)$ *and* $\text{Twist}(t_0) = \text{Twist}(t_1)$.

The theorem shows that tilings can be connected by flips and trits if we allow for (a sufficiently large number of) refinements. One might think that the refinements make connectivity almost too easy, namely, that with enough refinements everything is connected. Theorem 4.9 shows this is not the case. If two tilings have different values of the twist, then the two tilings cannot be flip connected regardless of how many refinements are made.

Question 4.10 *How many refinements are necessary in Theorem 4.9?*

Although the authors believe the number to be quite small, it remains an open problem to even bound the number of refinements needed. There are examples where at least one refinement is necessary, but there is no known example where more than one refinement is needed. Conversely, no upper bound on the number of refinements is known.

4.3 A second topological approach

Our second approach is also topological in nature. This approach uses cell complexes, homotopy theory, and fundamental groups.

As in the last section, we provide results given a little "extra space". Previously, we used refinements, here we consider adding extra vertical space to cylinders. A *cylinder* is a region $\mathcal{R} = \mathcal{D} \times [0, N]$, where $\mathcal{D}$ is a fixed two dimensional region. We assume $\mathcal{D}$ to be connected and simply connected, a topological disk. Extra vertical space is added by moving from $\mathcal{R} = \mathcal{D} \times [0, N]$ to $\mathcal{R}' = \mathcal{D} \times [0, N + M]$. A tiling of $\mathcal{R}$ is given space by adding all vertical dominoes in the top M floors.

We begin by defing the *domino complex* $\mathcal{C}_\mathcal{D}$ of a disk $\mathcal{D}$. $\mathcal{C}_\mathcal{D}$ is a 2-dimensional CW-complex, constructed such that tilings of $\mathcal{R} = \mathcal{D} \times [0, N]$ are closed paths in the complex. The vertices, edges, and two-cells correspond to tilings as follows. First, when considering a floor of a tiling, some tiles lie completely within the floor, while others intersect vertically, occupying only one square of the floor. Vertices correspond to the partial tilings of a floor, marking only vertical intersections. More formally, a *plug* is a balanced set of squares in $\mathcal{D}$, and there is one vertex in our CW-complex for each plug of $\mathcal{D}$. There is an edge between two disjoint plugs for each way the union can be completed to a full tiling of the floor. In this way, edges correspond to sets of disjoint horizontal tiles that lie within a floor. The two disjoint plugs correspond to vertical dominoes intersecting a given floor, one plug for dominoes using the next floor, and another plug for dominoes using the previous floor.

There are two types of 2-cells, both of which reflect flip moves. Glue a bigon for each flip within a floor, as in Figure 12; glue a quadrilateral for each vertical flip, as in Figure 13.

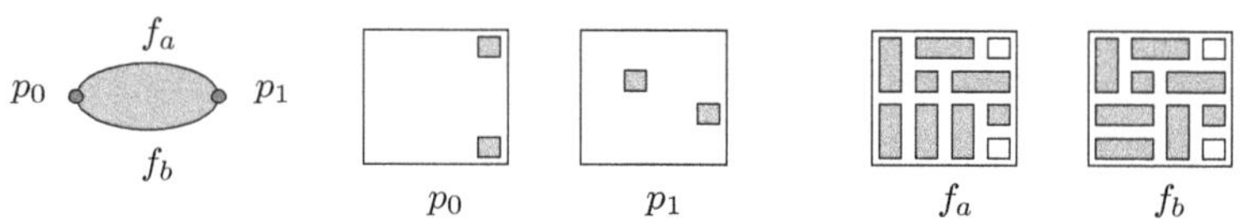

Figure 12: An example of a 2-cell, corresponding to a horizontal flip, see [27].

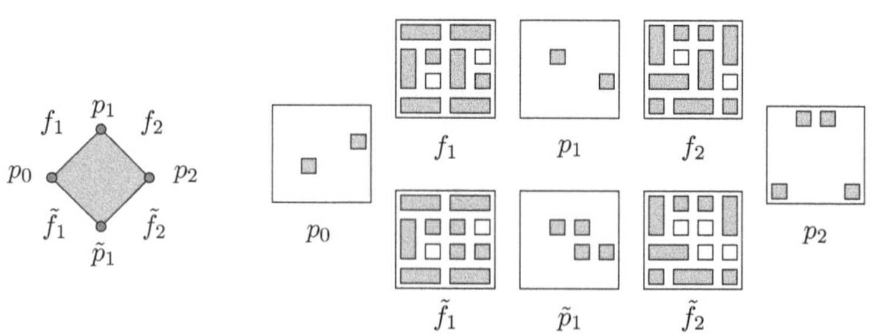

Figure 13: Another 2-cell, corresponding to a vertical flip, see also [27].

The importance of the construction is in the following result.

Proposition 4.11 *Two tilings of a cylinder are flip connected (with possible vertical space added) if and only if their paths in the domino complex are homotopic.*

Definition 4.12 The *domino group* $\mathcal{G}_\mathcal{D}$ is the fundamental group of the domino complex:

$$\mathcal{G}_\mathcal{D} = \pi_1(\mathcal{C}_\mathcal{D}).$$

The twist of a tiling defines a group homomorphism from the domino group to the integers. Define the homomorphism Tw by appending the parity of the height N of the region:

$$\mathrm{Tw} : \mathcal{G}_\mathcal{D} \to \mathbf{Z} \oplus \mathbf{Z}/2.$$

The homomorphism Tw assumes values in $\mathbf{Z} \oplus \mathbf{Z}/2$, while the twist, as defined above, assumes values in $\mathbf{Z}$: these are therefore closely related, but slightly different objects.

Proposition 4.13 *The homomorphism* Tw *is an isomorphism precisely when two tilings of the same twist are flip connected (with possible vertical space).*

In higher dimension, twist assumes values in $\mathbf{Z}/2$, and Tw is a group homomorphism of the form $\mathrm{Tw} : \mathcal{G}_\mathcal{D} \to \mathbf{Z}/2 \oplus \mathbf{Z}/2$. It has been shown that the twist is an isomorphism for a number of different regions as collected below, see [9, 10, 11, 20, 27].

Theorem 4.14 *The homomorphism* Tw *is an isomorphism for:*

1. *Two dimensional boxes with sides at least* 4*;*
2. *Higher dimensional boxes (except a few very small examples);*
3. *Hamiltonian disks without bottlenecks.*

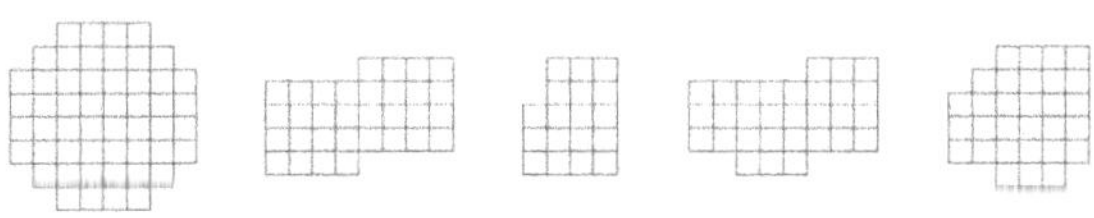

Figure 14: Various disks for which Tw is an isomorphism, see [11].

Figure 14 shows a number of disks $\mathcal{D}$ such that Tw is an isomorphism: these disks are Hamiltonian (i.e., the dual graph is Hamiltonian) and have no bottlenecks (see [10] for a definition of bottleneck). Therefore, for any of the cases listed in Theorem 4.14 or shown in Figure 14, any two tilings of the same twist of a cylinder above the region are flip connected, provided we are allowed to add a little extra vertical space.

It remains an open question to understand generally for which regions Tw is an isomorphism. This seems to be true for regions without small pinch points, or bottlenecks. However, current proofs rely on an additional niceness assumption, such as having a Hamiltonian path or cycle. Notice that the existence of a Hamiltonian path or cycle is sometimes a delicate question.

The amount M of extra vertical space required is a function of $\mathcal{D}$ only, that is, it does not depend on the height N of the region. The proof of this fact does not give a practical way to compute M. It would of course be interesting to obtain explicit estimates, at least for such examples as the square $\mathcal{D} = [0,4]^2$. Experimental evidence seems to indicate that a small value of M would already work.

For instance, a brute force computation verifies the following fact. For a tiling $\mathbf{t}$ of $\mathcal{R} = [0,4]^2 \times [0,4]$, let $\mathbf{t}'$ be the tiling of $\mathcal{R}' = [0,4]^2 \times [0,6]$ obtained from $\mathbf{t}$ by adding two floors of vertical dominoes. Notice that $\operatorname{Twist}(\mathbf{t}') = \operatorname{Twist}(\mathbf{t})$. If $\mathbf{t}_0, \mathbf{t}_1$ are tilings of $\mathcal{R}$ and $\operatorname{Twist}(\mathbf{t}_0) = \operatorname{Twist}(\mathbf{t}_1)$, then there is a finite sequence of flips taking $\mathbf{t}_0'$ to $\mathbf{t}_1'$ (in $\mathcal{R}'$).

4.4 A commutative algebra approach

In this section, we describe a different approach to higher dimensional tilings, involving techniques from commutative algebra, as introduced in Chin [5], Gross and Yazmon [13]. The central idea to this approach is that tilings are encoded as monomials in a polynomial ring. Local moves between tilings are then interpreted by constructing binomials. As is typical in commutative algebraic approaches in combinatorics, one forms ideals out of the objects of interest. In this way, the rich literature of monomial and binomial ideals can be applied to questions in domino tilings. We construct two such ideals – the tiling ideal and the flip ideal.

Let $\mathcal{R} \subset \mathbf{R}^N$ be a balanced cubiculated region, and let $\mathcal{G}_R = \mathcal{G}(R^*)$ be the graph dual of $\mathcal{R}$. In the commutative algebra setting, we will abuse notation slightly and consider domino tilings of $\mathcal{R}$ in terms of the corresponding perfect matching in $\mathcal{G}_R$. Associate a variable x_e to each edge $e \in E(\mathcal{G}_R)$, and consider the polynomial ring generated by the edge variables:

$$k[E] = k[x_e \,|\, e \in E(G)].$$

Let T be a tiling of the region $\mathcal{R}$. The monomial of T is the product of all edges in the tiling:

$$\mathbf{x}^T = \prod_{x_e \in T} x_e.$$

Let T_1 and T_2 be two tilings of a region $\mathcal{R}$. Define the difference of T_1 and T_2 as the binomial:

$$x^{T_1} - x^{T_2} = \prod_{x_e \in T_1} x_e - \prod_{x'_e \in T_2} x'_e.$$

As above, the difference of two tilings is a system of cycles in $\mathcal{G}_R$.

Definition 4.15 The *tiling ideal* of $\mathcal{R}$ is the binomial ideal

$$I_{\text{Tiling}} := \langle x^{T_1} - x^{T_2} \,|\, T_1, T_2 \text{ are tilings of } \mathcal{R} \rangle.$$

Definition 4.16 A *flip move* is (ab, cd), where a, b, c, d are edges which form a 4-cycle in $\mathcal{G}_R$ with edges a, b opposite each other, and edges c, d opposite each other.

Definition 4.17 The *flip ideal* of $\mathcal{R}$ is the ideal

$$I_{\text{Flip}} := \langle x_a x_b - x_c x_d \mid (ab, cd) \text{ is a flip move of } \mathcal{R} \rangle.$$

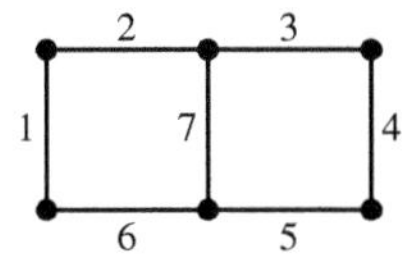

Figure 15: A small section of $\mathcal{G}_R$ with edge labels.

Example 4.18 Figure 15 shows two adjacent 4-cycles, which is $\mathcal{G}_R$ for the 2×3 rectangle. The term associated to each 4-cycle is an element of the flip ideal. For example,

$$x_4 x_7 - x_3 x_5 \in I_{\text{flip}},$$

reflecting that dominoes in positions corresponding to edges 4 and 7 could be flipped to positions 3 and 5.

The tiling ideal of the region in Figure 15 contains the term

$$x_1(x_4 x_7 - x_3 x_5) = x_1 x_4 x_7 - x_1 x_3 x_5 \in I_{\text{tiling}},$$

reflecting two tilings that differ only by the flip across the cycle on the right.

The connectivity of a region by flip moves can be determined by comparing the two ideals. As in the example, one always has that a flip binomial can be extended to an element of the tiling ideal by multiplying by all other variables held fixed by the flip move. We now state the main theorem of this approach, Theorem 3.12 [13].

Theorem 4.19 *A region $\mathcal{R}$ is flip connected if and only if*

$$I_{\textit{Tiling}} \subseteq I_{\textit{Flip}}.$$

Using Theorem 4.19, the authors are able to recover the connectivity result in two dimensions:

Theorem 4.20 *Let $\mathcal{R}$ be a 2-dimensional simply connected cubiculated region. Any binomial arising from two distinct tilings of $\mathcal{R}$ is generated by quadratics. In particular, $I_{\textit{Tiling}} \subseteq I_{\textit{Flip}}$.*

The work [5] explores primary decompositions of the flip ideal. Informally, a primary decomposition breaks an ideal into its irreducible components. The ideal is then perhaps better understood by understanding its building blocks.

A complete primary decomposition is given for the flip ideal of narrow grids, and a conjecture is given for the general grid case. The idea is that the prime components are reflected by neighborhoods of internal vertices and boundary conditions.

5 Random tilings

What does a random tiling look like? As we have already mentioned above, this is a very active area of research concerning domino tilings in dimension two. As an example, Cohn, Kenyon, and Propp showed a variational principle for domino tilings in the plane [8].

There is also some very interesting current work in higher dimensions. The approach and techniques are sufficiently different from the remainder of the survey that we cannot really do justice to this area of research here. We simply mention a recent achievement in the probability theory of higher dimensional tilings: a large deviation result by Chandgotia, Sheffield, and Wolfram [4].

Figure 16: A random tiling of the Aztec octahedron, see [4].

Figure 16 shows a random tiling of the "Aztec octahedron". The traditional algorithms used to generate such figures involve random walks in the space of tilings in order to uniformly sample a tiling. Given an initial tiling $\mathbf{t}_0$, we generate a random cycle as follows. Randomly select an initial vertex (or unit cube) v_0: the tiling $\mathbf{t}_0$ matches v_0 with a vertex v_1. Select a random neighbor $v_2 \neq v_0$ of v_1: the tiling $\mathbf{t}_0$ matches v_2 with a vertex v_3. More generally, given v_{2k-1} and v_{2k-2}, select a random neighbor $v_{2k} \neq v_{2k-2}$ of v_{2k-1}. If v_{2k} is not equal to any v_{2j}, $0 \leq j < k-1$, then we proceed: the tiling $\mathbf{t}_0$ matches v_{2k} with a vertex v_{2k+1}. If $v_{2k} = v_{2j}$, then the desired random cycle is $v_{2j}, v_{2j+1}, \ldots, v_{2k-1}, v_{2k} = v_{2j}$. We then move from tiling $\mathbf{t}_0$ to tiling $\mathbf{t}_1$ by deleting the dominoes $(v_{2j}, v_{2j+1}), \ldots, (v_{2k-2}, v_{2k-1})$ and placing the dominoes $(v_{2j+1}, v_{2j+2}), \ldots, (v_{2k-1}, v_{2k})$.

If the random cycle generated as above has length 4 (resp., 6), then it is a flip (resp., trit or combination of a small number of flips). The above algorithm uses cycles of unbounded length (meaning that there is no bound independent of the region). Placing an a priori bound on the length of cycles is equivalent to using local moves: this leads naturally to conjectures such as Question 4.3.

6 Twist

As we have seen already above and will be further motivated below, the twist is an important statistic for higher dimensional tilings. We note that although defined topologically, for sufficiently nice regions, the twist can be computed purely combinatorially. This allows us to gather some data on the twist.

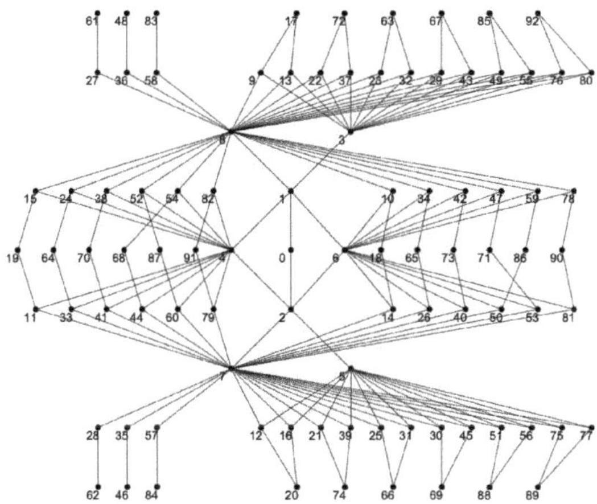

Figure 17: The graph of connected components under flips of the $4 \times 4 \times 4$ box.

Figure 17 shows the graph whose vertices are the connected components under the flip move for tilings of the $4 \times 4 \times 4$ box. The $4 \times 4 \times 4$ box has 93 flip connected components; they are listed with multiplicity in the table of Section 3. The largest component is labeled 0. Two flip connected components A and B are connected by an edge in the graph if there exists tilings $T_1 \in A$ and $T_2 \in B$ such that T_1 and T_2 are connected by a trit move. Note there may be many such tilings, but we only draw one edge.

The graph is laid out so that vertices in the same row have the same value of twist. Vertices corresponding to tilings of twist 0 are across the middle. Tilings one level up have twist 1. Tilings one level down have twist -1. Recall that a trit move changes the twist by exactly 1, hence, we only have edges between adjacent rows. The vertical symmetry of the graph is expected due to the symmetry of the region. Other aspects of the graph, such as the varying number of components per level or the varying degrees of components, are not easily anticipated.

In this example, the maximum and minimum values for the twist are ± 4. For a general box, the possible values of the twist can be bounded as follows, see [21, Proposition 6.1].

Proposition 6.1 *For* $\mathcal{R} = [0, L] \times [0, M] \times [0, N]$,

$$\frac{1}{162} \leq \frac{\mathrm{Tw}_{\max}(\mathcal{R})}{LMN \min(L, M, N)} \leq \frac{1}{16}.$$

Question 6.2 *What are the best constants for the bounds in Proposition 6.1?*

There is no conjecture for what the best constants are. The numbers in the statement can almost certainly be improved.

7 Helicity

Let $M \subset \mathbf{R}^3$ be a simply connected domain with smooth boundary ∂M. Let ξ be a divergence-free vector field in M, i.e., the divergence $\nabla \cdot \xi$ equals 0; equivalently, the flux of ξ through a small sphere is equal to zero. Assume that ξ is tangent to the boundary ∂M. The *helicity* of ξ is defined by

$$\mathrm{Hel}(\xi) = \int_M (\xi, \mathrm{curl}^{-1} \xi)\, dxdydz.$$

In this formula, $\mathrm{curl}^{-1} \xi$ stands for a vector field ζ, satisfying $\mathrm{curl}\, \zeta = \xi$; it is not hard to verify that if $\mathrm{curl}\, \zeta_0 = \mathrm{curl}\, \zeta_1 = \xi$, then

$$\int_M (\xi, \zeta_0)\, dxdydz = \int_M (\xi, \zeta_1)\, dxdydz;$$

here, (ξ, ζ) denotes the inner product. Thus, helicity is an invariant, and it can be interpreted as measuring the linkage and knottedness of orbits. More precisely, in [2] (see also [3]), Arnold proved that an asymptotic version of the Hopf invariant for a vector field is equal to the field's helicity.

If ξ is not tangent to the boundary, we can still define the *relative helicity* of ξ. In a nutshell, extend ξ to a thin tubular neighborhood of ∂M to define an *isolating shell* (a tubular neighborhood of ∂M is homeomorphic to the cartesian product $\partial M \times (-\epsilon, \epsilon)$). Among vector fields with the same behavior at the boundary, helicity is then well defined, up to an additive constant.

The similarity between the concepts of twist (for tilings) and helicity (for vector fields) has been informally remarked several times. An explicit construction relating the two concepts is the *five pipe construction*, see [19]. Consider a domino tiling of a 3d region $\mathcal{R}$. Inside each domino, construct five narrow pipes joining faces of the domino as in Figure 18.

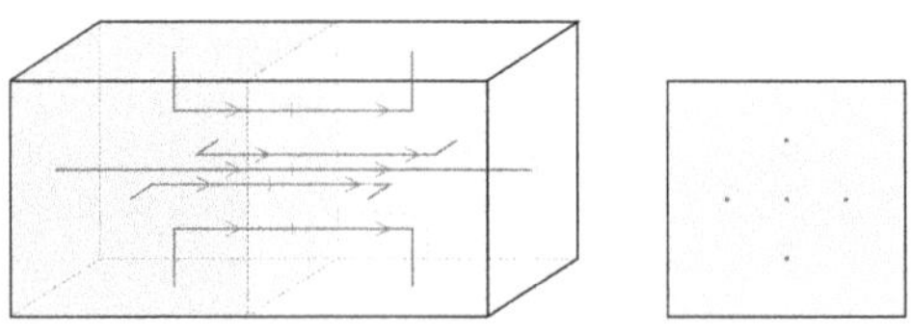

Figure 18: The 5 pipe construction (one domino), see [19].

Notice that, in the interior of $\mathcal{R}$, pipes connect nicely. At the boundary $\partial\mathcal{R}$, the pipe pattern is independent from the tiling. Turn the water on, with flux $\varphi = 1/6$ in each pipe. The water defines a vector field ξ, which is clearly divergence-free. We claim that the relative helicity agrees with the twist, which are both defined up to additive constant: this is the main result (Theorem 1.1) from [19], which we restate in a slightly simplified form.

Theorem 7.1 *Let $\mathcal{R} \subset \mathbf{R}^3$ be a cubiculated compact region. We assume that $\mathcal{R}$ is tileable and simply connected. Let $\mathbf{t}_0$ and $\mathbf{t}_1$ be tilings of $\mathcal{R}$: perform the 5 pipe construction with flux $\varphi = 1/6$ to obtain vector fields $\xi_{\mathbf{t}_0}$ and $\xi_{\mathbf{t}_1}$. We then have*

$$\mathrm{Hel}(\xi_{\mathbf{t}_1}) - \mathrm{Hel}(\xi_{\mathbf{t}_0}) = \mathrm{Tw}(\mathbf{t}_1) - \mathrm{Tw}(\mathbf{t}_0).$$

We call attention to the fact that the vector field ξ thus constructed is not tangent to the boundary. In other words: there are pipes arriving transversally at the boundary of $\mathcal{R}$. In Figure 19, the blue pipes are internal to $\mathcal{R}$. This is not a problem because we are talking about relative helicity: we can construct an isolating shell aroud $\mathcal{R}$ and pipes in the isolating shell so that this augmented pipe system does not have dead ends. The red pipes in Figure 19 show the isolating shell: they are external to the box $\mathcal{R}$ and independent of the tiling.

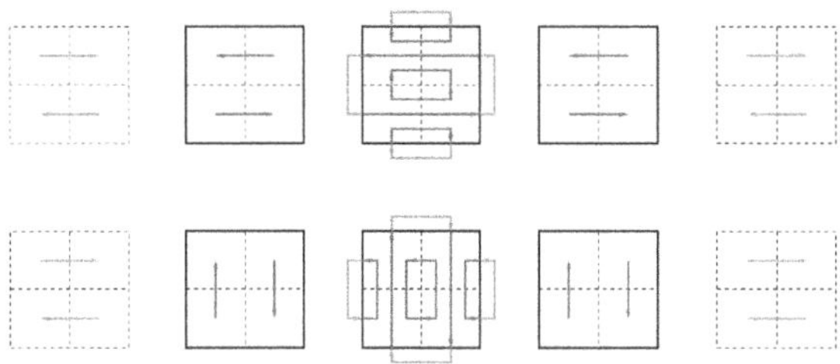

Figure 19: The pipe systems corresponding to a flip, see [19].

The proof of Theorem 7.1 relies on the results of [12]. More precisely, we need to compute the change of relative helicity when we perform a flip, a trit or a refinement. After many such figures, we verify that, as expected, relative helicity remains unchanged under flip or refinement and changes by ± 1 under a trit (with correct sign). As an example, we show in Figure 19 the pipe systems and vector fields corresponding to a flip: the relative helicity does not change. We verify that, in both cases, we have a family of unknotted and unlinked cycles with self-linking also equal to zero: the helicity is therefore zero in both cases. The curious reader will find far more complicated diagrams in [19]: also there, helicity is computed via linking and self-linking numbers.

8 Enumeration

A natural question is: how many domino tilings are there of a given region in 3-dimensions? The analogous question in 2-dimensions has been studied extensively both for domino and other types of tilings. Enumerating such tilings proves to be quite difficult and leads to fascinating connections and results. Recent research of Krattenthaler investigates the enumeration questions of tiling in 2-dimensions in great depth.

As one might imagine, there is less known about the case in higher dimensions. Recall the table of Section 3, giving the sizes of the flip connected components for small boxes.

The number of tilings of the $3 \times 3 \times 2$ box is 229.

The number of tilings of the $4 \times 4 \times 4$ box is $5,051,532,105$.

The number of tilings of the $4 \times 4 \times 8$ box is $175,220,727,982,196,365,632$.

Question 8.1 *How many tilings are there of the* $4 \times 6 \times 6$ *box? Of the* $6 \times 6 \times 6$ *box? Of the* $8 \times 8 \times 8$ *box?*

There are a tremendous number of tilings in higher dimensions, growing rapidly with the size of the box. In general, we do not have a simple formula for the number of tilings of a given region, even for a box: even the explicit and relatively small examples in Question 8.1 are open problems (the examples before the question show that a brute force approach is out of question). This situation should be contrasted with the corresponding problem in dimension 2, for which so much more is known: Kasteleyn's theorem about pfaffians yields an explicit formula for the number of tilings of a rectangle, the Aztec diamond has been thoroughly studied and there are many other results about counting perfect matchings of planar graphs.

Question 8.2 *How many tilings are there of the d-dimensional box?*

Unfortunately, due to the next result, it is unlikely that a nice formula can be found for the number of domino tilings in higher dimensions. First proved by Valiant [31] and later strengthened by Pak and Yang [23], we have the following complexity result.

Theorem 8.3 *The number of domino tilings of a region in 3-dimensions is #P-complete.*

Although a perfect count may be out of reach, understanding the approximate number of domino tilings is an interesting question and important in applied physical contexts. A number of works have proved upper and lower bounds on the number of tilings of the d-dimensional box. We will not discuss this line of inquiry here, but refer the reader to [6, 14, 22].

In two dimensions, many of the classical enumerative techniques are linear-algebraic in nature. For example, one constructs a signed weighted adjacency matrix, known as the Kasteleyn matrix, and then takes a determinant in order to enumerate tilings of rectangular regions. Such a construction can be mimicked for 3-dimensional boxes [20]. The result is not an exact count, but it is an interesting statistic on tilings. For a box $\mathcal{R}$, let $K = K(\mathcal{R})$ be the generalized Kasteleyn matrix. We give an indirect description of the matrix $K \in \mathbf{Z}^{N\times N}$. Rows and columns correspond to black and white unit cubes, respectively. Every entry belongs to $\{0, \pm 1\} \subset \mathbf{Z}$. The entry K_{ij} is nonzero if and only if the i-th black unit cube b_i and the j-th white unit cube w_j share a common face. If the two black unit cubes b_{i_0} and b_{i_1} and the two white unit cubes w_{j_0} and w_{j_1} share a common edge, then $K_{i_0 j_0} K_{i_0 j_1} K_{i_1 j_0} K_{i_1 j_1} = -1$.

Theorem 8.4 *The determinant of the matrix* $K(\mathcal{R})$ *gives an alternating sum of tilings by the parity of the twist:*

$$\det(K(\mathcal{R})) = \sum_{\mathbf{t}} (-1)^{\mathrm{Twist}(\mathbf{t})}.$$

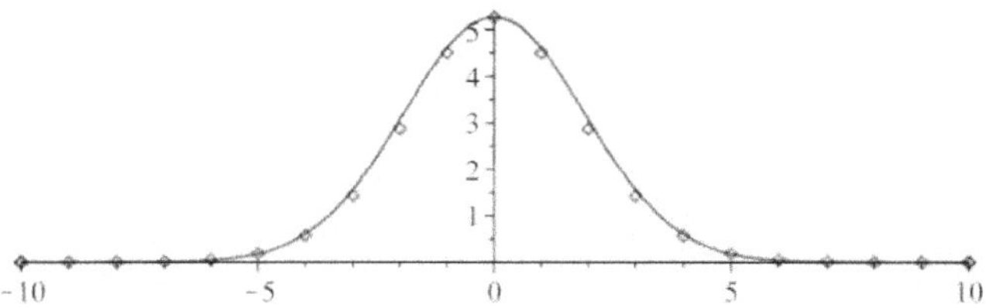

Figure 20: #Tilings per value of twist of the $4 \times 4 \times 60$ box.

There is a similar statement using pfaffians (instead of determinants) and a slightly different matrix.

Theorem 8.4 reinforces a natural question refining Question 8.2: How many tilings are there of the d-dimensional box for a fixed value of the twist? Given the difficulty of exact counting and the large number of tilings, we reconsider the problem as understanding the *distribution* of twist for a random tiling of the box.

Again, looking at the data of the table in Section 3, one might be tempted to conjecture that there is a giant component in the graph of flip connected components, and that almost all tilings of the box have twist equal to 0. However, we know that this is not the case.

Figure 20 plots the number of tilings by twist of the $4 \times 4 \times 60$ box. The vertical axis is scaled by a factor of 10^{156}, and the curve drawn in red is the true Gaussian.

For cylinders, i.e., regions of the form $\mathcal{R} = D \times [0, N]$, where D is a fixed two dimensional disk, it is known that the twist is normally distributed as $N \to \infty$, see [26, Theorem 3]. The distribution of the twist is unknown (but conjectured to approach a normal distribution) for arbitrary boxes when more than one dimension goes to infinity.

Question 8.5 *Is the twist normally distributed for boxes* $\mathbf{R} = [0, N] \times [0, M] \times [0, L]$ *as* $\min N, M, L \to \infty$ *?*

9 Slabs

Very little is known about tilings by other higher dimensional tiles, but there are a few results about tilings of boxes by *slabs*, $2 \times 2 \times 1$ blocks. We present a small sample of such results, see [1]. Figure 21 shows an example of a slab tiling.

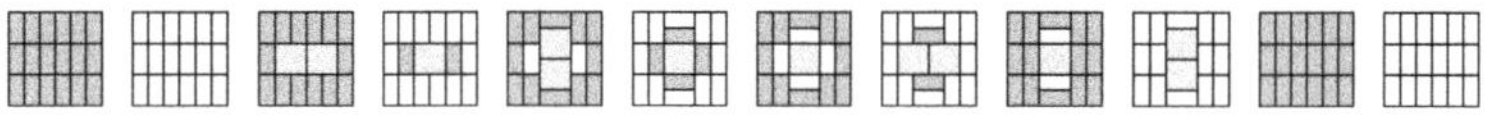

Figure 21: A slab tiling of the $6 \times 6 \times 12$ box, see [1].

A *flip* is a local move, involving two slabs. Is the tiling in Figure 21 equivalent via a finite sequence of flips to a horizontal tiling? No! There exists an invariant, the *triple twist*, assuming values in $\mathbf{Z}^3$. Begin by painting the unit cubes in four colors, red, yellow, green, and blue, as in Figure 22. More formally: label the cubes by $\mathbf{Z}^3$, and let $L \subset \mathbf{Z}^3$ be the additive group generated by $(2,0,0)$, $(0,2,0)$, $(0,0,2)$ and $(1,1,1)$. Each color corresponds to a coset $v + L \subset \mathbf{Z}^3$.

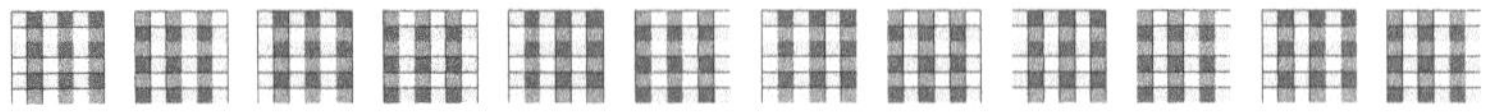

Figure 22: Painting unit cubes using four colors, see [1].

Choose two colors, say red and green. Inflate the red and green cubes; deflate the yellow and blue; the result is shown in Figure 23. The slab tiling in Figure 21 becomes a domino tiling, as in Figure 24.

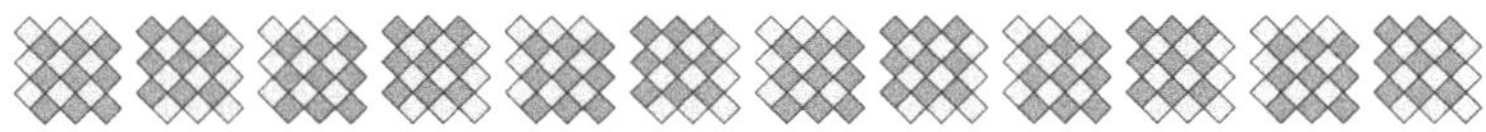

Figure 23: Inflate red and green cubes, see [1].

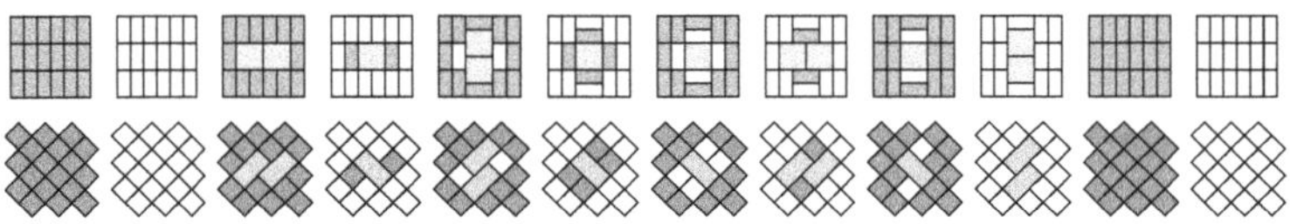

Figure 24: The slab tiling becomes a domino tiling, see [1].

Compute the twist of the resulting domino tiling: the value of the twist is invariant by flips. Notice that the horizontal tiling yields twist zero, while the example in the figures yields twist $+2$, proving the claim above. One easy way to verify that the twist equals to $+2$ is to rotate the figure counterclockwise by $\pi/4$, complete a box with vertical dominoes, and use the method described in Figure 11. Perform the same construction for other pairs of colors to define a total of six related invariants. They turn out not to be linearly independent, but we still have three dimensions, thus defining the triple twist.

Theorem 9.1 *There is a map, called the triple twist, taking slab tilings to elements of $\mathbf{Z}^3$. The triple twist is invariant under flips. For a cubical box $N \times N \times N$, where N is large and even, the triple twist assumes approximately CN^{12} different values.*

10 Four dimensions and above

The two dimensional theory of domino tilings is well studied, and we have seen above how many ideas can be extended to the three dimensional case, although with far more open questions than answers. There is a distinct shift when we move to four dimensions and above. Borrowing from Section 4, for example, we might expect a new topological invariant and corresponding local move as we go up in dimension. This is, however, not the case.

In dimensions $n \geq 4$, a domino is a $2 \times 1 \times \cdots \times 1$ block. In the paper [20], the authors establish the following theory. The *twist* of a tiling is defined; it is no longer an integer but an element of $\mathbf{Z}/(2)$. On the one hand, this is simpler than the three dimensional case, on the other hand, it collapses much of the interesting behavior seen above. The *flip* and *trit* moves remain the same, involving either two or three dominoes, and take place in a two or three dimensional subspace of the tiling region. The flip move does not alter the value of the twist, a trit move changes the value of the twist to the only other possible value.

In dimensions $n \geq 4$, there is both a connectedness-by-local-moves theorem and an enumerative theorem. The connectedness result is Theorem 4.14, already stated in Section 4.3. In particular, the result uses the theory of domino complexes and the domino group for cylinders $\mathcal{R}_\mathcal{D} = \mathcal{D} \times [0, N]$, where $\mathcal{D}$ is a region in $\mathbf{R}^{n-1}$. Reiterating more precisely:

Theorem 10.1 *Consider $n \geq 4$ and a box $\mathcal{D} = [0, N_1] \times \cdots \times [0, N_{n-1}]$. Assume that $N_i \geq 2$ (for all i), and that $N_1 > 2$ or $n > 4$. Any two tilings of $\mathcal{R}_\mathcal{D}$ with the same twist can be connected by a sequence of flip moves, with possible vertical space added.*

In this case, the amount of extra vertical space needed can be bounded in terms of the disk $\mathcal{D}$. These results then allow us to understand the distribution of the twist in higher dimensions. We give a simplified version here, see [20, Corollary 1.2] for the full statement.

Theorem 10.2 *Let $\mathcal{D}$ be as in Theorem 10.1. The set of flip connected components of tilings of the cylinder $\mathcal{R}_\mathcal{D}$ has two twin giant components, one for each value of the twist. The ratio between the sizes of the two giant components tends to 1, when the height N goes to infinity.*

Acknowledgements

The first author acknowledges support from the NSF. The second author acknowledges support from CNPq, CAPES, and Faperj. The authors thank the referee for a very careful and helpful report.

References

[1] George L. D. Alencar, Nicolau C. Saldanha, and Arthur M. M. Vieira, *Slab tilings, flips and the triple twist*, Discrete and Computational Geometry (2025).

[2] Vladimir I. Arnold, *The asymptotic hopf invariant and its applications*, Vladimir I. Arnold - Collected Works: Hydrodynamics, Bifurcation Theory, and Algebraic Geometry 1965-1972 (Alexander B. Givental, Boris A. Khesin, Alexander N. Varchenko, Victor A. Vassiliev, and Oleg Ya. Viro, eds.), Springer Berlin Heidelberg, 2014, pp. 357–375.

[3] Vladimir I. Arnold and Boris A. Khesin, *Topological methods in hydrodynamics*, 2nd ed., Applied Mathematical Sciences, vol. 125, Springer, 2021.

[4] N. Chandgotia, S. Sheffield, and C. Wolfram, *Large deviations for the 3D dimer model*, Preprint arXiv:math/2304.08468 (2023).

[5] Tracy Chin, *A computational commutative algebra approach to tilings*, Bachelor's Thesis, Brown University, 2019.

[6] Mihai Ciucu, *An improved upper bound for the 3-dimensional dimer problem*, Duke Math. J. **94** (1998), no. 1, 1–11.

[7] Henry Cohn, Noam Elkies, and James Propp, *Local statistics for random domino tilings of the Aztec diamond*, Duke Math. J. **85** (1996), no. 1, 117–166.

[8] Henry Cohn, Richard Kenyon, and James Propp, *A variational principle for domino tilings*, J. Amer. Math. Soc. **14** (2001), no. 2, 297–346.

[9] Raphael de Marreiros, *3D domino tilings: irregular disks and connected components under flips*, Preprint arXiv:2209.03109 (2022).

[10] ———, *Domino tilings of three-dimensional cylinders: regularity of hamiltonian disks*, Preprint arXiv:2406.11777 (2024).

[11] ———, *Domino tilings of 3-dimensional cylinders*, PhD Thesis, Pontificia Universidade Catolica do Rio de Janeiro, 2025.

[12] Juliana Freire, Caroline J. Klivans, Pedro H. Milet, and Nicolau C. Saldanha, *On the connectivity of spaces of three-dimensional domino tilings*, Trans. Amer. Math. Soc. **375** (2022), no. 3, 1579–1605.

[13] Elizabeth Gross and Nicole Yamzon, *Binomial ideals of domino tilings*, Discrete Math. **344** (2021), no. 11.

[14] J. M. Hammersley, *An improved lower bound for the multidimensional dimer problem*, Proc. Cambridge Philos. Soc. **64** (1968), 455–463.

[15] W. Jockusch, J. Propp, and P. Shor, *Random domino tilings and the arctic circle theorem*, Preprint arXiv:math/9801068 (1998).

[16] P.W. Kasteleyn, *The statistics of dimers on a lattice: I. The number of dimer arrangements on a quadratic lattice*, Physica **27** (1961), no. 12, 1209–1225.

[17] R. Kenyon, *An introduction to the dimer model*, Preprint arXiv:math/0310326 (2003).

[18] Richard Kenyon, Andrei Okounkov, and Scott Sheffield, *Dimers and amoebae*, Ann. of Math. (2) **163** (2006), no. 3, 1019–1056.

[19] B. Khesin and N. Saldanha, *Relative helicity and tiling twist*, Transactions of the American Mathematical Society (2025).

[20] Caroline J. Klivans and Nicolau C. Saldanha, *Domino tilings and flips in dimensions 4 and higher*, Algebr. Comb. **5** (2022), no. 1, 163–185.

[21] Pedro Milet, *Domino tilings of three-dimensional regions*, PhD Thesis, Pontificia Universidade Catolica do Rio de Janeiro, 2015.

[22] Henryk Minc, *An asymptotic solution of the multidimensional dimer problem*, Linear and Multilinear Algebra **8** (1979/80), no. 3, 235–239.

[23] Igor Pak and Jed Yang, *The complexity of generalized domino tilings*, Electron. J. Combin. **20** (2013), no. 4.

[24] James Propp and the MIT Tilings Research Group, *My life with Aztec diamonds*, 2021, Mathematical Enchantments Blog, https://mathenchant.wordpress.com/2021/01/16/my-life-with-aztec-diamonds/.

[25] N. C. Saldanha, C. Tomei, M. A. Casarin, Jr., and D. Romualdo, *Spaces of domino tilings*, Discrete Comput. Geom. **14** (1995), no. 2, 207–233.

[26] Nicolau C. Saldanha, *Domino tilings of cylinders: connected components under flips and normal distribution of the twist*, Electron. J. Combin. **28** (2021), no. 1, Paper No. 1.28, 23.

[27] ———, *Domino tilings of cylinders: the domino group and connected components under flips*, Indiana Univ. Math. J. **71** (2022), no. 3, 965–1002.

[28] Nicolau C. Saldanha and Carlos Tomei, *An overview of domino and lozenge tilings*, Resenhas **2** (1995), no. 2, 239–252, Combinatorics Week (São Paulo, 1994).

[29] H. N. V. Temperley and Michael E. Fisher, *Dimer problem in statistical mechanics—an exact result*, Philos. Mag. (8) **6** (1961), 1061–1063.

[30] William P. Thurston, *Conway's tiling groups*, Amer. Math. Monthly **97** (1990), no. 8, 757–773.

[31] L. G. Valiant, *Completeness classes in algebra*, Conference Record of the 11th Annual ACM Symposium on Theory of Computing (Atlanta, Ga., 1979), ACM, New York, 1979, pp. 249–261.

Division of Applied Mathematics
Brown University
Box F
Providence, RI 02906
USA
klivans@brown.edu

Departamento de Matemática, PUC-Rio
R Mq S Vicente 225, Gávea
Rio de Janeiro, RJ 22451-900
Brazil
saldanha@puc-rio.br

For EU product safety concerns, contact us at Calle de José Abascal, 56–1°,
28003 Madrid, Spain or eugpsr@cambridge.org.

ww.ingramcontent.com/pod-product-compliance
ram Content Group UK Ltd.
ld, Milton Keynes, MK11 3LW, UK
V020201290926
UKWH00005B/994

8 1 0 0 9 7 6 6 0 4 3 *